Polymeric Materials for Wastewater Treatment Applications

Polymeric Materials for Wastewater Treatment Applications

Guest Editors

Marta Otero
Ricardo N. Coimbra

Basel • Beijing • Wuhan • Barcelona • Belgrade • Novi Sad • Cluj • Manchester

Guest Editors

Marta Otero
Department of Applied
Chemistry and Physics
University of León
León
Spain

Ricardo N. Coimbra
Department of Environment
and Planning
University of Aveiro
Aveiro
Portugal

Editorial Office
MDPI AG
Grosspeteranlage 5
4052 Basel, Switzerland

This is a reprint of the Special Issue, published open access by the journal *Polymers* (ISSN 2073-4360), freely accessible at: www.mdpi.com/journal/polymers/special_issues/Polymeric_Materials_Wastewater_Treatment_Applications.

For citation purposes, cite each article independently as indicated on the article page online and using the guide below:

Lastname, A.A.; Lastname, B.B. Article Title. *Journal Name* **Year**, *Volume Number*, Page Range.

ISBN 978-3-7258-3460-0 (Hbk)
ISBN 978-3-7258-3459-4 (PDF)
https://doi.org/10.3390/books978-3-7258-3459-4

© 2025 by the authors. Articles in this book are Open Access and distributed under the Creative Commons Attribution (CC BY) license. The book as a whole is distributed by MDPI under the terms and conditions of the Creative Commons Attribution-NonCommercial-NoDerivs (CC BY-NC-ND) license (https://creativecommons.org/licenses/by-nc-nd/4.0/).

Contents

Preface vii

Marta Otero and Ricardo N. Coimbra
Polymeric Materials for Wastewater Treatment Applications
Reprinted from: *Polymers* 2025, 17, 552, https://doi.org/10.3390/polym17040552 1

Adenike A. Akinsemolu and Helen Onyeaka
Advances in Hydrogel Polymers for Microbial Control in Water Systems
Reprinted from: *Polymers* 2024, 16, 2205, https://doi.org/10.3390/polym16152205 10

Haipeng Yan, Pan Wang, Lingsha Li, Zixin Zhao, Yang Xiang and Haoqian Guo et al.
Development Status of Solar-Driven Interfacial Steam Generation Support Layer Based on Polymers and Biomaterials: A Review
Reprinted from: *Polymers* 2024, 16, 2427, https://doi.org/10.3390/polym16172427 25

Haneen Abdelrazeq and Majeda Khraisheh
Porosity Effect of Polystyrene Membranes on Desalination Performance: A Combined Experimental and Numerical Heat and Mass Transfer Study in Direct Contact Membrane Distillation
Reprinted from: *Polymers* 2023, 15, 1821, https://doi.org/10.3390/polym15081821 43

Alexis López-Borrell, Jaime Lora-García, Salvador C. Cardona, María-Fernanda López-Pérez and Vicent Fombuena
Vapor Pressure and Evaporation Studies of Saline Solutions on Natural and Synthetic Fabrics for Industrial Water Treatment
Reprinted from: *Polymers* 2024, 16, 2335, https://doi.org/10.3390/polym16162335 63

Fan Li, Thomas Shean Yaw Choong, Luqman Chuah Abdullah, Siti Nurul Ain Md. Jamil and Nurul Nazihah Amerhaider Nuar
Effective Removal of Glyphosate from Aqueous Systems Using Synthesized PEG-Coated Calcium Peroxide Nanoparticles: Kinetics Study, H_2O_2 Release Performance and Degradation Pathways
Reprinted from: *Polymers* 2023, 15, 775, https://doi.org/10.3390/polym15030775 91

Hao Luo, Hao Liu and Chengwu Sun
Removal of Sulfide Ions from Kraft Washing Effluents by Photocatalysis with N and Fe Codoped Carbon Dots
Reprinted from: *Polymers* 2023, 15, 679, https://doi.org/10.3390/polym15030679 108

Zirun Wang, Chao Jia, Hengxue Xiang and Meifang Zhu
Palladium Nanoparticle-Loaded Mesostructural Natural Woods for Efficient Water Treatment
Reprinted from: *Polymers* 2023, 15, 658, https://doi.org/10.3390/polym15030658 121

Yassine EL-Ghoul and Salman Alsamani
Highly Efficient Biosorption of Cationic Dyes via Biopolymeric Adsorbent-Material-Based Pectin Extract Polysaccharide and Carrageenan Grafted to Cellulosic Nonwoven Textile
Reprinted from: *Polymers* 2024, 16, 585, https://doi.org/10.3390/polym16050585 132

Rachid El Kaim Billah, Ikrame Ayouch, Youness Abdellaoui, Zineb Kassab, Moonis Ali Khan and Mahfoud Agunaou et al.
A Novel Chitosan/Nano-Hydroxyapatite Composite for the Adsorptive Removal of Cd(II) from Aqueous Solution
Reprinted from: *Polymers* 2023, 15, 1524, https://doi.org/10.3390/polym15061524 151

Maria Victoria X. Klaus, Angela M. Gutierrez and J. Zach Hilt
Development of Poly(acrylamide)-Based Hydrogel Composites with Powdered Activated Carbon for Controlled Sorption of PFOA and PFOS in Aqueous Systems
Reprinted from: *Polymers* **2023**, *15*, 4384, https://doi.org/10.3390/polym15224384 . . . . **168**

Francisco de Borja Ojembarrena, Jose Luis Sánchez-Salvador, Sergio Mateo, Ana Balea, Angeles Blanco and Noemí Merayo et al.
Modeling of Hexavalent Chromium Removal with Hydrophobically Modified Cellulose Nanofibers
Reprinted from: *Polymers* **2022**, *14*, 3425, https://doi.org/10.3390/polym14163425 . . . . **183**

Cuizhu Ye, Ziyan Pan and Yi Shen
Facile Conversion of Polystyrene Waste into an Efficient Sorbent for Water Purification
Reprinted from: *Polymers* **2022**, *14*, 4477, https://doi.org/10.3390/polym14214477 . . . . **211**

Preface

Water is indispensable to the functioning of most known life forms, and good-quality water is essential to human health, social and economic development, and ecosystem functioning. Nonetheless, population growth has led to the degradation and depletion of freshwater resources around the world. Because of these circumstances, ensuring sufficient and safe water supplies for everyone was one of the Sustainable Development Goals (SDGs) set by the United Nations General Assembly in 2015 for the year 2030. For this goal to be achieved, we must ensure the development and implementation of appropriate and efficient wastewater treatments that make it possible to reduce water pollution, which is a major challenge.

The application of polymers and polymeric materials in wastewater treatment is a research field that has developed considerably in recent years. Conventional and novel approaches have been used by researchers from different areas, who have demonstrated that polymers and polymeric materials may have an important role in the removal of pollutants of different origins and natures from wastewater, in the disposal of sludge, in the recycling of materials, in the improved efficiency and economy of wastewater, etc.

In view of the relevant contributions that polymers and polymeric materials may make in the conservation of the aquatic environment—namely, through their use in wastewater treatment—a Special Issue (SI) entitled "Polymeric Materials for Wastewater Treatment Applications" aimed to publish original research and review papers within this field. Scientific contributions related to any aspect of the utilization of polymers and polymeric materials—whether synthetic or natural—in the treatment or purification of wastewater were welcomed. Given the large spectrum of materials, types of pollution, and treatment processes, this was a widely inclusive SI, which succeeded in gathering very relevant scientific contributions. Dr. Marta Otero and Dr. Ricardo N. Coimbra, as Guest Editors of this SI, acknowledge all contributors and peer-reviewers that allowed for such a success and are happy to share a compilation of the published papers in this book.

Marta Otero and Ricardo N. Coimbra
Guest Editors

Editorial

Polymeric Materials for Wastewater Treatment Applications

Marta Otero [1,*] and Ricardo N. Coimbra [2,*]

1. Departamento de Química y Física Aplicadas, Universidad de León, Campus de Vegazana s/n, 24071 León, Spain
2. Department of Environment and Planning, University of Aveiro, Santiago Campus, 3810-193 Aveiro, Portugal
* Correspondence: marta.otero@unileon.es (M.O.); ricardo.coimbra@ua.pt (R.N.C.)

1. Introduction

Water of adequate quality is crucial for the survival of most life forms, playing a key role in human health, social and economic progress, and the functioning of ecosystems [1–4]. Still, only 2.5% of the Earth's water is freshwater, and less than 1% of that is easily accessible [5,6]. This availability is further threatened by climate change and human activity [7–9]. Additionally, the growth of populations and industries continues to drive the ongoing deterioration of freshwater quality worldwide, putting its sustainability at risk [10–12]. Given these challenges, ensuring safe and sufficient water supplies for all is one of the Sustainable Development Goals (SDGs) set by the United Nations General Assembly in 2015 with a target for 2030 [13–15]. Achieving this goal requires the development and implementation of effective wastewater treatment solutions [16–18].

Urban wastewater is a major contributor to water pollution if not properly collected and treated [19]. It is often contaminated with bacteria, viruses, harmful chemicals—including micropollutants—and excessive nutrients, which, if left untreated and released into the environment, pose risks to human health and harm rivers, lakes, and coastal waters [20–23]. Taking the example of the European Union (EU), the adoption of the Urban Wastewater Treatment Directive in 1991 (Directive 91/271/EEC) allowed for the improvement of the quality of European rivers, lakes, and seas [24–27]. Despite its success, more than 30 years later, an update built on those achievements was necessary [28–30]. An updated Urban Wastewater Treatment Directive (UWWTD) [31], which entered into force on 1 January 2025, addresses ongoing pollution issues, as well as new challenges in managing urban wastewater [32]. This UWWTD [31] aligns with European environmental strategies, including the European Green Deal [33], the Zero Pollution Action Plan [34], and the Circular Economy Plan [35]. Among the new regulations, the UWWTD [31] mandates the following: (i) collection and treatment of wastewater in all urban areas of more than 1000 inhabitants; (ii) more exigent limits on the removal of nutrients with appropriate tertiary treatment; (iii) removal of micropollutants with quaternary treatment, financed through extended producer responsibility by the sectors responsible for the pollution; (iv) ensuring that treatment plants are energy-neutral and reducing their greenhouse gas emissions by 2045. Similar regulations aiming to protect the environment and human health have already been or will soon be in force across different world regions.

Compliance with increasingly demanding discharge limits makes necessary the implementation of advanced water treatments and the development of efficient processes and materials [36–38]. In this sense, polymers offer remarkable properties and capacities so their applications in water treatment are a growing research topic [39–42]. Researchers from various fields have explored and/or developed traditional and innovative processes,

Received: 31 January 2025
Accepted: 14 February 2025
Published: 19 February 2025

Citation: Otero, M.; Coimbra, R.N. Polymeric Materials for Wastewater Treatment Applications. *Polymers* **2025**, *17*, 552. https://doi.org/10.3390/polym17040552

Copyright: © 2025 by the authors. Licensee MDPI, Basel, Switzerland. This article is an open access article distributed under the terms and conditions of the Creative Commons Attribution (CC BY) license (https://creativecommons.org/licenses/by/4.0/).

highlighting the significant potential of polymeric materials in removing undesired substances, organisms, and pollutants of diverse origins from wastewater, managing sludge disposal, recycling materials, and enhancing the efficiency and cost-effectiveness of water treatment processes.

In the above-referred context, the published Special Issue (SI) and the present editorial aim to highlight the relevant role that polymers and polymeric materials may play in the conservation of the aquatic environment, namely, by their application in water treatment. Nineteen manuscripts were submitted to this Special Issue and twelve were finally published. The following section is an overview of the latter, summarizing their main contributions and novelties.

2. Overview of Published Articles

Among the published manuscripts in the SI, there were ten research works and two review papers. The latter included a review on the removal of microbial contamination [43] and another on a support layer for solar-driven interfacial steam generation (SISG) application to water treatment [44]:

The work by Akinsemolu and Onyeaka [43] reviewed the published literature on the application of hydrogel polymers, which are networks of insoluble hydrophilic polymers (either natural or synthetic) that can hold large water volumes without dissolving, for microbial control in water treatment systems. The synthesis, structure, and properties of hydrogel polymers received special attention together with the mechanisms for the removal of pathogens from wastewater. It was evidenced that apart from antimicrobial capacity, biodegradability, living cell compatibility, sensitivity to environmental stimuli, and mechanical strength are important properties for the increased overall performance of hydrogels in wastewater treatment.

Yan et al. [44] presented a literature review on the existing progress in the field of multi-layer interface evaporators based on various polymers and biomaterials, along with their advantages and disadvantages for SISG water treatment processes, which has gained attention due to its low energy consumption, simple operation, and eco-friendliness. The typical multi-layer SISG evaporator consists of a photothermal layer and a support layer, with the latter playing a key role in thermal management, stability, and water transport to the evaporation interface. Although less studied than the photothermal layer, the significance of the support layer is highlighted in this review, which summarizes advancements in materials like polymers (foams and gels) and biomaterials (natural plants and carbonized materials) and discusses the structural design strategies for the support layer, emphasizing its role in enhancing the efficiency of the SISG. It is pointed out that unmodified polymers and biomaterials are simple to prepare but lack durability in harsh conditions, while modified support layers show improved performance, though scaling them for large-scale applications remains challenging. The review concludes with potential future research directions and applications of support layer materials to address global water challenges.

Regarding the research papers, two studied desalination water treatments, either by membrane filtration [45] or by evaporation on natural and synthetic fabrics [46]:

Abdelrazeq and Khraisheh [45] investigated the effect of polystyrene membrane porosity on desalination performance by direct contact membrane distillation (MD), which is a thermal-based process with high potential for water treatment. For such a purpose, the relationship between permeate flux and feed temperature in MD using polystyrene membranes with varying porosities (77%, 89%, and 94%) was assessed under pilot-scale operation. Increasing membrane porosity by 15% was found to result in a 14.6% rise in thermal efficiency and a 5% increase in evaporation efficiency. Indeed, efficiency in the rejection of dissolved solutes was significantly reduced for porosities below 89%. The work included mathematical validation and computational predictions, providing insights into the impact of membrane porosity on thermal and evaporation efficiencies in DCMD.

López-Borrell et al. [46] examined evaporation behavior from saturated NaCl and $CuSO_4 \cdot 5H_2O$ solutions using natural fabrics (jute (Jut), bamboo (Bam), linen composed of 50% polylactic acid (LPLA), and a non-woven fabric composed of 70% palm prunings, 20% lyocell, and 10% PLA (WL-T)) and synthetic fabrics (non-woven polyester (PES) and a fabric with an aramid taffeta structure (Ara)) in view of brackish water treatment under zero liquid discharge (ZLD) conditions. For NaCl, the evaporation rate increased with the number of cycles, as salt deposits on the fabrics enhanced evaporation, with polymeric fabrics, namely PES and LPLA, being the most promising for long-term use. In contrast, $CuSO_4 \cdot 5H_2O$ reduced fabric absorption capacity over cycles due to salt precipitation, which blocked the fabric structure, decreased interaction with the solution, and ultimately slowed evaporation. These findings serve as support for the selection of fabrics for ZLD wastewater treatment, offering potential benefits for industrial-scale applications.

Advanced oxidation processes (AOPs) for wastewater treatment and, specifically, the development, characterization, and application of novel oxidants or catalysts, were the focus of three of the research papers published in the SI [47–49]:

In the study by Li et al. [47], polyethylene glycol (PEG)-coated calcium peroxide nanoparticles (nCPs) were synthesized and used as oxidants to generate hydroxyl free radicals (•OH) in the presence of Fe^{2+} for the removal of glyphosate from water in a Fenton-based advanced oxidation process. A co-precipitation method using calcium chloride ($CaCl_2$) as a precursor and polyethylene glycol 200 (PEG 200) as a surface stabilizer was used to obtain nCPs (40.88 nm) with high surface area (28.09 m^2/g), which were characterized using various techniques like Fourier-transform infrared spectroscopy (FTIR), X-ray diffraction (XRD), Brunauer–Emmett–Teller (BET) surface area, dynamic light scattering (DLS), and field emission (FE) scanning electron microscopy (SEM). Under optimized conditions (pH 3.0, 0.2 g nCPs, Ca^{2+}/Fe^{2+} molar ratio of 6, initial glyphosate concentration of 50 mg/L), 99.6% glyphosate and 75.1% chemical oxygen demand removal were achieved within just 75 min. The degradation of glyphosate was fast and followed the Behnajady–Modirshahla–Ghanbery (BMG) kinetic model. The results demonstrated that nCPs are a much more effective oxidant than commercial calcium peroxide (CaO_2 (CP)) for removing glyphosate from water and promising for application in advanced wastewater treatment.

Kraft washing effluents (KWEs), whose major constituent is lignin biopolymer, usually contain high concentrations of sulfide ions (S(II)) since the kraft process uses sodium sulfide (Na_2S). Luo et al. [48] synthesized N and Fe co-doped carbon dots (N,Fe-CDs) from citric acid, L-glutamic acid, and ferric chloride via a hydrothermal method to photocatalytically remove S^{2-} from KWE. The photocatalytic degradation of S^{2-} followed first-order kinetics, with an activation energy of 33.77 kJ/mol. The N,Fe-CDs (fluorescent nanoparticles with an average 3.8 nm diameter) were stable at temperatures up to 80 °C and could be reused for at least four cycles, retaining over 90% of their catalytic activity. In the treatment of KWE, the N,Fe-CDs reduced S^{2-} concentration from 1.19 to 0.41 mmol/L in 6 h, achieving near-complete remediation in 24 h. The photocatalyst also removed half of the chemical oxygen demand and was found to be safe at concentrations up to 200 mg/L, supporting its potential for cleaner kraft pulping processes.

Natural wood is primarily made up of cellulose, hemicellulose, and lignin, which together constitute over 90% of its total weight. Five types of natural wood of two types, namely hardwood (bass, beech, and balsa) and softwood (pine and fir) were impregnated by Wang et al. [49] with palladium nanoparticles to obtain different palladium nanoparticle-loaded natural wood membranes (PNNW membranes). These PNNW membranes were tested for their catalytic performance in the removal of dyes (methylene blue (MB), methyl orange (MO), and 4-nitrophenol (4-N)) from their single and mixed aqueous solutions. The softwoods, which have small-diameter channels and therefore water flux through them is relatively slow, exhibited higher degradation efficiency for single 4-NP compared to hardwood due to the longer contact time between the dye and catalyst. Both the hardwood and softwood showed high degradation efficiency for single MB and MO, but mixed pollutants interfered with each other, reducing catalytic performance. All the wood membranes demonstrated good degradation properties for different concentrations of MB, though high concentrations of MO and 4-NP resulted in lower efficiency.

Finally, adsorption was the most approached wastewater treatment in this SI, with five research papers [50–54] on the synthesis, characterization, and utilization of novel adsorbent materials for the removal of pollutants from wastewater:

EL-Ghoul and Alsamani [50] presented the design of a new adsorbent material made from a cellulosic non-woven textile grafted with two extracted biopolymers using a layer-by-layer grafting technique. The study detailed the extraction of Suaeda fruticosa polysaccharide (SFP), which was confirmed to have a pectin-like structure, followed by grafting SFP with carrageenan crosslinked using 1,2,3,4-butanetetracarboxylic acid. Characterization through FTIR and SEM revealed successful chemical grafting, the total filling of material micro-spaces with layers of grafted biopolymers, and a rough surface morphology of the synthesized adsorbent. The cationic dye methylene blue (MB) was selected as the target pollutant to test the synthesized adsorbent, which provided an adsorptive removal as high as 803 mg/g in batch experiments under agitation. The adsorption of MB followed pseudo-first-order kinetics and Langmuir/Temkin adsorption isotherms, indicating that the developed biosorbent has great potential for low-cost, efficient wastewater treatment.

El Kaim Billah et al. [51] produced a novel polymer bio-composite, CS/n-Hap, to remove toxic cadmium ions (Cd^{2+}) from water, combining chitosan (CS) and nano-hydroxyapatite (n-Hap). The structure and composition of CS/n-Hap were confirmed through XRD and FT-IR analysis, which showed that both CS and n-Hap retained their characteristic features. Batch adsorption experiments under agitation revealed that CS/n-Hap outperformed pristine CS in removing Cd^{2+}, with a maximum uptake of 127 mg/g under optimized conditions. The adsorption of Cd^{2+} onto CS/n-Hap was endothermic and spontaneous and followed pseudo-second-order kinetics and Freundlich isotherm equilibrium, suggesting the formation of chemical bonds between Cd^{2+} and CS/n-Hap and multi-layer adsorption. Electrostatic interactions combined with chelation, ion exchange, and surface complexation were underneath the Cd^{2+} adsorption. Additionally, CS/n-Hap was successfully regenerated and reused with only a 3% reduction in Cd^{2+} removal after five cycles, highlighting its potential for eco-friendly and cost-effective water treatment.

Klaus et al. [52] prepared a series of acrylamide-based hydrogels and then hydrogel composites with powdered activated carbon (PAC), characterized the obtained materials, and assessed their ability to adsorb per- and polyfluoroalkyl substances (PFASs). Physicochemical characterization using FTIR, thermogravimetric analysis (TGA), and swelling studies confirmed successful PAC incorporation in the composites and revealed how crosslinking density influenced the swelling ratio. Also, surface analysis showed carbon-rich areas in the composite. The removal of PFAS at relevant concentrations was analyzed by liquid chromatography–mass spectrometry (LC-MS/MS), with the hydrogel composite achieving up to 98% removal for PFOS and 96% for PFOA. The adsorption capacity of hydrogel composites was larger than that of hydrogels but lower than that of PAC. However, contrary to PAC, hydrogel composites showed high selectivity and tunable properties to optimize their performance, pointing to their significant potential as advanced materials for selective PFAS removal from water.

The study by Ojembarrena et al. [53] proved that the adsorption performance of cellulose nanofibers (CNF), which are already known for their high specific surface area and active groups, may be enhanced by hydrophobization. In this work, CNFs obtained by TEMPO-mediated oxidation followed by mechanical disintegration was hydrophobized with methyl trimethoxysilane (MTMS) and tested for the adsorptive removal of hexavalent chromium (Cr^{6+}) from wastewater. Optimization was performed by adjusting the MTMS dosage for hydrophobization and operational parameters, namely contact time, pH, the initial concentration of Cr^{6+}, and adsorbent dosage. Hydrophobized CNFs with 1.5 mmol MTMS/g at dosages above 500 mg/L achieved nearly complete Cr^{6+} removal (<97%), with the best performance at pH 3. The batch adsorption results obtained under stirring fitted pseudo-second-order kinetics at low concentrations and intraparticle diffusion at higher concentrations and Freundlich equilibrium.

Ye et al. [54] explored the conversion of polystyrene (PS) plastic waste into sorbent foam via a simple sulfonation process, producing sulfonated polystyrene (SPS) foams with a porous, hydrophilic surface and abundant sulfonic groups for effective pollutant binding. SPS was tested for the adsorption of Pb^{2+}, lysozyme, and MB in batch, filtration, and cyclic experiments, with Pb^{2+} and lysozyme fitting the Langmuir model and maximum adsorption capacities of 10.5 mg/g and 15.7 mg/g, respectively, while MB adsorption followed the Freundlich isotherm. Despite lower adsorption capacities compared to PS-based sorbents in the literature, SPS's bulky self-standing morphology allowed easy post-use separation. SPS was easily regenerated by acid washing and showed excellent cyclic performance. Due to its abundant feedstock, simple preparation, and effective regeneration, SPS foams were highlighted as promising sorbents for wastewater treatment.

3. Challenges and Future Perspectives

The published papers in this SI provide a thorough understanding of the application of polymeric materials in wastewater treatment. A wide range of important topics was covered, and the latest research findings on different types of treatments, from desalination membranes to hydrogels, catalysts, absorbers, and adsorbents for the removal of different types of contaminants were assessed. These assessments were derived either from the comprehensive literature reviews or from practical materials development and/or engineering applications, reflecting concern about the decontamination of water.

Apart from the relevant information, findings, technical basis, or practical guidance provided in this SI, challenges and necessities for future research were raised.

Akinsemolu and Onyeaka [43] emphasized in their review the need for further research to optimize hydrogel polymers for pathogen removal from wastewater, focusing on assessing their efficiency against specific microorganisms and comparing the performance and cost of reused versus new polymers. Overall, refining hydrogel formulations to improve selectivity, long-term stability, and scalability, thereby contributing to cleaner water and more sustainable environmental management, are key aims for future research on hydrogels [55]. For its part, the review by Yan et al. [44] highlighted the challenges in designing support layers for photothermal evaporators, stressing the need for broader material options, enhanced performance, and commercial viability to improve SISG technology for sustainable water purification. For this purpose, and in view of synergistic applications in water purification, exploring hybrid photothermal materials should be considered [56].

The findings by Abdelrazeq and Khraisheh [45] constitute a relevant basis for the optimization of MD processes for wastewater treatment, but further studies are needed to increase energy efficiency for large-scale applications. Despite the advantages of polymeric membranes, namely enhanced processability, low cost, polymer variability, and availability, challenges remain for MD's large-scale application, including high energy consumption, flux decay related to fouling, and economic feasibility [57]. The study by López-Borrel et al. [46] proved the applicability of polymeric fabrics for brackish ZLD water treatment. In line with recent advances [58], more salts apart from the studied ones (NaCl and $CuSO_4 \cdot 5H_2O$) need to be tested, operational conditions should be optimized, and scaling assessed to support practical application in view of ZLD.

Li et al. [47] developed an efficient oxidant for advanced wastewater treatment via the Fenton process, achieving nearly complete glyphosate removal with 75% mineralization. Besides the analysis of transformation products, further testing on other organic micropollutants and real matrices is needed for practical implementation. Luo et al. [48] assessed the applicability of their N,Fe-CDs for the removal of S^{2-} from real matrices (KFE) at the laboratory scale, with upscale studies needed for real applications. The authors highlighted that more research on N,Fe-CDs could help in the catalytic removal of odorous sulfur-containing organic compounds produced during kraft pulping (e.g., CH_3SCH_3 or CH_3SSCH_3). On the other hand, the results obtained by Wang et al. [49] on their PNNW membranes highlighted the need for tailored wood-based membranes depending on the pollutants in water, with the efficient catalytic degradation of mixed pollutants from real complex matrices being a main challenge. In general, AOP applications in water treatment face several challenges, such as the short lifespan of free radicals, high reagent consumption, and secondary pollution risks [57]. On the other hand, the efficiency of AOP is highly affected by operational parameters and matrix effects, which need to be assessed for an optimized performance [59].

A simple and novel low-cost polymeric biosorbent material was designed by EL-Ghoul and Alsamani [50] and tested for MB cationic dye adsorptions. An efficient bio-composite was developed by El Kaim Billah et al. [51], which was shown to be efficient for the adsorptive removal of Cd^{2+}. The acrylamide-based hydrogel composites with PAC developed by Klaus et al. [52] are very efficient in the removal of environmentally relevant concentrations of PFOs and PFAS. Hydrophobized CNFs hydrogel produced by Ojembarrena et al. [53] provided better adsorption than CNFs and cellulose nanomaterials for the removal of Cr^{6+}. Along with that, Ye et al. [54] obtained interesting results on the utilization of SPS, whose production is in line with the principles of circular economy, for the adsorption of Pb^{2+}, lysozyme, and MB. Future challenges to expand the application of polymeric advanced adsorbent materials include the following [41,60,61]: (i) the assessment of their potential in the removal of different toxic metals, dyes, pesticides, pharmaceuticals, etc.; (ii) exploring their selectivity and competitive effects in complex real matrices; (iii) carrying out pilot and full-scale experiments to support their practical applicability; (iv) developing protocols for their synthesis and performance evaluation; (v) assessing reusability; (vi) developing multifunctional polymeric materials; and (vii) integrating adsorption processes with advanced remediation technologies. On the basis of research and innovations devoted to upgrading performance, selectivity, and sustainability, polymeric adsorbents are set to play a key role in sustainable water treatment [41].

Author Contributions: Conceptualization: M.O. and R.N.C.; writing—original draft preparation: M.O. and R.N.C.; writing—review and editing: M.O. and R.N.C. All authors have read and agreed to the published version of the manuscript.

Acknowledgments: The authors would like to acknowledge the novel and relevant contributions to the Special Issue (SI) entitled "Polymeric Materials for Wastewater Treatment Applications", which has been published in *Polymers*. The authors of these contributions are thanked for considering this SI for the publication of their outstanding works. Moreover, we would like to recognize the important role of the peer-reviewers in supporting the robustness of the scientific process. The reviewers who kindly sent their relevant comments, corrections, and suggestions on the published and rejected manuscripts in this SI are acknowledged with thanks.

Conflicts of Interest: The authors declare no conflicts of interest.

References

1. Vardon, M.J.; Le, T.H.L.; Martinez-Lagunes, R.; Pule, O.B.; Schenau, S.; May, S.; Grafton, R.Q. Accounting for Water: A Global Review and Indicators of Best Practice for Improved Water Governance. *Ecol. Econ.* **2025**, *227*, 1–10. [CrossRef]
2. Yan, T.; Shen, S.L.; Zhou, A. Indices and Models of Surface Water Quality Assessment: Review and Perspectives. *Environ. Pollut.* **2022**, *308*, 1–15. [CrossRef] [PubMed]
3. Ngatia, M.; Kithiia, S.M.; Voda, M.; Ssembajwe, R. Spatial and Temporal Variations in Surface Water Quality: A Continental Review. *Geogr. Tech.* **2024**, *19*, 185. [CrossRef]
4. Elmadani, M.; Kasmai Kiptulon, E.; Klára, S.; Orsolya, M. Systematic Review of the Impact of Natural Resource Management on Public Health Outcomes: Focus on Water Quality. *Resources* **2024**, *13*, 122. [CrossRef]
5. Suresh Isravel, R.; Karunamurthy, K. A Review of Solar Thermal Desalination Using Nanotechnology. In *Materials Today: Proceedings*; Elsevier, Ltd.: Amsterdam, The Netherlands, 2023; Volume 84, pp. 74–82.
6. Mishra, R.K.; Dubey, S.C. Fresh Water Availability and It's Global Challenge. *J. Mar. Sci. Res.* **2023**, *2*. [CrossRef]
7. Islam, M.S.; Mostafa, M.G. Impacts of Climate Change on Global Freshwater Quality and Availability: A Comprehensive Review. *J. Water Env. Technol.* **2024**, *22*, 1–26. [CrossRef]
8. Kulionis, V.; Pfister, S. A Planetary Boundary-Based Method to Assess Freshwater Use at the Global and Local Scales. *Environ. Res. Lett.* **2022**, *17*, 094021. [CrossRef]
9. Khondoker, M.; Mandal, S.; Gurav, R.; Hwang, S. Freshwater Shortage, Salinity Increase, and Global Food Production: A Need for Sustainable Irrigation Water Desalination—A Scoping Review. *Earth* **2023**, *4*, 223–240. [CrossRef]
10. Hekmatnia, M.; Isanezhad, A.; Ardakani, A.F.; Ghojghar, M.A.; Ghaleno, N.D. An Attempt to Develop a Policy Framework for the Global Sustainability of Freshwater Resources in the Virtual Water Trade. *Sustain. Prod. Consum.* **2023**, *39*, 311–325. [CrossRef]
11. Syeed, M.M.M.; Hossain, M.S.; Karim, M.R.; Uddin, M.F.; Hasan, M.; Khan, R.H. Surface Water Quality Profiling Using the Water Quality Index, Pollution Index and Statistical Methods: A Critical Review. *Environ. Sustain. Indic.* **2023**, *18*, 100145. [CrossRef]
12. Rahman, A.; Kumar, P.; Dominguez, F. Increasing Freshwater Supply to Sustainably Address Global Water Security at Scale. *Sci. Rep.* **2022**, *12*, 24314. [CrossRef] [PubMed]
13. Asante-Darko, D.; Dadzie, S.A.; Kwarteng, A.; Agbodjah, S.; Aryee, T.E. Effects of Circular Economy Practices and Accounting Innovations on Sustainable Development Goals. *Circ. Econ. Sustain.* **2024**, *4*, 103961. [CrossRef]
14. Garcia, C.; López-Jiménez, P.A.; Pérez-Sánchez, M.; Sanchis, R. Methodology for Assessing Progress in Sustainable Development Goals Indicators in Urban Water Systems. How Far Are We from the 2030 Targets? *Sustain. Cities Soc.* **2024**, *112*, 105616. [CrossRef]
15. Ojeda-Matos, G.; Jones-Crank, J.L. The Potential of the Water-Energy-Food Nexus Approach in Advancing the Sustainable Development Goals: A PRISMA-Based Systematic Review. *Env. Sci. Policy* **2025**, *163*, 103961. [CrossRef]
16. Jones, E.R.; Bierkens, M.F.P.; Wanders, N.; Sutanudjaja, E.H.; van Beek, L.P.H.; van Vliet, M.T.H. Current Wastewater Treatment Targets Are Insufficient to Protect Surface Water Quality. *Commun. Earth Env.* **2022**, *3*, 554. [CrossRef]
17. Saadatinavaz, F.; Alomari, M.A.; Ali, M.; Saikaly, P.E. Striking a Balance: Decentralized and Centralized Wastewater Treatment Systems for Advancing Sustainable Development Goal 6. *Adv. Energy Sustain. Res.* **2024**, *2024*, 100097. [CrossRef]
18. Sanei, E.; Gómez-Gallegos, M.A.; Márquez, I. Chapter 4: Advances in Wastewater Treatment Technologies as Enablers to Reach Sustainable Development Goal 6. In *Contributions of Chemical Engineering to Sustainability*; Segovia-Hernández, J.G., Ramírez-Corona, N., Aristizábal-Marulanda, V., Eds.; Springer: Gewerbestrasse, Switzerland, 2024; pp. 101–124.
19. Villez, K.; Aguado, D.; Alferes, J.; Plana, Q.; Ruano, M.V.; Samuelsson, O. *Metadata Collection and Organization in Wastewater Treatment and Wastewater Resource Recovery Systems*; IWA Publishing: London, UK, 2024; ISBN 9781789061154.
20. Carneiro, R.B.; Nika, M.C.; Gil-Solsona, R.; Diamanti, K.S.; Thomaidis, N.S.; Corominas, L.; Gago-Ferrero, P. A Critical Review of Wastewater-Based Epidemiology as a Tool to Evaluate the Unintentional Human Exposure to Potentially Harmful Chemicals. *Anal. Bioanal. Chem.* **2024**, *2024*, 100. [CrossRef]
21. Elehinafe, F.B.; Agboola, O.; Vershima, A.D.; Bamigboye, G.O. Insights on the Advanced Separation Processes in Water Pollution Analyses and Wastewater Treatment—A Review. *S. Afr. J. Chem. Eng.* **2022**, *42*, 188–200. [CrossRef]
22. Mohammadi, S.A.; Najafi, H.; Zolgharnian, S.; Sharifian, S.; Asasian-Kolur, N. Biological Oxidation Methods for the Removal of Organic and Inorganic Contaminants from Wastewater: A Comprehensive Review. *Sci. Total Environ.* **2022**, *843*, 161053. [CrossRef] [PubMed]
23. Alshami, A.; Ali, E.; Elsayed, M.; Eltoukhy, A.E.E.; Zayed, T. IoT Innovations in Sustainable Water and Wastewater Management and Water Quality Monitoring: A Comprehensive Review of Advancements, Implications, and Future Directions. *IEEE Access* **2024**, *12*, 58427–58453. [CrossRef]

24. Källqvist, T.; Molver, J.; Oug, E.; Berge, D.; Tjomsland, T.; Johansen, S.S. *Implementation of the Urban Waste Water Treatment Directive in Norway—An Evaluation of the Norwegian Approach Regarding Wastewater Treatment*; Norwegian Institute for Water Research (NIVA): Oslo, Norway, 2002.
25. Pistocchi, A.; Dorati, C.; Grizzetti, B.; Udias, A.; Vigiak, O.; Zanni, M. Water quality in Europe: Effects of the Urban Wastewater Treatment Directive. A retrospective and scenario analysis of Dir. 91/271/EEC, EUR 30003 EN, Publications Office of the European Union, Luxembourg. 2019. Available online: https://publications.jrc.ec.europa.eu/repository/handle/JRC115607 (accessed on 13 February 2025).
26. Preisner, M.; Smol, M.; Szołdrowska, D. Trends, Insights and Effects of the Urban Wastewater Treatment Directive (91/271/EEC) Implementation in the Light of the Polish Coastal Zone Eutrophication. *Env. Manag.* **2021**, *67*, 342–354. [CrossRef] [PubMed]
27. Smith, A. Fitting in with Brussels: Implementing the Urban Waste Water Treatment Directive in England and Wales. *J. Environ. Policy Plann.* **2000**, *2*, 115–134. [CrossRef]
28. Prochaska, C.; Zouboulis, A. A Mini-Review of Urban Wastewater Treatment in Greece: History, Development and Future Challenges. *Sustainability* **2020**, *12*, 3206. [CrossRef]
29. Badhofer, A. Quaternary Treatment According to the Proposal for a Recast of the EU Urban Wastewater Treatment Directive and Its Monetary Implications for Candidate Countries (e.g., Serbia). Master's Thesis, Technische Universität Wien, Vienna, Austria, 2023.
30. Commission Staff. *Working Document Impact Assessment Accompanying the Document "Proposal for a Directive of the European Parliament and of the Council Concerning Urban Wastewater Treatment (Recast)"*; European Commission: Brussels, Belgium, 2022.
31. The European Parliament and the Council of the European Union. *Directive (EU) 2024/3019 of the European Parliament and of the Council of 27 November 2024 Concerning Urban Wastewater Treatment (Recast) (Text with EEA Relevance)*; European Commission: Brussels, Belgium, 2024.
32. Kardos, M.K.; Patzier, M.; Jolánkai, Z.; Clement, A. The New Urban Wastewater Treatment Directive from the Perspective of the Receiving Rivers' Quality. *Env. Sci. Eur.* **2025**, *37*, 10. [CrossRef]
33. European Commission. *The European Green Deal*; European Commission: Brussels, Belgium, 2019; pp. 1–25.
34. European Commission. *Pathway to a Healthy Planet for All Action Plan: "Towards Zero Pollution for Air, Water and Soil"*; European Commission: Brussels, Belgium, 2021.
35. European Commission. *A New Circular Economy Action Plan: For a Cleaner and More Competitive Europe*; European Commission: Brussels, Belgium, 2020.
36. Shamshad, J.; Ur Rehman, R. Innovative Approaches to Sustainable Wastewater Treatment: A Comprehensive Exploration of Conventional and Emerging Technologies. *Environ. Sci. Adv.* **2024**, *5*, 112. [CrossRef]
37. Gedda, G.; Balakrishn, K.; Devi, R.U.; Shah, K.J. Introduction to Conventional Wastewater Treatment Technologies: Limitations and Recent Advances. In *Advances in Wastewater Treatment I. Materials Research Foundations*; Materials Research Forum, LLC: Millersville, PA, USA, 2021; Volume 91, pp. 1–36.
38. Bairagi, S.; Wazed Ali, S. Conventional and Advanced Technologies for Wastewater Treatment. In *Environmental Nanotechnology for Water Purification*; Ul Islam, S., Ed.; Willey & Sons, Ltd.: Hoboken, NJ, USA, 2020; pp. 33–56. ISBN 978-1-119-64045-5.
39. Ahmaruzzaman, M.; Roy, P.; Bonilla-Petriciolet, A.; Badawi, M.; Ganachari, S.V.; Shetti, N.P.; Aminabhavi, T.M. Polymeric Hydrogels-Based Materials for Wastewater Treatment. *Chemosphere* **2023**, *331*, 138766. [CrossRef]
40. Saravanan, A.; Thamarai, P.; Kumar, P.S.; Rangasamy, G. Recent Advances in Polymer Composite, Extraction, and Their Application for Wastewater Treatment: A Review. *Chemosphere* **2022**, *308*, 136368. [CrossRef]
41. Alkhaldi, H.; Alharthi, S.; Alharthi, S.; AlGhamdi, H.A.; AlZahrani, Y.M.; Mahmoud, S.A.; Amin, L.G.; Al-Shaalan, N.H.; Boraie, W.E.; Attia, M.S.; et al. Sustainable Polymeric Adsorbents for Adsorption-Based Water Remediation and Pathogen Deactivation: A Review. *RSC Adv.* **2024**, *14*, 33143–33190. [CrossRef]
42. Lakkimsetty, N.R.; Feroz, S.; Karunya, S.; Motilal, L.; Saidireddy, P.; Suman, G. Synthesis, Characterization and Application of Polymer Composite Materials in Wastewater Treatment. In *Materials Today: Proceedings*; Elsevier, Ltd.: Amsterdam, The Netherlands, 2022; Volume 59, pp. 1726–1734.
43. Akinsemolu, A.A.; Onyeaka, H. Advances in Hydrogel Polymers for Microbial Control in Water Systems. *Polymers* **2024**, *16*, 314. [CrossRef]
44. Yan, H.; Wang, P.; Li, L.; Zhao, Z.; Xiang, Y.; Guo, H.; Yang, B.; Yang, X.; Li, K.; Li, Y.; et al. Development Status of Solar-Driven Interfacial Steam Generation Support Layer Based on Polymers and Biomaterials: A Review. *Polymers* **2024**, *16*, 540. [CrossRef] [PubMed]
45. Abdelrazeq, H.; Khraisheh, M. Porosity Effect of Polystyrene Membranes on Desalination Performance: A Combined Experimental and Numerical Heat and Mass Transfer Study in Direct Contact Membrane Distillation. *Polymers* **2023**, *15*, 1821. [CrossRef] [PubMed]
46. López-Borrell, A.; Lora-García, J.; Cardona, S.C.; López-Pérez, M.F.; Fombuena, V. Vapor Pressure and Evaporation Studies of Saline Solutions on Natural and Synthetic Fabrics for Industrial Water Treatment. *Polymers* **2024**, *16*, 2335. [CrossRef] [PubMed]

47. Li, F.; Choong, T.S.Y.; Abdullah, L.C.; Siti Nurul, S.N.A.; Amerhaider Nuar, N.N. Effective Removal of Glyphosate from Aqueous Systems Using Synthesized PEG-Coated Calcium Peroxide Nanoparticles: Kinetics Study, H_2O_2 Release Performance and Degradation Pathways. *Polymers* **2023**, *15*, 775. [CrossRef]
48. Luo, H.; Liu, H.; Sun, C. Removal of Sulfide Ions from Kraft Washing Effluents by Photocatalysis with N and Fe Codoped Carbon Dots. *Polymers* **2023**, *15*, 679. [CrossRef]
49. Wang, Z.; Jia, C.; Xiang, H.; Zhu, M. Palladium Nanoparticle-Loaded Mesostructural Natural Woods for Efficient Water Treatment. *Polymers* **2023**, *15*, 658. [CrossRef]
50. EL-Ghoul, Y.; Alsamani, S. Highly Efficient Biosorption of Cationic Dyes via Biopolymeric Adsorbent-Material-Based Pectin Extract Polysaccharide and Carrageenan Grafted to Cellulosic Nonwoven Textile. *Polymers* **2024**, *16*, 585. [CrossRef]
51. El Kaim Billah, R.; Ayouch, I.; Abdellaoui, Y.; Kassab, Z.; Khan, M.A.; Agunaou, M.; Soufiane, A.; Otero, M.; Jeon, B.H. A Novel Chitosan/Nano-Hydroxyapatite Composite for the Adsorptive Removal of Cd(II) from Aqueous Solution. *Polymers* **2023**, *15*, 1524. [CrossRef]
52. Klaus, M.V.X.; Gutierrez, A.M.; Hilt, J.Z. Development of Poly(Acrylamide)-Based Hydrogel Composites with Powdered Activated Carbon for Controlled Sorption of PFOA and PFOS in Aqueous Systems. *Polymers* **2023**, *15*, 4384. [CrossRef]
53. Ojembarrena, F.d.B.; Sánchez-Salvador, J.L.; Mateo, S.; Balea, A.; Blanco, A.; Merayo, N.; Negro, C. Modeling of Hexavalent Chromium Removal with Hydrophobically Modified Cellulose Nanofibers. *Polymers* **2022**, *14*, 3425. [CrossRef]
54. Ye, C.; Pan, Z.; Shen, Y. Facile Conversion of Polystyrene Waste into an Efficient Sorbent for Water Purification. *Polymers* **2022**, *14*, 4477. [CrossRef]
55. Visan, A.I.; Negut, I. Environmental and Wastewater Treatment Applications of Stimulus-Responsive Hydrogels. *Gels* **2025**, *11*, 15. [CrossRef] [PubMed]
56. Tessema, A.A.; Wu, C.M.; Motora, K.G.; Lee, W.H.; Peng, Y.T. A Review on State of Art of Photothermal Nanomaterials for Interfacial Solar Water Evaporation and Their Applications. *Desalination* **2024**, *591*, 1–15. [CrossRef]
57. Kakkar, R.; Kaur, D.P.; Raj, S. Membrane Distillation for Sustainable Water Desalination: A Review of Principles, Materials, and Applications. *Water Air Soil. Pollut.* **2025**, *236*, 76. [CrossRef]
58. Scelfo, G.; Trezzi, A.; Vassallo, F.; Cipollina, A.; Landi, V.; Xenogianni, C.; Tamburini, A.; Xevgenos, D.; Micale, G. Demonstration of Ultra-High-Water Recovery and Brine Concentration in a Prototype Evaporation Unit: Towards Zero Liquid Discharge Desalination. *Sep. Purif. Technol.* **2025**, *354*, 129427. [CrossRef]
59. Khan, Q.; Sayed, M.; Khan, J.A.; Rehman, F.; Noreen, S.; Sohni, S.; Gul, I. Advanced Oxidation/Reduction Processes (AO/RPs) for Wastewater Treatment, Current Challenges, and Future Perspectives: A Review. *Environ. Sci. Pollut. Res. Int.* **2024**, *31*, 1863–1889. [CrossRef]
60. Mazumder, M.A.J. Polymeric Adsorbents: Innovative Materials for Water Treatments. *Curr. Anal. Chem.* **2023**, *19*, 105–110. [CrossRef]
61. Saleh, T.A. Protocols for Synthesis of Nanomaterials, Polymers, and Green Materials as Adsorbents for Water Treatment Technologies. *Environ. Technol. Innov.* **2021**, *24*, 101346. [CrossRef]

Disclaimer/Publisher's Note: The statements, opinions and data contained in all publications are solely those of the individual author(s) and contributor(s) and not of MDPI and/or the editor(s). MDPI and/or the editor(s) disclaim responsibility for any injury to people or property resulting from any ideas, methods, instructions or products referred to in the content.

Review

Advances in Hydrogel Polymers for Microbial Control in Water Systems

Adenike A. Akinsemolu [1,2,*] and Helen Onyeaka [2,*]

1 Department of Integrated Science, Adeyemi Federal University of Education, Ondo 351101, Ondo, Nigeria
2 School of Chemical Engineering, University of Birmingham, Birmingham B15 2TT, UK
* Correspondence: a.akinsemolu@bham.ac.uk (A.A.A.); h.onyeaka@bham.ac.uk (H.O.)

Abstract: Every year, contaminated water is responsible for over one million deaths globally. Microbiology leads other fields in the development of solutions to water contamination to reduce these deaths while advancing the achievement of SDG 6, which aims to ensure universal access to water and sanitation. This article explores hydrogel polymers as a solution to water contamination through microbial control. Using a systematic approach, this study collects, reviews, analyzes, and synthesizes the findings of studies on the structure, properties, and mechanisms used by hydrogel polymers in pathogen control in water systems, emphasizing recent advances in microbiology that have improved the antimicrobial properties of hydrogel polymers, enhanced their synthetic properties, and improved their overall ability to control the spread of pathogens in water. Other additional notable findings, including the applications of hydrogel polymers in water systems, the environmental implications of using the method to decontaminate and purify water for various purposes, and the regulatory standards needed to reinforce the viability and effectiveness of the adaptation of hydrogel polymers for the control of harmful or unwanted microorganisms in water systems, inform the presented inferences on the future of hydrogel technologies and new opportunities for the expansion of their commercial use.

Keywords: hydrogel polymers; microbial control; pathogens; microorganisms; antimicrobial

Citation: Akinsemolu, A.A.; Onyeaka, H. Advances in Hydrogel Polymers for Microbial Control in Water Systems. *Polymers* **2024**, *16*, 2205. https://doi.org/10.3390/polym16152205

Academic Editors: Marta Otero and Ricardo N. Coimbra

Received: 11 June 2024
Revised: 26 July 2024
Accepted: 31 July 2024
Published: 2 August 2024

Copyright: © 2024 by the authors. Licensee MDPI, Basel, Switzerland. This article is an open access article distributed under the terms and conditions of the Creative Commons Attribution (CC BY) license (https://creativecommons.org/licenses/by/4.0/).

1. Introduction

According to the World Health Organization (WHO), half of the global population lacks access to clean and safe drinking water [1]. Consequently, they use dirty and contaminated water, which is associated with a high incidence of preventable diseases. The WHO links contaminated water with at least 1.4 million preventable deaths every year, namely owing to diarrhea, cholera, and other diseases that are spread through contaminated water [1]. In 2021, diarrhea alone caused 829,000 deaths globally, including 300,000 deaths of children below the age of five [2]. These deaths could have been prevented if the affected families and communities had access to clean water for drinking, sanitation, and hand hygiene. The main causes of water contamination include agricultural chemicals, chemicals from household and industrial waste, sewage, fuels, pharmaceutical waste, and microorganisms [3]. Harmful microorganisms enter water sources through sewage containing human fecal matter, waste from hospitals and pharmaceutical companies, agricultural waste containing animal waste, and industrial waste [4]. The most common microorganisms associated with water contamination are *Escherichia coli*, *Shigella dysenteriae*, *Shigella flexneri*, *Shigella boydii*, *Shigella sonnei*, *Salmonella enterica*, and *Vibrio cholerae*, which are responsible for diseases such as cholera, dysentery, acute diarrhea, gastroenteritis, and typhoid [5]. The death toll from these diseases and their impact on the quality of life of affected individuals, households, communities, nations, and regions warrant the development and implementation of microbial control approaches to reduce water contamination by disease-causing pathogens.

Hydrogel polymers present an effective, environmentally friendly, and affordable method of microbial control in contaminated water [6]. Hydrogel polymers are insoluble polymers that are made from either natural or synthetic polymers and can hold large volumes of water without dissolving [7]. Their ability to absorb large volumes of water makes hydrogel polymers suitable for various applications, including the treatment of contaminated or polluted water. The application of polymers in wastewater treatment harnesses their adsorption abilities, through which contaminants adhere to their surface to remove them from the contaminated water [8]. Currently, hydrogel polymers are used to remove both organic and inorganic contaminants such as metal ions and dyes from polluted water. They can be adapted to remove microbial contaminants as well as to leverage microorganisms' adherence to hydrogel surfaces.

Recent and past reviews on the topic have explored various dimensions of the adaptation of hydrogel polymers for the control of harmful or unwanted microorganisms in water systems. A recent study explored the application of hydrogel polymers to the treatment of industrial effluent, focusing on applications such as dye removal, the adsorption of heavy metal ions, and the viability of different types of hydrogels for the treatment of industrial waste [9]. However, the study does not address hydrogels with antimicrobial abilities or the integrations of remediating microorganisms with hydrogel polymers to remediate microbial pollutants. A second study evaluated the use of natural hydrogel polymers with antimicrobial abilities in the remediation of environmental pollutants [10]. One of the applications explored in the study is the use of hydrogels in water treatment, highlighting mechanisms such as adsorption, the use of hydrogels as carriers of remediating microorganisms, and the use of hydrogels with natural antimicrobial abilities. While the study highlights these mechanisms, it does not identify the hydrogel polymers with antimicrobial abilities nor does it discuss their adaptation for different water systems. A third study explores the use of cellulose-based hydrogels in the treatment of wastewater [11]. Its focus is on the remediation of metal ions and its discussion of wastewater remediation relative to the use of microorganisms is limited to an identification of the environmental conditions affecting the stability and performance of cellulose hydrogels. Evidently, most studies on the use of hydrogel polymers in water treatment focus on other pollutants such as metal ions and fail to address microbial pollution comprehensively. Furthermore, the integration of antimicrobials with hydrogel polymers as well as polymers with natural antimicrobial properties and their application in controlling microbial pollution in the treatment of water are not addressed comprehensively in recent studies and reviews. To this end, this article explores how historic and recent advances in hydrogel polymers can and are being adapted to control microorganisms in water systems with the overarching goal of evaluating the efficiency, cost-effectiveness, sustainability, and replicability of their use for microbial control in water systems.

2. Materials and Methods

A broad search was conducted using the keywords and prompts "hydrogel polymers", "the applications of hydrogel polymers", "hydrogel polymers and water treatment", "hydrogel polymers and microbial control", and hydrogel polymers and microbial control in water systems". Relevant research studies, case studies, and books were identified, schemed through, and screened using inclusive criteria that isolated peer-reviewed sources, factual case studies describing relevant applications of hydrogel polymers, and recent publications published between 2010 and 2024 for relevance and timeliness. The research articles, studies, books, and case studies that met the inclusion criteria were studied extensively and analyzed and information was extracted for use in this study.

3. Results

3.1. Fundamentals of Hydrogel Polymers

Hydrogel polymers, also known as polymeric hydrogels, are polymeric materials with hydrophilic qualities and structures that enhance their ability to hold large volumes of water [7]. The polymers have porous cross-linked three-dimensional structures that can be synthesized from one or more monomers. They are classified into five categories as shown in Figure 1. The first group of polymers, which are categorized by source, include natural and synthetic polymers. Natural hydrogel polymers are formulated from materials extracted from nature, such as chitosan and cellulose, while synthetic hydrogel polymers are formulated from petroleum and include poly (hydroxyethyl methacryla te) PHEMA and polyacrylic acid (PAA) [12,13]. The second classification characterizes hydrogel polymers by their method of preparation, yielding polymers produced through solution casting, free radical mechanism, and interpenetrating networks [14]. Third, the cross-linking method used in the formulation of hydrogel polymers is used to categorize them into chemical and physical cross-linked polymers based on the type of reaction used in their formulation. Due to the differences in these reactions, chemically cross-linked hydrogels have permanent links while their physically linked counterparts have temporary connections [13]. Fourth, hydrogel polymers exhibit different responses to physical or chemical stimuli such as pH, temperature, and magnetic and electric fields. Finally, hydrogel polymers fall under three distinct categories based on network electrical charge, which determines their ability to swell and shrink. These are neutral, cationic, and anionic hydrogel polymers [13].

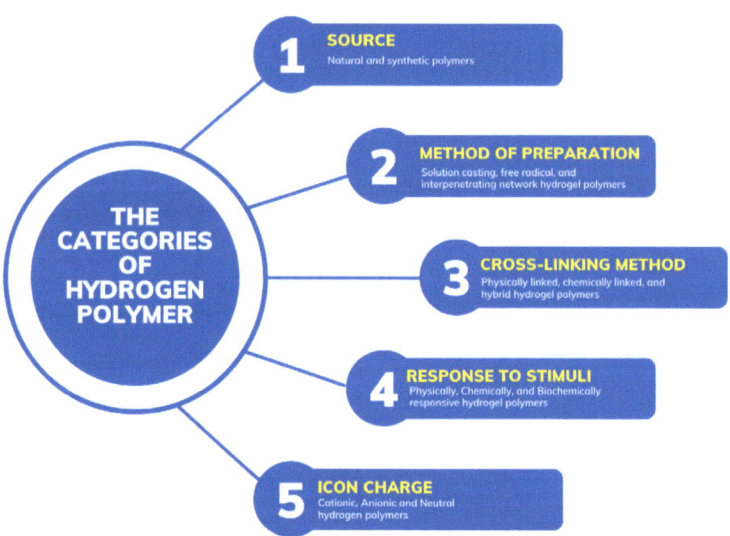

Figure 1. The categories of hydrogel polymers.

Due to their significant hydrophilic qualities, hydrogel polymers can hold water within their interstices without affecting their network structure [14]. Furthermore, since the polymers can be synthesized from one or more monomers, their ability to hold water can be modified and both their physicochemical and mechanical properties enhanced [10]. The properties that make hydrogels suitable for water purification include their porosity, hydrophilicity, separation capability, switchable wettability, and the presence of different states of water in their structure [15]. Hydrogel polymers have pores or interstices through which they absorb water. This quality of the polymers can be manipulated to improve their ability to absorb higher amounts of liquid [16]. Hydrophilicity refers to the ability of polymeric hydrogels to absorb water and moisture. The hydrogels are adapted for water absorption by their crosslinks, which prevent them from dissolving in water. Third, the water absorbed by the polymers can exist in three states, namely polymer-bound water, intermediate water, and free water. The polymers' ability to absorb and store some of these states such as free water can be enhanced by increasing porosity. Additionally, the storage of intermediate water requires less energy while the mechanisms involved in the absorption and presence of bulk water enable the concentration of reactive agents such as bactericides [15]. Finally, switchable wettability allows hydrogel polymers to release absorbed water, enhancing their recyclability. These qualities make hydrogel polymers effective water purifiers and decontaminants.

Some hydrogels have natural microbial abilities that naturally adapt them for microbial control in wastewater. These hydrogels are classified based on the microorganisms against which they possess antimicrobial activity. The first category of hydrogel polymers with antimicrobial properties is antibacterial polymers, which gain their inherent antibacterial qualities from the materials used in their formulation. These materials include antibacterial peptides, chitosan, and organic acids, which are both biocompatible and biodegradable [17]. Some of the polymers that fall under this category target bacterial strains that have developed resistance against other antimicrobial agents. For instance, polymers with crude ethyl acetate extract (CEE) are particularly effective against Methicillin-resistant Staphylococcus aureus (MRSA) and other multi-drug resistant strains of bacteria [18]. A second category of polymers have natural antifungal properties that enable them to kill fungi on contact. These include hydrogels such as amphogel, which has desirable qualities such as prompt action against fungi, reusability, and biocompatibility, Poly (hexamethylene biguanide) hydrochloride (PHMB), and poly-ε-lysine (ε-PL) [19]. These hydrogels, together with others without antimicrobial properties, have been used successfully for the remediation of wastewater, particularly in the control of microbial contaminants. Some of the hydrogels that have been used for this purpose include macroporous antimicrobial polymeric gel (MAPG), polyacrylamide (PAAm) hydrogels, and polyethylene glycol (PEG) [20]. See Figure 1.

3.2. Synthesis of Hydrogel Polymers for Microbial Control

Polymeric hydrogels are synthesized through chemical, physical, and hybrid bonding through various routes, including solution polymerization, interpenetrating network formation, solution casting and mixing, the radiation method, and free radical mechanism [14]. Solution polymerization yields highly porous gels using a monomer solution, a chemical initiator, and a cross-linking agent [21]. The initiator is activated using a redox initiator system, temperature, or UV radiation. Solution polymerization is ideal for the large-scale production of hydrogel polymers for various purposes, including water treatment and decontamination, due to its low cost, fast production of polymers, and the significant capacity of the resulting hydrogel polymer to absorb water [22]). Comparatively, free radical polymerization, a type of chain growth polymerization, is regarded as the most suitable technique for the synthesis of hydrogels based on amides, acrylates, and other monomers. It produces hydrogels from natural polymers, yielding polymerized polymers that do not allow impurities to pass through. Consequently, polymers produced through the synthesis pathway are ideal for microbial control in water systems as well as the removal of other impurities from water [22]. Like hydrogels produced through solution polymerization, the products of free radical polymerization are formed using crosslinking agents, initiators, and monomers. A third method of hydrogel preparation uses ionizing high-energy radiation such as electron beams and gamma rays to form cross-linked polymers [7]. The polymers have a significantly high absorption capacity, their particle size and porosity can be adjusted to meet desired levels, and their re-wetting capability is adequate. Furthermore, the polymers are durable and stable in various stages of absorption and adsorption and are broken down easily upon disposal since they are biodegradable. These qualities, which make the polymers easily adaptable for microbial control in contaminated water, have led to their extensive application in the biomedical field [23]. Ultimately, the synthesis method used to produce hydrogel polymers influences the viability of their use in controlling microbial contamination in water by having direct implications on their cost, porosity, stability during water absorption and storage, and structural integrity.

The applications of hydrogel polymers beyond water decontamination and microbial control in water systems include tissue engineering, drug delivery, and the production of medical dressing, biosensors, and contact lenses [14]. The magnitude of these uses has necessitated the dedication of research, funding, and resources to synthesis techniques to yield new applications, enhance the ability of hydrogel polymers to carry out existing uses, reduce the cost of their production, and improve their recyclability and biodegradability to make them more environmentally friendly. These efforts and resources have yielded several advances in synthesis techniques. One such advancement is the emergence of 3D bioprinting technology, which has made the customization of hydrogel polymers to specific uses relatively easier [24]. A second advancement in the synthesis of hydrogel polymers is the combination of cross-linked polymeric hydrogel (HDG) networks with nanoparticles to yield polymerized hydrogels with high water content and flexibility [25]. This has been found to enhance their mechanical properties and make them more suitable for specific applications such as water decontamination. These and other advances in the synthesis of polymers enhance their usability and applicability to the treatment and purification of contaminated water. However, these uses of hydrogel polymers raise some challenges. For instance, regardless of the synthetic pathway followed in their creation, hydrogels based on synthetic monomers lack antimicrobial activity [26]. This limits their applicability for microbial control in water systems and raises the need for the consideration of natural monomers in the synthesis of polymerized hydrogels for use in water treatment. Second, some synthesis methods are not environmentally friendly and use hazardous solvents [26]. This raises the need for greener synthesis approaches that use less energy and rely on less hazardous solvents. Third, some synthesis methods are not cost-effective and require specialized equipment [14]. This raises the issue of the economic viability of the use of hydrogel polymers produced using these methods for microbial control.

3.3. Mechanisms of Microbial Control by Hydrogel Polymers

Hydrogel polymers use various techniques to purify water and remove pathogenic contaminants in dirty water or wastewater. First, they erect barriers that encapsulate and immobilize microorganisms. Hydrogels are the ideal material for the encapsulation of living cells [27]. Traditionally, this ability was used to protect the encapsulated living cells, such as bacteria and other microorganisms, from damage during their delivery into the body as drugs to fight pathogens. The ability has since been adapted for the removal of harmful microorganisms from water systems using the same mechanisms. Essentially, the polymers' high moisture and water content provide an ideal environment for microorganisms to thrive [28]. Once the microorganisms are bound to the polymers, the polymers' cross-linked nature encapsulates and immobilizes them, controlling their spread until the proper conditions for their release are achieved. The mechanism used by hydrogel polymers in encapsulation and immobilization mimics the natural biofilm formed by microorganisms. In addition to attracting microorganisms, holding them using cross-linked polymer networks, and serving as separation media to draw them from contaminated water, hydrogels can inhibit microbial growth by depriving microorganisms of nutrients. Finally, some hydrogel polymers produce antimicrobial agents that eliminate microorganisms. An example of such polymers is chitosan-based hydrogels, which contain antiviral, antibacterial, and antifungal qualities. Antimicrobial hydrogel polymers use various mechanisms to kill the microorganisms against which they are deployed. Some damage the membrane of the microorganisms and release the dead microbes to prevent their accumulation in the gel [29]. This solves a critical problem of the retention of dead microorganisms by hydrogel polymers, freeing them up for reuse. Others use electrical stimulation moderated by external stimuli such as light and heat to release high concentrations of active oxygen, which kills microorganisms such as bacteria [17]. A third category of antimicrobial physical hydrogels, particularly those formulated through physical crosslinking with hyaluronic acid, form hydroxyl free radicals that destroy the cells of bacteria, effectively killing them [30]. Notably, some antimicrobial hydrogel polymers have selective properties and can target specific species of microorganisms. This solves the problem of the elimination of good and beneficial microorganisms alongside pathogenic and potentially invasive microorganisms that would upset biodiversity in the water bodies to which wastewater flows after treatment. To demonstrate the antimicrobial efficacy of hydrogel polymers, the diameter of inhibition zones for various hydrogel polymers tested against different bacterial strains is shown in Table 1 below.

Table 1. Diameter of the inhibition zones of various hydrogel polymers against different bacteria.

Hydrogel Polymer	Bacteria	Diameter of Inhibition Zone (mm)	References
Poly(acrylamide) Hydrogel	Escherichia coli	13	[31]
Poly(vinyl alcohol) Hydrogel	Staphylococcus aureus	20	[32]
Poly(ethylene glycol) Hydrogel	Pseudomonas aeruginosa	18	[33]
Chitosan Hydrogel	Bacillus subtilis	22.4	[34]
Alginate Hydrogel	Klebsiella pneumoniae	14	[35]
Polyvinyl Alcohol/Carboxymethyl Cellulose (PVA/CMC) Hydrogel	Proteus vulgaris	19	[36]
WO3NPs-hydroxyethyl Cellulose Hydrogel	Bacillus cereus	17	[37]
Benzyl Isothiocyanate-β-cyclodextrin-κ-carrageenan (BITC- βCD-KC) Hydrogel	Listeria monocytogenes	16.7	[38]
Gelatin methacryloyl/Hyaluronic acid Hydrogel	Enterococcus faecalis	5.5	[39]
Hyaluronic Acid Hydrogel	Escherichia coli	20.7	[40]

3.4. Applications of Hydrogel Polymers in Water Systems

Hydrogel polymers are used as adsorbent materials for the purification of drinking water. Chitosan-based hydrogels are particularly preferred for the absorption of contaminants to produce quality drinking water due to their desirable qualities. They include the sustainability of the polymers since chitosan is a natural polymer that is biodegradable and has low levels of toxicity, the antiviral, antifungal, and antimicrobial qualities of chitosan, which enable the polymers to target, encapsulate, and eliminate specific microorganisms, and the low cost of production of chitosan-based polymers, which makes their use in drinking water system affordable and economical [41]. For a detailed summary of the characteristics, mechanisms, and applications of hydrogel polymers, see Table 2. Another common trend in microbial control in drinking water systems, in addition to the use of biodegradable polymers with antimicrobial qualities such as chitosan, is the use of a combination of renewable sources of energy together with natural hydrogel polymers to purify water contaminated with pathogens [42]. One such hydrogel polymer was developed by researchers at Princeton University in 2023 [43]. The researchers developed a solar absorber gel that uses sunlight to eliminate water contaminants, including bacteria. The water purification system uses a hydrogel made from poly(N-isopropylacrylamide) or PNIPAm to absorb and release water at set temperatures. A FESEM image of the polymer is shown in Figure 2 below. When temperatures fall below 33 degrees Celsius or 91 degrees Fahrenheit, the hydrogel absorbs water from its designated source of water, which can be any water body including rivers and lakes. Once the gel is taken out of the water and heated by sunlight to temperatures above 33 degrees Celsius, it begins to release the water while filtering pollutants and contaminants such as microorganisms, microplastics, and heavy metals [43]. The gel has been tested on large water videos and the large-scale production of clean drinking water researchers are currently working to develop a smaller size model for use in households. A similar hydrogel polymer developed to purify water contaminated with microplastics, oil, metals, and small molecules was found to be effective in the removal of bacteria, particularly Escherichia coli, from contaminated water as well [44]. The hydrogel uses solar energy to purify water. It eliminates bacteria by leveraging their adhesive capabilities to attach them to their surface and the generation of a surface layer to contain the microorganisms upon the release of the now-purified water [44].

In addition to their application in the purification of drinking water, hydrogel polymers are used extensively in the treatment of wastewater, particularly in the removal of hard metals, dyes, and other organic and inorganic pollutants. The removal of these pollutants is primarily performed through adsorption by leveraging various interactions between the polymers and the pollutants or contaminants they target. Some of the common interactions used to eliminate these unwanted substances in wastewater include hydrophobic interactions, which prevent the gels from dissolving in water, hydrophilic interactions through which the hydrogel polymers absorb water, chelation, through which the ions of pollutant metals are bonded to the polymers, and ion exchange [15]. These interactions have been adapted to the use of polymers produced from various organic and inorganic materials in wastewater treatment. For instance, a polyacrylic acid hydrogel polymer, a synthetic and high-molecular-weight polymer, has been found to be highly effective in wastewater treatment. The polymer has most of the suitable qualities, including a high capacity for adsorption and reusability, which enhances the polymer's sustainability for the application, cost-effectiveness since the polymer is cheap and easy to produce, and rapid removal of kinetics [6]. The hydrogel polymer has been found to be effective in the treatment of industrial wastewater. One recent development in the adaptation of hydrogel polymers for wastewater treatment is the use of cellulose-based polymers in water decontamination systems. The polymers are low-cost and biodegradable and their adsorption capabilities can be enhanced [45]. They have been successfully integrated into the treatment of wastewater in hybrid water systems [45]. The typical wastewater treatment process is a five-step process. It begins with the pretreatment stage, in which particles and dirt are bound to form larger particles that settle at the bottom of the wastewater. In the second stage, further coagulation and flocculation occur. Polymer flocculants are introduced at this stage to separate more particles from the water before the third stage [15]. The integration of hydrogel polymers begins in the third stage, in which purification occurs. In this stage, hydrogel polymers are incorporated to remove pollutants and contaminants such as pathogens, hard metals, and dyes in wastewater through adsorption before the water proceeds to the second-last stage of tertiary treatment [46]. In the fourth stage, oxidation, membrane filtration, and other chemical processes occur before the water is directed to clarifiers as the last step in its treatment. Therefore, hydrogel polymers are integrated into the secondary treatment stage of wastewater treatment, preceding tertiary treatment to complete a series of physical, chemical, and mechanical processes that rid the water of solid and microscopic contaminants and pollutants.

The applications of hydrogel polymers in both drinking water systems and wastewater treatment are enhanced and scaled for use in industrial water systems. While evaluating the applicability and effectiveness of the polymers in the treatment of industrial effluent, Flores-Valenzuela, Gonzalez-Fernandez, and Carranza-Oropeza identified some of the factors considered before their integration in industrial water treatment plants, which include their regeneration costs, maintenance needs, and capacity for adsorption [9]. In industrial water systems, hydrogels play more than one role. In addition to the treatment of wastewater and sludge using the five-step process described above, the polymers are adapted for use in cooling systems such as industrial cooling towers. This application leverages their hydrophilic and hydrophobic properties, which are part of the polymers' self-adaptive cooling systems. The two abilities allow them to lose water through evaporation in high temperatures and regenerate when temperatures fall, creating a cooling effect during the day or in instances of high heat [47]. In addition to their integration into industrial cooling systems and their role in the treatment of industrial effluent, hydrogel polymers are used widely in the food, pharmaceutical, and agriculture industries [48]. In the food industry, their ability to absorb water is leveraged and adapted for the development of food packaging that controls humidity and microbial activity [49]. In the pharmaceutical industry, hydrogels are used in drug delivery, wound dressing, and wound management, which leverage their ability to absorb moisture to keep wounds dry [50]. In agriculture, hydrogel polymers enhance the capacity of the soil to hold water in addition to serving as nutrient conduits and reservoirs [51]. All these applications leverage the ability of hydrogel polymers to absorb, hold, and release water, which are the same qualities upon which their application in drinking water purification, wastewater treatment, and the treatment of industrial effluent are founded.

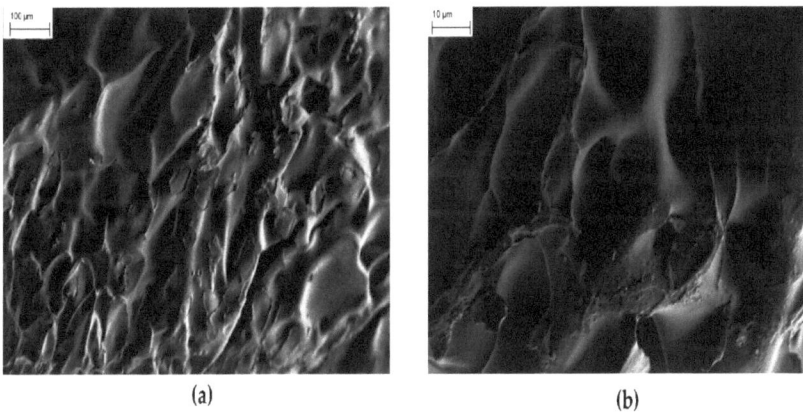

Figure 2. (a) Field emission scanning electron microscopy (FESEM) image of a pure PNIPAM hydrogel. (b) FESEM image of the same pure PNIPAM hydrogel at higher magnification [52].

Table 2. Overview of hydrogel polymer characteristics, mechanisms, and applications in water microbial control.

	Description	Applications	Examples	Reference
Absorption Capacity	Ability to hold large volumes of water without dissolving, due to porous cross-linked structures.	Water purification, removal of organic and inorganic contaminants	Polyacrylamide (PAAm), Polyethylene glycol (PEG)	[20]
Synthesis Methods	Includes solution polymerization, free radical polymerization, and radiation methods.	Tailoring properties for specific water treatment needs	Solution polymerization for large-scale water decontamination	[14]
Mechanisms of Action	Encapsulation and immobilization of microorganisms, release of antimicrobial agents.	Controlling spread of pathogens in drinking and wastewater systems	Chitosan-based hydrogels for their antiviral and antibacterial properties	[27,28,31,32]
Environmental Impact	Biodegradability, toxicity, and impact on non-target organisms.	Ensuring sustainability and safety in water treatment	Biodegradable chitosan versus synthetic hydrogels	[14,24,31,32]
Technological Advances	Smart hydrogels responsive to environmental changes, integration with nanotechnology.	Enhancing effectiveness and application range in water systems	Smart hydrogels for adaptive water purification systems	[48,52,53]

4. Discussion

4.1. Performance Evaluation and Optimization

A comparison of the effectiveness of hydrogel polymers with other methods of microbial water treatment sets the polymers apart as one of the most effective means of controlling microorganisms in water systems. Other methods of microbial control in water systems include boiling, filtration, chlorination, sedimentation, and coagulation [54]. While effective in microbial control in drinking water systems, chlorination requires supplementation with other methods such as boiling as traces of pathogens have been isolated in chlorinated water [55]. Similarly, traces of microorganisms such as *Bacillus* spp. have been found to survive boiling, particularly if an open container is used for the process [53,56]. Filtration, on the other hand, is highly effective, removing nearly all traces of harmful microorganisms such as *Coliforms bacteria*, *Clostridium perfringens*, and *Enterococcus faecalis* [57]. However, the method is costly, has a high potential for toxicity to both human beings and the environment, and is incompatible with some water systems [58]. The drawbacks of these methods of microbial control in water systems warrant an evaluation of the effectiveness of hydrogel polymers, a viable alternative, general water treatment and purification, the environmental impact of their use, and the potential for integration with current and emerging technologies to enhance their effectiveness.

The fundamental property of hydrogel polymers is their hydrophilic nature and ability to hold large volumes of water [59]. This property, together with its characteristics, synthetic routes, mechanisms for microbial control, and applications in water treatment has been discussed extensively in the literature. These parameters can be used to evaluate their effectiveness in controlling microorganisms to eliminate pathogens in drinking water, wastewater, and industrial effluent. One of the most effective methods to this end is the Water Quality Index, which evaluates the quality of drinking water using parameters established by drinking water standards [60]. One of these parameters is the absence of microorganisms in the water [60]. This method could be applied by determining the species and population of microorganisms in water before its treatment or purification using hydrogel polymers. The drinking water, wastewater, or industrial effluent would be tested again for microorganisms after its purification to determine the effectiveness of the polymers as well as the quality and safety of the water for its intended purpose. A second method focuses on the activities of the polymers rather than the outcome of their applications. Typically, hydrogel polymers have inherent qualities that make them suitable for microbial control. For instance, chitosan, a natural polymer that is biodegradable and has low levels of toxicity, produces hydrogels with antiviral, antifungal, and antimicrobial qualities that enable them to target, encapsulate, and eliminate specific microorganisms [41]. Such abilities enhance the effectiveness of the polymers that possess them in microbial control. Third, the mechanical interactions between the cells of microorganisms and polymeric substances are good indicators of the effectiveness of hydrogel polymers in microbial control. For instance, the stiffness of hydrogels is a key determinant of their ability to encapsulate microorganisms [61]. However, the characteristics of the microorganisms, such as their ability to adhere to the surfaces of the hydrogel polymers, also influence encapsulation and affect the effectiveness of the gels in microbial control. Fortunately, the qualities of hydrogel polymers can be enhanced and adapted for the control of different microorganisms.

While the qualities and characteristics of both polymers and microorganisms are crucial determinants of the effectiveness of the hydrogels in microbial control, some parameters beyond both elements influence the performance of hydrogels as well as the outcome of their application. First, polymer composition determines the degree of adherence of microorganisms to hydrogel polymers. Polymer coating, for instance, affects the adherence of microorganisms such as bacteria [62]. The resulting differences in the level of adhesion to hydrogel polymers affect their ability to remove pathogens from drinking or wastewater. Another dimension of polymer composition is the material used in their formulation. Some polymers are made from materials that have natural antimicrobial qualities, which make them highly effective in microbial control while the rest rely on other mechanisms to remove microorganisms and other contaminants from water. Hydrogels that contain chitosan, for instance, have natural antimicrobial properties. Inorganic hydrogels can also be enhanced to contain antimicrobial qualities through their augmentation with nanoparticles that damage the cells of microorganisms [63]. Second, the properties of individual polymers influence their ability to eliminate or control microorganisms in water systems. The properties of hydrogel polymers differ based on the material from which they are produced and their adaptable characteristics, such as porosity. These characteristics determine the performance and effectiveness of the polymers. For instance, the hydrophilic structure of hydrogels is an indicator of their capacity to hold water, which influences their ability to treat and purify water by volume. Similarly, their foundational material, which is either natural or synthetic, influences their mechanical strength, durability, and rate of degradation, which affect both their performance and the length of time for which they can be used and reused in water treatment [7]. Third, the environmental conditions under which hydrogel polymers are used for microbial control in water systems influence their performance. Different polymeric hydrogels have different optimum levels of performance at different conditions such as temperature. Furthermore, they demonstrate different responses to chemical stimuli such as pH, physical stimuli such as light, and biological stimuli such as enzymes [64]. Therefore, these conditions will influence the performance of the polymers and aid in the determination of the suitability of different polymers for different environmental conditions. These determinants of the performance of hydrogel polymers can be used to optimize it in different water systems. Organic hydrogels can be augmented with inorganic elements in their networks to enhance durability and reusability, reduce costs, and simplify their production [15]. Second, hydrogel polymers can be adapted for different environmental conditions to enhance their performance and increase their effectiveness in microbial control. This can be conducted simply through the use of materials that are suitable for the conditions of different water systems or the adoption of smart hydrogel technology, which produces hydrogels that can respond to and adapt to environmental stimuli [65]. Finally, the integration of hydrogel polymers with existing water treatment systems optimizes their effectiveness in different water systems. A proposed solution to water turbidity demonstrates this strategy. The proposed strategy involves incorporating hydrogel polymers in set chambers in municipal water supply systems to purify water as it flows through the pipes to households and commercial premises. Integrating the polymers with existing water systems is inexpensive since it relies on existing systems.

4.2. Environmental and Health Impacts

Whilst hydrogel polymers are effective in the elimination of microorganisms from water, inexpensive, and environmentally friendly, particularly when the polymers used are organic and biodegradable, their use raises the issues of toxicity and the potential impacts on non-target organisms. Hydrogels that are synthesized from natural polymers are non-toxic [14]. However, some hydrogel polymers, particularly those that are synthesized from synthetic polymers, are toxic, cause extensive damage to the environment, and are incompatible with live cells. This raises the dilemma of the safety of their use to remove microbial and other contaminants from water, particularly drinking water. On the one hand, they are highly effective in microbial control, durable and reusable, easily and cheaply produced, and effective in the removal of other harmful and toxic contaminants such as metal ions from water. Furthermore, their properties can be enhanced and adapted to specific environmental conditions; but, on the other hand, they could be toxic [66]. Fortunately, their toxicity can be addressed by changing the crosslinkers used in their synthesis, which are often the source of toxicity [67]. A second challenge that arises in the use of hydrogel polymers in microbial control is their elimination of all microorganisms, including beneficial microbes. This can be addressed through the augmentation of hydrogel polymers with specific microorganism recognition elements that bind select microorganisms to their surface without capturing other microorganisms [68]. However, this could have some implications on the cost, effectiveness, and ease of production of the polymers.

In addition to the inherent lack of selectivity relative to microorganisms and the toxicity of some hydrogel polymers, their use in microbial control in water systems raises the issues of environmental impact. Natural polymers are naturally biodegradable and have little impact on their environment during use and upon disposal. However, synthetic polymers, which are often more effective than their organic counterparts, can be non-biodegradable [69]. Furthermore, they are made from petroleum, whose production and use are detrimental to the environment. Fortunately, some conditions can be established to degrade synthetic hydrogels. These include the use of microorganisms such as fungi to degrade the polymers [69]. This addresses the problem of the environmental impact of hydrogel polymers and justifies their continued use in microbial control in water systems. However, their use requires the establishment of and adherence to strict regulatory and health safety standards. These include restrictions on the types and materials of hydrogels that can be used to treat drinking water, water quality indices to determine the level of purification reached by the polymers, and the maximum levels of toxicity above which polymers cannot be used in water treatment or purification.

4.3. Future Perspectives and Innovations

Hydrogel technology is evolving and advancing to yield new applications of hydrogels, improve the effectiveness of existing applications, and improve the synthesis of hydrogels to make them more suitable for both types of applications. One of the emerging trends in hydrogel technology is smart hydrogels. The hydrogels are responsive to stimuli and environmental conditions such as temperature, pH, and light [70]. Their responsiveness to these triggers is reversible such that they can adapt quickly to changes in their environment and go back to their initial form just as promptly. They adapt by changing their shape, through interactions with solvents such as water, and via their hydrophilicity. In the pharmaceutical industry, these qualities of smart hydrogels are used to control dosage and dosing frequency, ensure the drug is only delivered to target tissues by minimizing its accumulation in non-target tissues, and prolong the drug release period [70]. In water systems, the qualities of smart hydrogels can be used to adapt them to changes in conditions such as acidity in wastewater, temperature in industrial water treatment plants, and pressure from piped water systems, to which they can respond by changing their shape. The viability of the use of smart hydrogels in water treatment has been evaluated, validated, and found to yield significant benefits, including easy synthesis, biocompatibility, and low costs [6]. A second emerging trend in hydrogel technology is responsive systems. These can be a feature of smart hydrogels. The gels can swell in high temperatures and contract in cold temperatures while others respond to light, pressure, and molecules such as glucose [71]. Some of the advantages of responsive hydrogels include fixed timing and precision [71]. These qualities enhance their performance in the treatment and purification of water by allowing the setting of specific timelines for their application. Once determined, timelines can be set automatically to determine the optimum purification. A third emerging trend, which is particularly relevant to microbial control, is the integration of sensing techniques in hydrogel polymer synthesis to allow them to capture microorganisms. This innovation has been adapted for the successful capture and elimination of bacteria [69]. Fourth, hydrogel polymers are being integrated with other technologies to enhance their efficacy in microbial control in water systems and other applications. Their integration with nanotechnology, for instance, yields nanohydrogels, which have superior benefits such as biodegradability, selective abilities, and a wide range of applications [72]. These emerging trends and technologies yield highly effective hydrogel polymers with improved attributes, which can be commercialized for use in the treatment of water across a variety of water systems. However, they raise the challenges of cost and the need for specialized knowledge and equipment.

5. Conclusions

In conclusion, hydrogel polymers are networks of polymers with hydrophilic qualities. Their properties, which include porosity, the ability to absorb large volumes of water without dissolving, and their ability to swell allow the polymer's adaptation for the treatment of water through the removal of contaminants, including pathogens and unwanted microorganisms. Some hydrogel polymers have additional properties that are leveraged for microbial control in water systems such as antimicrobial abilities. Other qualities that enhance the overall performance of the polymers in water treatment and purification include biodegradability, compatibility with live cells, sensitivity to environmental stimuli, and mechanical strength. The use of hydrogel polymers for microbial control in water systems benefits from significant inventions and technological advancements in related fields, which have yielded trends such as improved synthesis techniques that enhance the performance and properties of the polymers. However, a few areas remain understudied and warrant further investigation to fully harness the abilities of the polymers. These include the suitability of different hydrogel polymers for the elimination of specific types and species of microorganisms and cost comparisons of reuse versus the synthesis of new polymers.

Author Contributions: Conceptualization, A.A.A.; methodology, A.A.A.; resources, H.O.; writing—original draft preparation, A.A.A.; writing—review and editing, A.A.A. and H.O.; visualization, A.A.A.; supervision, A.A.A. All authors have read and agreed to the published version of the manuscript.

Funding: This research received no external funding.

Institutional Review Board Statement: Not applicable.

Data Availability Statement: Data are contained within the article.

Acknowledgments: We would like to thank V.W. Ogundero for his unwavering support.

Conflicts of Interest: The authors declare no conflicts of interest.

References

1. World Health Organization. Improving Access to Water, Sanitation and Hygiene Can Save 1.4 Million Lives per Year, Says New WHO Report. WHO. 2023. Available online: https://www.who.int/news/item/28-06-2023-improving-access-to-water--sanitation-and-hygiene-can-save-1.4-million-lives-per-year--says-new-who-report (accessed on 23 March 2024).
2. Lin, L.; Yang, H.; Xu, X. Effects of water pollution on human health and disease heterogeneity: A review. *Front. Environ. Sci.* **2022**, *10*, 880246. [CrossRef]
3. Li, P.; Karunanidhi, D.; Subramani, T.; Srinivasamoorthy, K. Sources and consequences of groundwater contamination. *Arch. Environ. Contam. Toxicol.* **2021**, *80*, 1–10. [CrossRef] [PubMed]
4. Some, S.; Mondal, R.; Mitra, D.; Jain, D.; Verma, D.; Das, S. Microbial pollution of water with special reference to coliform bacteria and their nexus with environment. *Energy Nexus* **2021**, *1*, 100008. [CrossRef]
5. Cabral, J.P. Water microbiology. Bacterial pathogens and water. *Int. J. Environ. Res. Public Health* **2010**, *7*, 3657–3703. [CrossRef] [PubMed]
6. Shah, L.A.; Khan, S.A. Polymer hydrogels for wastewater treatment. In *Environmental Chemistry and Recent Pollution Control Approaches*; IntechOpen: Rijeka, Croatia, 2019.
7. Ahmed, E.M. Hydrogel: Preparation, characterization, and applications: A review. *J. Adv. Res.* **2015**, *6*, 105–121. [CrossRef] [PubMed]
8. Akter, M.; Bhattacharjee, M.; Dhar, A.K.; Rahman FB, A.; Haque, S.; Rashid, T.U.; Kabir, S.F. Cellulose-based hydrogels for wastewater treatment: A concise review. *Gels* **2021**, *7*, 30. [CrossRef] [PubMed]
9. Flores-Valenzuela, L.E.; González-Fernández, J.V.; Carranza-Oropeza, M.V. Hydrogel Applicability for the Industrial Effluent Treatment: A Systematic Review and Bibliometric Analysis. *Polymers* **2023**, *15*, 2417. [CrossRef] [PubMed]
10. Ahmadi, S.; Pourebrahimi, S.; Malloum, A.; Pirooz, M.; Osagie, C.; Ghosh, S.; Dehghani, M.H. Hydrogel-based materials as super adsorbents and antibacterial agents for the remediation of emerging pollutants: A comprehensive review. *Emerg. Contam.* **2024**, *10*, 100336. [CrossRef]
11. Persano, F.; Malitesta, C.; Mazzotta, E. Cellulose-Based Hydrogels for Wastewater Treatment: A Focus on Metal Ions Removal. *Polymers* **2024**, *16*, 1292. [CrossRef]
12. Zhao, L.; Zhou, Y.; Zhang, J.; Liang, H.; Chen, X.; Tan, H. Natural polymer-based hydrogels: From polymer to biomedical applications. *Pharmaceutics* **2023**, *15*, 2514. [CrossRef] [PubMed]

13. Ahmad, Z.; Salman, S.; Khan, S.A.; Amin, A.; Rahman, Z.U.; Al-Ghamdi, Y.O.; Khan, S.B. Versatility of hydrogels: From synthetic strategies, classification, and properties to biomedical applications. *Gels* **2022**, *8*, 167. [CrossRef] [PubMed]
14. Bashir, S.; Hina, M.; Iqbal, J.; Rajpar, A.H.; Mujtaba, M.A.; Alghamdi, N.A.; Ramesh, S. Fundamental concepts of hydrogels: Synthesis, properties, and their applications. *Polymers* **2020**, *12*, 2702. [CrossRef] [PubMed]
15. Ahmaruzzaman, M.; Roy, P.; Bonilla-Petriciolet, A.; Badawi, M.; Ganachari, S.V.; Shetti, N.P.; Aminabhavi, T.M. Polymeric hydrogels-based materials for wastewater treatment. *Chemosphere* **2023**, *331*, 138743. [CrossRef] [PubMed]
16. Loo, S.L.; Vásquez, L.; Athanassiou, A.; Fragouli, D. Polymeric hydrogels—A promising platform in enhancing water security for a sustainable future. *Adv. Mater. Interfaces* **2021**, *8*, 2100580. [CrossRef]
17. Jia, B.; Li, G.; Cao, E.; Luo, J.; Zhao, X.; Huang, H. Recent progress of antibacterial hydrogels in wound dressings. *Mater. Today Bio* **2023**, *19*, 100582. [CrossRef]
18. Narayanan, K.B.; Park, G.T.; Han, S.S. Biocompatible, antibacterial, polymeric hydrogels active against multidrug-resistant Staphylococcus aureus strains for food packaging applications. *Food Control* **2021**, *123*, 107695. [CrossRef]
19. Zumbuehl, A.; Ferreira, L.; Kuhn, D.; Astashkina, A.; Long, L.; Yeo, Y.; Kohane, D.S. Antifungal hydrogels. *Proc. Natl. Acad. Sci. USA* **2007**, *104*, 12994–12998. [CrossRef] [PubMed]
20. Pinthong, T.; Yooyod, M.; Daengmankhong, J.; Tuancharoensri, N.; Mahasaranon, S.; Viyoch, J.; Ross, G.M. Development of Natural Active Agent-Containing Porous Hydrogel Sheets with High Water Content for Wound Dressings. *Gels* **2023**, *9*, 459. [CrossRef]
21. Thang, N.H.; Chien, T.B.; Cuong, D.X. Polymer-based hydrogels applied in drug delivery: An overview. *Gels* **2023**, *9*, 523. [CrossRef] [PubMed]
22. Madduma-Bandarage, U.S.; Madihally, S.V. Synthetic hydrogels: Synthesis, novel trends, and applications. *J. Appl. Polym. Sci.* **2021**, *138*, 50376. [CrossRef]
23. El Sayed, M.M. Production of polymer hydrogel composites and their applications. *J. Polym. Environ.* **2023**, *31*, 2855–2879. [CrossRef]
24. Yang, J.; Rao, L.; Wang, Y.; Zhao, Y.; Liu, D.; Wang, Z.; Liu, Y. Recent advances in smart hydrogels prepared by ionizing radiation technology for biomedical applications. *Polymers* **2022**, *14*, 4377. [CrossRef] [PubMed]
25. Sánchez, E.; Gómez-Blanco, J.C.; López Nieto, E.; Casado, J.G.; Macías-García, A.; Díaz Díez, M.A.; Pagador, J.B. Hydrogels for bioprinting: A systematic review of hydrogels synthesis, bioprinting parameters, and bioprinted structures behavior. *Front. Bioeng. Biotechnol.* **2020**, *8*, 776. [CrossRef] [PubMed]
26. Idumah, C.I.; Nwuzor, I.C.; Odera, R.S. Recent advances in polymer hydrogel nanoarchitectures and applications. *Curr. Res. Green Sustain. Chem.* **2021**, *4*, 100143. [CrossRef]
27. Sánchez-Cid, P.; Jiménez-Rosado, M.; Romero, A.; Pérez-Puyana, V. Novel trends in hydrogel development for biomedical applications: A review. *Polymers* **2022**, *14*, 3023. [CrossRef] [PubMed]
28. Tang, T.C.; Tham, E.; Liu, X.; Yehl, K.; Rovner, A.J.; Yuk, H.; Lu, T.K. Hydrogel-based biocontainment of bacteria for continuous sensing and computation. *Nat. Chem. Biol.* **2021**, *17*, 724–731. [CrossRef]
29. Salomé Veiga, A.; Schneider, J.P. Antimicrobial hydrogels for the treatment of infection. *Pept. Sci.* **2013**, *100*, 637–644. [CrossRef] [PubMed]
30. Liu, J.; Jiang, W.; Xu, Q.; Zheng, Y. Progress in antibacterial hydrogel dressing. *Gels* **2022**, *8*, 503. [CrossRef] [PubMed]
31. Agili, F.A.; Mohamed, S.F. Synthesis and Characterization of a Self-Crosslinked Organic Copolymer Kappa-Carrageenan/Polyacrylamide/Cetrimide (κ-CAR/PAAm/CI) Hydrogel with Antimicrobial and Anti-Inflammatory Activities for Wound Healing. *Chemistry* **2023**, *5*, 2273–2287. [CrossRef]
32. Lungoci, C.; Rîmbu, C.M.; Motrescu, I.; Serbezeanu, D.; Horhogea, C.E.; Vlad-Bubulac, T.; Ghiţău, C.S.; Puiu, I.; Neculai-Văleanu, A.-S.; Robu, T. Evaluation of the Antibacterial Properties of Polyvinyl Alcohol-Pullulan Scaffolds Loaded with Nepeta racemosa Lam. Essential Oil and Perspectives for Possible Applications. *Plants* **2023**, *12*, 898. [CrossRef] [PubMed]
33. Ermawati, D.E.; Surya, A.P.; Setyawati, R.; Niswah, S.U. The effect of glycerin and polyethylene glycol 400 as humectant on stability and antibacterial activity of nanosilver biosynthetic peel-off mask. *J. Appl. Pharm. Sci.* **2022**, *12*, 080–089. [CrossRef]
34. Mohamed, N.A.; Fahmy, M.M. Synthesis and antimicrobial activity of some novel cross-linked chitosan hydrogels. *Int. J. Mol. Sci.* **2012**, *13*, 11194–11209. [CrossRef] [PubMed]
35. Aydin, G.; Zorlu, E.B. Characterisation and antibacterial properties of novel biodegradable films based on alginate and roselle (*Hibiscus sabdariffa* L.) extract. *Waste Biomass Valorization* **2022**, *13*, 2991–3002. [CrossRef] [PubMed]
36. Ruiz, Y.P.; de Almeida Campos, L.A.; Alves Agreles, M.A.; Galembeck, A.; Macário Ferro Cavalcanti, I. Advanced hydrogels combined with silver and gold nanoparticles against antimicrobial resistance. *Antibiotics* **2023**, *12*, 104. [CrossRef] [PubMed]
37. Bao, Y.; He, J.; Song, K.; Guo, J.; Zhou, X.; Liu, S. Functionalization and antibacterial applications of cellulose-based composite hydrogels. *Polymers* **2022**, *14*, 769. [CrossRef]
38. Huang, Y.; Liu, J.; Li, Z.; Cao, Z.; Hao, H.; Bi, J.; Hou, H.; Wu, H.; Zhang, G. Antibacterial film based on κ-carrageenan with benzyl isothiocyanate-β-cyclodextrin inclusion complex: Characterization and application in chicken preservation. *Food Hydrocoll.* **2023**, *145*, 109063. [CrossRef]
39. Dorterler, O.C.; Akgun, B.; Alper, M.; Ayhan, F. Improving Antimicrobial Properties of GelMA Biocomposite Hydrogels for Regenerative Endodontic Treatment. *Polymers* **2024**, *16*, 1675. [CrossRef] [PubMed]

40. Salma-Ancane, K.; Sceglovs, A.; Tracuma, E.; Wychowaniec, J.K.; Aunina, K.; Ramata-Stunda, A.; Loca, D. Effect of crosslinking strategy on the biological, antibacterial and physicochemical performance of hyaluronic acid and ϵ-polylysine based hydrogels. *Int. J. Biol. Macromol.* **2022**, *208*, 995–1008. [CrossRef] [PubMed]
41. Pérez-Luna, V.H.; González-Reynoso, O. Encapsulation of biological agents in hydrogels for therapeutic applications. *Gels* **2018**, *4*, 61. [CrossRef] [PubMed]
42. Chelu, M.; Musuc, A.M.; Popa, M.; Calderon Moreno, J.M. Chitosan Hydrogels for Water Purification Applications. *Gels* **2023**, *9*, 664. [CrossRef] [PubMed]
43. Poore, C. Solar-Powered Gel Produces Enough Clean Water to Meet Daily Needs. Princeton University. 2023. Available online: https://research.princeton.edu/news/solar-powered-gel-produces-enough-clean-water-meet-daily-needs (accessed on 23 March 2024).
44. Xu, X.; Guillomaitre, N.; Christie, K.S.; Bay, R.K.; Bizmark, N.; Datta, S.S.; Priestley, R.D. Quick-release antifouling hydrogels for solar-driven water purification. *ACS Cent. Sci.* **2023**, *9*, 177–185. [CrossRef] [PubMed]
45. Radoor, S.; Karayil, J.; Jayakumar, A.; Kandel, D.R.; Kim, J.T.; Siengchin, S.; Lee, J. Recent advances in cellulose-and alginate-based hydrogels for water and wastewater treatment: A review. *Carbohydr. Polym.* **2023**, *323*, 121339. [CrossRef] [PubMed]
46. Maćczak, P.; Kaczmarek, H.; Ziegler-Borowska, M. Recent achievements in polymer bio-based flocculants for water treatment. *Materials* **2020**, *13*, 3951. [CrossRef] [PubMed]
47. Hossain, M.S.; Hossain, M.M.; Khatun, M.K.; Hossain, K.R. Hydrogel-based super adsorbents for efficient removal of heavy metals in industrial wastewater treatment and environmental conservation. *Environ. Funct. Mater.* **2023**, *2*, 142–158. [CrossRef]
48. Feng, C.; Yang, P.; Liu, H.; Mao, M.; Liu, Y.; Xue, T.; Liu, K. Bilayer porous polymer for efficient passive building cooling. *Nano Energy* **2021**, *85*, 105971. [CrossRef]
49. Rehman, T.U.; Shah, L.A. Rheological investigation of polymer hydrogels for industrial application: A review. *Int. J. Polym. Anal. Charact.* **2022**, *27*, 430–445. [CrossRef]
50. Ćorković, I.; Pichler, A.; Šimunović, J.; Kopjar, M. Hydrogels: Characteristics and application as delivery systems of phenolic and aroma compounds. *Foods* **2021**, *10*, 1252. [CrossRef] [PubMed]
51. Jacob, S.; Nair, A.B.; Shah, J.; Sreeharsha, N.; Gupta, S.; Shinu, P. Emerging role of hydrogels in drug delivery systems, tissue engineering and wound management. *Pharmaceutics* **2021**, *13*, 357. [CrossRef] [PubMed]
52. Uva, M.; Atrei, A. Surface morphology at the microscopic scale, swelling/deswelling, and the magnetic properties of PNIPAM/CMC and PNIPAM/CMC/Fe$_3$O$_4$ hydrogels. *Gels* **2016**, *2*, 30. [CrossRef]
53. Clasen, T.F.; Thao, D.H.; Boisson, S.; Shipin, O. Microbiological effectiveness and cost of boiling to disinfect drinking water in rural Vietnam. *Environ. Sci. Technol.* **2008**, *42*, 4255–4260. [CrossRef] [PubMed]
54. Mustapha, S.; Oladejo, T.J.; Muhammed, N.M.; Saka, A.A.; Oluwabunmi, A.A.; Abdulkabir, M.; Joel, O.O. Fabrication of porous ceramic pot filters for adsorptive removal of pollutants in tannery wastewater. *Sci. Afr.* **2021**, *11*, e00705. [CrossRef]
55. Karikari, A.Y.; Ampofo, J.A. Chlorine treatment effectiveness and physico-chemical and bacteriological characteristics of treated water supplies in distribution networks of Accra-Tema Metropolis, Ghana. *Appl. Water Sci.* **2013**, *3*, 535–543. [CrossRef]
56. Rice, E.W.; Rose, L.J.; Johnson, C.H.; Boczek, L.A.; Arduino, M.J.; Reasoner, D.J. Boiling and Bacillus spores. *Emerg. Infect. Dis.* **2004**, *10*, 1887. [CrossRef] [PubMed]
57. Wysowska, E.; Wiewiórska, I.; Kicińska, A. The impact of different stages of water treatment process on the number of selected bacteria. *Water Resour. Ind.* **2021**, *26*, 100167. [CrossRef]
58. Fahimirad, S.; Fahimirad, Z.; Sillanpää, M. Efficient removal of water bacteria and viruses using electrospun nanofibers. *Sci. Total Environ.* **2021**, *751*, 141673. [CrossRef] [PubMed]
59. Tariq, Z.; Iqbal, D.N.; Rizwan, M.; Ahmad, M.; Faheem, M.; Ahmed, M. Significance of biopolymer-based hydrogels and their applications in agriculture: A review in perspective of synthesis and their degree of swelling for water holding. *RSC Adv.* **2023**, *13*, 24731–24754. [CrossRef] [PubMed]
60. Zhang, Z.; Zhang, W.; Hu, X.; Li, K.; Luo, P.; Li, X.; Duan, C. Evaluating the efficacy of point-of-use water treatment systems using the water quality index in rural southwest China. *Water* **2020**, *12*, 867. [CrossRef]
61. Kandemir, N.; Vollmer, W.; Jakubovics, N.S.; Chen, J. Mechanical interactions between bacteria and hydrogels. *Sci. Rep.* **2018**, *8*, 10893. [CrossRef] [PubMed]
62. Al-Ahmad, A.; Wollensak, K.; Rau, S.; Guevara Solarte, D.L.; Paschke, S.; Lienkamp, K.; Staszewski, O. How do polymer coatings affect the growth and bacterial population of a biofilm formed by total human salivary bacteria?—A study by 16S-rna sequencing. *Microorganisms* **2021**, *9*, 1427. [CrossRef] [PubMed]
63. Li, S.; Dong, S.; Xu, W.; Tu, S.; Yan, L.; Zhao, C.; Chen, X. Antibacterial hydrogels. *Adv. Sci.* **2018**, *5*, 1700527. [CrossRef] [PubMed]
64. Qureshi, D.; Nayak, S.K.; Maji, S.; Anis, A.; Kim, D.; Pal, K. Environment sensitive hydrogels for drug delivery applications. *Eur. Polym. J.* **2019**, *120*, 109220. [CrossRef]
65. El-Husseiny, H.M.; Mady, E.A.; Hamabe, L.; Abugomaa, A.; Shimada, K.; Yoshida, T.; Tanaka, R. Smart/stimuli-responsive hydrogels: Cutting-edge platforms for tissue engineering and other biomedical applications. *Mater. Today Bio* **2022**, *13*, 100186. [CrossRef] [PubMed]
66. Khan, F.; Atif, M.; Haseen, M.; Kamal, S.; Khan, M.S.; Shahid, S.; Nami, S.A. Synthesis, classification and properties of hydrogels: Their applications in drug delivery and agriculture. *J. Mater. Chem. B* **2022**, *10*, 170–203. [CrossRef] [PubMed]

67. Dsouza, A.; Constantinidou, C.; Arvanitis, T.N.; Haddleton, D.M.; Charmet, J.; Hand, R.A. Multifunctional composite hydrogels for bacterial capture, growth/elimination, and sensing applications. *ACS Appl. Mater. Interfaces* **2022**, *14*, 47323–47344. [CrossRef] [PubMed]
68. Adjuik, T.A.; Nokes, S.E.; Montross, M.D. Biodegradability of bio-based and synthetic hydrogels as sustainable soil amendments: A review. *J. Appl. Polym. Sci.* **2023**, *140*, e53655. [CrossRef]
69. Bordbar-Khiabani, A.; Gasik, M. Smart hydrogels for advanced drug delivery systems. *Int. J. Mol. Sci.* **2022**, *23*, 3665. [CrossRef] [PubMed]
70. Roy, A.; Manna, K.; Pal, S. Recent advances in various stimuli-responsive hydrogels: From synthetic designs to emerging healthcare applications. *Mater. Chem. Front.* **2022**, *6*, 2338–2385. [CrossRef]
71. Hu, C.; Yang, L.; Wang, Y. Recent advances in smart-responsive hydrogels for tissue repairing. *MedComm–Biomater. Appl.* **2022**, *1*, e23. [CrossRef]
72. Quazi, M.Z.; Park, N. Nanohydrogels: Advanced polymeric nanomaterials in the era of nanotechnology for robust functionalization and cumulative applications. *Int. J. Mol. Sci.* **2022**, *23*, 1943. [CrossRef] [PubMed]

Disclaimer/Publisher's Note: The statements, opinions and data contained in all publications are solely those of the individual author(s) and contributor(s) and not of MDPI and/or the editor(s). MDPI and/or the editor(s) disclaim responsibility for any injury to people or property resulting from any ideas, methods, instructions or products referred to in the content.

Review

Development Status of Solar-Driven Interfacial Steam Generation Support Layer Based on Polymers and Biomaterials: A Review

Haipeng Yan [1,†], Pan Wang [1,*,†], Lingsha Li [1], Zixin Zhao [1], Yang Xiang [1], Haoqian Guo [1], Boli Yang [1], Xulin Yang [1], Kui Li [1], Ying Li [1], Xiaohong He [2] and Yong You [3]

1. School of Mechanical Engineering, Chengdu University, Chengdu 610106, China; yhp200519@foxmail.com (H.Y.); lilingsha@stu.cdu.edu.cn (L.L.); zhao19913609107@163.com (Z.Z.); xiangyang@stu.cdu.cn (Y.X.); jingyu264452@foxmail.com (H.G.); itjbsyday16245791@163.com (B.Y.); yangxulin@cdu.edu.cn (X.Y.); likui@cdu.edu.cn (K.L.); liying@cdu.edu.cn (Y.L.)
2. School of Automation, Chengdu University of Information Technology, Chengdu 610225, China; hexiaohong@cuit.edu.cn
3. Key Laboratory of General Chemistry of the National Ethnic Affairs Commission, School of Chemistry and Environment, Southwest Minzu University, Chengdu 610041, China; youyong@swun.edu.cn
* Correspondence: wangpan@cdu.edu.cn
† These authors contributed equally to this work.

Citation: Yan, H.; Wang, P.; Li, L.; Zhao, Z.; Xiang, Y.; Guo, H.; Yang, B.; Yang, X.; Li, K.; Li, Y.; et al. Development Status of Solar-Driven Interfacial Steam Generation Support Layer Based on Polymers and Biomaterials: A Review. *Polymers* **2024**, *16*, 2427. https://doi.org/10.3390/polym16172427

Academic Editor: George Z. Kyzas

Received: 5 July 2024
Revised: 23 August 2024
Accepted: 23 August 2024
Published: 27 August 2024

Copyright: © 2024 by the authors. Licensee MDPI, Basel, Switzerland. This article is an open access article distributed under the terms and conditions of the Creative Commons Attribution (CC BY) license (https://creativecommons.org/licenses/by/4.0/).

Abstract: With the increasing shortage of water resources and the aggravation of water pollution, solar-driven interfacial steam generation (SISG) technology has garnered considerable attention because of its low energy consumption, simple operation, and environmental friendliness. The popular multi-layer SISG evaporator is composed of two basic structures: a photothermal layer and a support layer. Herein, the support layer underlies the photothermal layer and carries out thermal management, supports the photothermal layer, and transports water to the evaporation interface to improve the stability of the evaporator. While most research focuses on the photothermal layer, the support layer is typically viewed as a supporting object for the photothermal layer. This review focuses on the support layer, which is relatively neglected in evaporator development. It summarizes existing progress in the field of multi-layer interface evaporators, based on various polymers and biomaterials, along with their advantages and disadvantages. Specifically, mainly polymer-based support layers are reviewed, including polymer foams, gels, and their corresponding functional materials, while biomaterial support layers, including natural plants, carbonized biomaterials, and other innovation biomaterials are not. Additionally, the corresponding structure design strategies for the support layer were also involved. It was found that the selection and optimal design of the substrate also played an important role in the efficient operation of the whole steam generation system. Their evolution and refinement are vital for advancing the sustainability and effectiveness of interfacial evaporation technology. The corresponding potential future research direction and application prospects of support layer materials are carefully presented to enable effective responses to global water challenges.

Keywords: water treatment; solar evaporation; support layer; polymer; biomaterial

1. Introduction

Solar-driven interfacial steam generation (SISG) has emerged as a highly promising strategy for seawater desalination [1]. Different from traditional approaches like reverse osmosis [2–4], distillation [5–7], and nanofiltration [8,9], photothermal evaporation employs solar energy by converting sunlight irradiation into heat energy, which drives the vapor; the condensed water is collected after evaporation to complete the desalination of sea water. Solar energy is renewable and environmentally friendly, and there is no requirement for large energy consumption, expensive equipment, or high maintenance costs. Generally, as shown in Figure 1, there are three strategies for unitizing solar energy for evaporation [1]: (1) evaporation based on internal heating, in which uniformly dispersed solar absorbers convert incident solar photons into thermal energy to heat the liquid; (2) evaporation based on bottom heating, in which solar energy is absorbed by a solar absorber and converted into thermal energy to heat a large amount of liquid at the bottom of the evaporator; (3) evaporation based on interface heating, where solar thermal conversion and heating are limited to the gas–liquid interface. In recent decades, solar-driven interface evaporation technology has been widely studied; interface water evaporation technology uses solar energy to concentrate heat at the water–air interface for local water evaporation, thus minimizing heat loss in the desalination process [10]. The advanced fabrication of photothermal evaporators, including the employment of reasonable materials and optimized structure design, is the key to realizing efficient interfacial water evaporation.

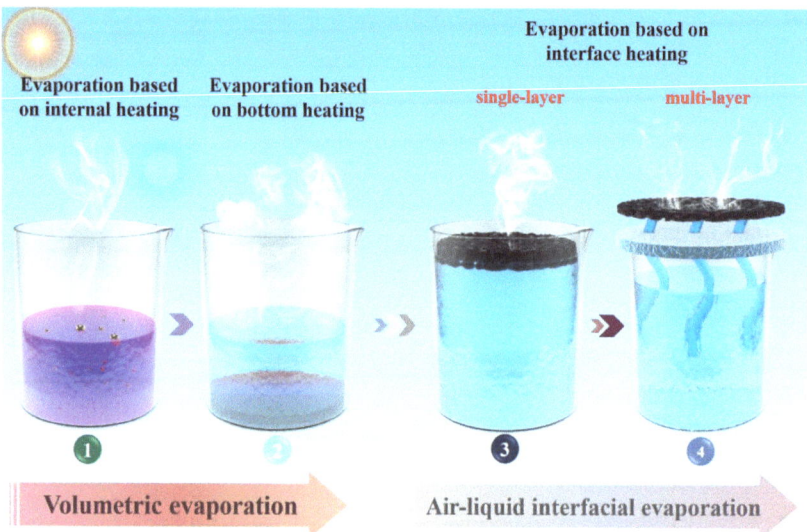

Figure 1. The overall material/structural scheme of an interfacial solar evaporation system [11].

Common SISG evaporators can be classified into single-layer and multi-layer evaporators. The single-layer evaporator structure typically consists of a lightweight floating layer with photothermal conversion materials directly placed on the water's surface. Since photothermal conversion materials usually need to be loaded onto a supporting substrate in order to float on the water's surface, the substrate is usually a porous membrane material, typically a polymer fiber membrane. However, this simplicity often leads to poor durability, heat loss, and other problems (Figure 1) [11–14].

The multi-layer evaporator typically combines at least a photothermal conversion layer and a support layer. The support layer is usually located at the bottom, providing stable structural support and preventing direct contact between bulk water and the thermal layer, reducing heat loss. It is typically constructed with lightweight materials that can float on the water [15]. The working principle of the multi-layer evaporator is that the top photothermal layer absorbs the incident sunlight and converts it into heat energy. Through capillary action, water gradually infiltrates the photothermal layer from bottom to the top through the water vapor transport channel arranged in the support layer under the photothermal layer, and accumulates into a thin surface water layer at the top of the photothermal layer. The photothermal layer heats the water in the surface water layer, causing it to evaporate into steam. The water in the surface water layer gradually evaporates, and the impurities in the water are retained in the photothermal absorption layer because they cannot evaporate, thus achieving desalination. Multi-layer interfacial evaporators for efficient interfacial evaporation technology are attracting increased attention from researchers for the scientifically based optimization of their parameters such as nature, morphology, thickness, interface, and structure. For example, after years of research, carbon-based materials [16], metal materials [17,18], and semiconductor materials have emerged as good choices for photothermal layer materials. They usually exhibit excellent light absorption and efficient evaporation characteristics. Compared with the core photothermal layer, the support layer has received less attention in many studies. However, with the deepening of research on the support layer, improving the support layer has gradually become an indispensable way to improve the overall performance of an SISG evaporator.

From the increasing number of research reports on SISG evaporators in recent years, it has been found that reasonable material selection and structural design of the support layer can significantly improve evaporation efficiency and structural stability. However, current research often relegates the support layer to a supporting role in discussions, and there remains a lack of comprehensive summarization in this field. This review mainly explores the support layer materials, including various porous polymer materials, biomaterials, and their corresponding composites. In addition, the common optimization of the structural design of the support layer, specifically the new 3D structure (vertebrae and arch), is also presented here. This work aims to describe the development status of the support layer in multi-layer evaporators, and finally propose future research directions for advancing support layer construction in the field of SISG technology for water treatment.

2. Support Layer Materials

In order to manage the heat and material loss from direct water surface contact in photothermal evaporators, the typical design involves separating the photothermal layer and support layer. This structure provides thermal insulation protection and structural support. Thus, the thermal conductivity and water absorption characteristics should be carefully considered when choosing the support layer material. At present, most common support layer materials are based on polymers. Among the family of polymer-based support layers, commercial polymer foams are characterized by their excellent thermal insulation and buoyancy [19]. These foams can also be enhanced through modifications to form composite materials, which often exhibit superior properties [20]. Aerogels and hydrogels are also promising support layer materials, and they stand out because of their high polymer crosslinking, facilitating efficient water transport and thermal conductivity [21,22]. At the same time, a support layer fabricated using biomaterials possesses the advantages of a wide range of sources, low cost, and eco-friendliness [23]. Biomaterials offer natural structural advantages, such as porosity and micro-channels, which ensure effective water transport. Some biomaterials even have natural beam tube structures, enhancing their utility as support layer materials. Moreover, reasonable treatments of biomaterials, including drying, carbonization, and modification with functional components, enable them to achieve high evaporation efficiency and suitability for diverse practical applications

2.1. Polymer Foam

2.1.1. Commercial Polymer Foam

Polymer foam is a common candidate for support layer fabrication, offering a particularly wide array of choices, such as polyurethane (PU) foams [24,25], polystyrene (PS) foams [3,26], and polydimethylsiloxane (PDMS) foam. These foams are characterized by their light weight, ability to float, and low thermal conductivity, as well as their cost-effectiveness, making them applicable as support layer materials. Their light weight structure allows them to float freely on the water's surface, which is a critical factor in evaporator design. In addition, foams ensure the stability and reliability of the evaporator. At the same time, foams' low thermal conductivity effectively insulates against heat loss, thereby preventing the heat generated from sunlight absorption by the photothermal layer from spreading the foam layer. This insulation is responsible for effectively maintaining high evaporation efficiency in the evaporator.

The porous structure of commercial plastic foams is another important feature that confers several advantages during application in SISG technology. The porosity of the foam allows water to easily enter the evaporator and reach the evaporation interface, thereby improving the evaporation efficiency. At the same time, the porous structure provides a larger surface area, allowing more water to absorb and diffuse into capillary water flow, further improving evaporation efficiency. Shi et al. [27] chose polystyrene (PS) foam as the support layer material and covered it with porous graphene oxide (rGO) film in their evaporator design. The top layer of porous rGO acted as a light absorber, efficiently capturing sunlight and converting it into thermal energy. Meanwhile, the bottom PS foam layer effectively breaks down water transport channels and acts as an excellent thermal barrier, minimizing heat transfer to the bulk water. The designed SISG evaporator showed efficient water transfer and a good evaporation efficiency: 1.31 kg m^{-2} h^{-1}.

The durability of the polymer foam deteriorates with an increase in the serving time, and the long-term accumulation of heat causes overflow at some point. To address this weakness, Nguyen et al. [28] designed an efficient SISG evaporator device with a polyethylene foam support layer. The durability of the designed evaporator was improved by wrapping polyethylene foam with cotton gauze, thereby minimizing the area of contact with bulk water (Figure 2a). The cotton gauze's ability to quickly absorb water also ensures the water supply during the evaporation process. Additionally, Kiriarachchi et al. [29] employed a cotton rod structure to raise the foam layer to the surface of the water, preventing direct contact between the water and polymer foam, and facilitating the transfer of water to the top layer for evaporation (Figure 2b).

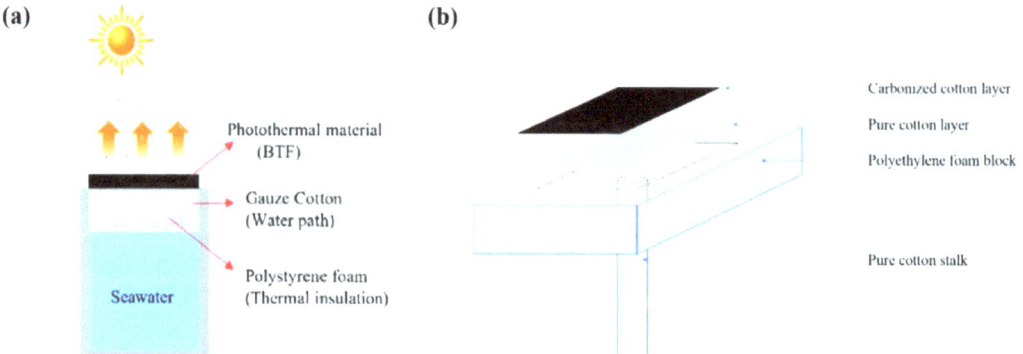

Figure 2. The schematic of the seawater evaporation experiment designed by Nguyen (a) [28] and schematic representation of the CC solar steam-generation device designed by Kiriarachchi (b) [29].

Due to the increases in evaporation rate achieved in many recent studies, this rate has long surpassed the rate of the support layer water supply, so the latter limits the performance of the SISG evaporator. Thus, some restructuring of the support layer using commercial polymer foam has been performed. Wang et al. [30] solved this problem by adding highly absorbent polymer particles (SAP) to ordinary plastic floating foam. SAP (highly absorbent polymer) is a functional polymer material that becomes a hydrogel after absorbing water, and has good thermal insulation properties and abundant water channels. The support layer consists of cylindrical hollow foam with a diameter of 32 mm. SAP particles are added to the hollow foam. The experiment involved related tests on the non-woven hollow foam with 0.1 g of SAP particles added. The test results showed that, compared to the foam without SAP particles, under the same illumination intensity and at the same time, the temperature change in the foam with SAP particles added decreased slightly, mainly because the increased water evaporation resulted in some heat leaving the surface of the photothermal layer. This shows that the addition of SAP particles achieves a fast water supply.

2.1.2. Functional Polymer Foam

The service environment of the evaporator affects its efficiency and service life in terms of various factors. For example, high salinity or pollution affect the durability of the support layer material [31]. Due to the limitations of traditional polymer foam in extreme environments, there is increasing research in developing functional composite foams. These include modified foams that are antibacterial, corrosion-resistant, etc. [32]. At the same time, recent studies have aimed to integrate functionalities into composite foams, combining capabilities for light and heat absorption with effective heat insulation [33,34]. Additionally, functional coating treatment and composite fabrication prior to the foaming process are also involved.

Coating Functionalized Polymer Foam

The high-performance and multi-functional coating material is configured and deposited on the surface of the polymer foam. The coated composite foam can be used as the support layer material to improve durability, functionality, etc. For surface coating treatment, one approach is to use a polymer component to coat the foam. This method usually improves the corrosion resistance and antibacterial properties of the foam [35]. Another method is adding functional nanomaterials to the surface coating to enhance its properties. It is noteworthy that, when employing surface-coating methods to treat foam, in addition to the usual polyurethane foam, melamine foam is also considered a good, original skeleton. This is because the open porous surface of melamine foam enables adhesion to any photothermal layer and is easy to combine with various coatings, without any surface degradation; compared with other polymer foams, melamine foam also has good thermal insulation properties and good pore orientation [36,37].

Pinto et al. [36] developed highly effective functional foams with antimicrobial properties (Figure 3a), explored the possibility of using them as antibacterial filters for water treatment, and decorated melamine foams with silver nanoparticles (ME/Ag). A uniform coating of silver nanoparticles (Ag NPs) with a diameter of less than 10 nm was formed in situ directly on the prop surface of the foam and then immersed in a $AgNO_3$ solution. It was demonstrated that the nanoparticles adhered stably to the foam, and when filtered at flow rates of up to 100 mL/h·cm^2, they were able to completely remove E. coli from the water, with the foam releasing less than 1 ppm of silver ions. After further dilution of the treated water, no bacterial regeneration was observed, and the level of mercury fell below the safety threshold for drinking water (0.1 ppm); subsequent combination with the photothermal layer can produce a multi-layer interfacial evaporator with an antibacterial support layer. Sometimes, it is necessary to ensure that the bottom support layer has a certain capacity to absorb heavy oil, under the premise of not destroying the photothermal layer, to achieve the effect of removing oil and ensuring interface evaporation at the photothermal layer. Traditional electric/solar heating absorption methods have low efficiency and high fire risk when heating high, easy fuel oil, which necessitates good flame retardancy in the support material. Shi et al. [38] developed a novel, highly efficient nanocoating for heavy oil absorption; the main component of the coating was rGO/Fe_3O_4, the coating showed both photothermal conversion ability and non-flammability, and it was loaded on the skeleton of melamine foam through a simple co-precipitation and dip-coating process. Thus, the modified foam could quickly and effectively remove high-viscosity heavy oil (3000 Pa·s). With the help of simulated sunlight, an absorption capacity for heavy oil of up to 75.1 g/g was achieved, and the absorption rate reached 9000 g m^{-2} min^{-1}. At the same time, thanks to the dual photothermal conversion ability of functional rGO/Fe_3O_4 coating, the fabricated foam also has a certain photothermal evaporation ability.

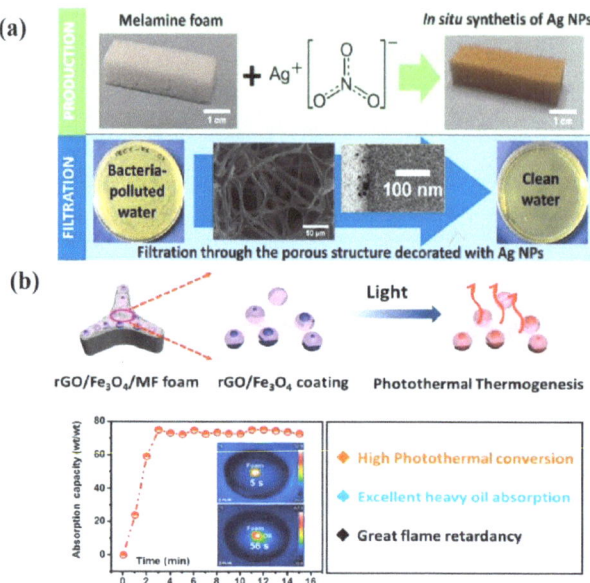

Figure 3. Photographs of the treated ME/Ag foams (**a**) [36], and schematic illustration for the preparation of the rGO/Fe$_3$O$_4$/MF foam (**b**) [38].

Substrate Modified Functional Polymer Foam

After adding modified materials to the raw material to form a composite material, foaming treatment was performed, for example by adding nanoparticles to the base material. In this way, the raw material can enhance the structural strength and durability of the foam. Wang et al. [39] used this method to prepare nano-Ag-based photothermal foams with good light absorption and low thermal conductivity. The photothermal Ag NPs were added to the polyurethane base material and stirred evenly with an electric mixer, and then a certain amount of isocyanate was added. The resulting mixture was stirred at high speed, and then quickly poured into a mold and foamed at room temperature. Due to the introduction of nanosilver, the newly developed foam had good photothermal properties. Polyacrylamide absorbent material (PAM) was also added to the foam pores to provide a fast water supply channel for rapid water supply.

Hitherto, there has been much research on the preparation of commercial foam and modified foam as the support layers of a photothermal evaporator to take and optimize the advantages of polymer foam. However, at the same time, a polymer foam-based support layer still presents the disadvantages of complex processes, pollutant discharge, and difficulty in degradation, making it unfriendly to the environment.

2.2. Gel

2.2.1. Hydrogel

A gel is a kind of flexible, solid polymer material with a unique three-dimensional structure, which is lightweight and floating, has good thermal conductivity, and is an ideal choice for support layers in a photothermal evaporator [40]. As common polymer gels, hydrogels are usually prepared with water as a dispersive medium and are crosslinked into a three-dimensional polymer network with acrylic polymer, acrylamide polymer, polyvinyl alcohol, etc. [41]. There are abundant hydrophilic groups such as hydroxyl, amino, and carboxyl groups in the polymer chains of the polymerized hydrogels. These functional groups are good at attracting water molecules into the three-dimensional networks of hydrogels, ensuring that they can quickly transport water to the photothermal layer when used as a support layer in a photothermal evaporator [21,40,42]. Excellent thermal conductivity can also prevent significant heat loss during light and heat evaporation. Evaporators using hydrogels as support layers have been used in water treatment fields such as seawater desalination and wastewater treatment.

Shi et al. [43] developed an interpenetrating network hydrogel evaporator with high mechanical strength, a high water evaporation rate, and good photoelectric conversion performance. Polyacrylamide (PAM) and crosslinking agents were used to form a three-dimensional crosslinked network structure, and the cis-hydroxyl group on the branch chain of galactomannan (GG) mannose formed a borate ester crosslinked network that enhanced the mechanical properties of the system. A photothermal layer composed of graphene oxide (GO) and Fe_3O_4 nanoparticles (Fe_3O_4 NPs) provides excellent photothermal transfer capability (Figure 4a). Under one sun irradiation, the GFCP evaporation rate reached 2.3 kg m^{-2} h^{-1}, and the conversion efficiency was 90.7%. Wang et al. [44] also prepared a new type of reduced graphene oxide (rGO)/$Cu_{7.2}S_4$ polyvinyl alcohol (PVA)/polyacrylamide (PAM) composite hydrogel evaporator. The two-component pore structure of PVA/PAM hydrogel can improve the influence of water supply and drainage. The rGO/$Cu_{7.2}S_4$ provides photothermal conversion capability and achieved a high evaporation rate (2.19 kg m^{-2} h^{-1}) (Figure 4b).

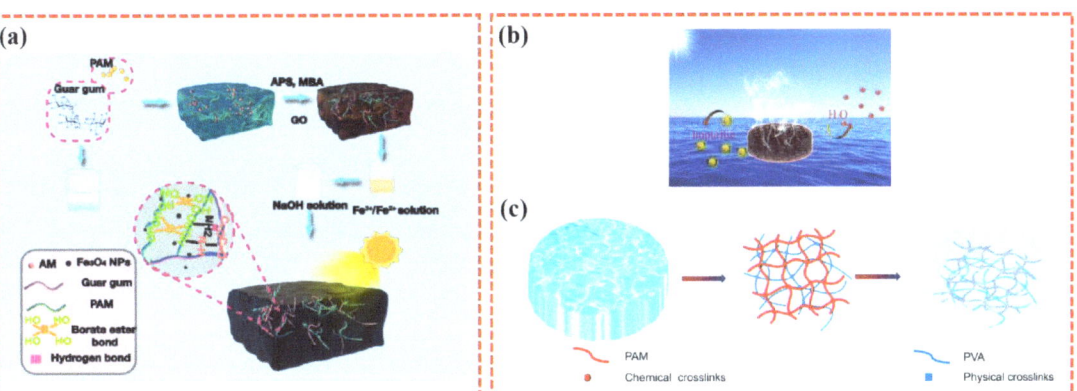

Figure 4. Flowchart of GFGP hydrogel evaporator preparation (**a**) [43]. A schematic diagram of the hydrogel mixing device generated through solar steam generation (**b**). A schematic diagram of the PVA/PAM hydrogel polymer network's formation (**c**) [44].

As an emerging green material, hydrogels show many unique advantages, excellent water absorption, heat insulation, biocompatibility, and flexibility, making hydrogels an ideal choice for the support layers of photothermal evaporators. However, hydrogels have limited durability, and prolonged exposure to sunlight can affect the evaporation rate. Therefore, when preparing hydrogels, consideration should be given to improving their durability. Some researchers are also trying to modify hydrogels' properties further, such as their stain resistance [45,46], salt resistance [47,48], and bacteriostatic effects [49,50], to provide hydrogels with more prospective uses.

2.2.2. Aerogel

Aerogels are materials that trap air molecules in porous (up to 99.8%) filled solids that prevent air from escaping. There are certain differences between aerogels and hydrogels in terms of the solvent selection, gelation method, special treatment, and drying method. Hydrogels are prepared by using water as the main solvent. Aerogels usually use organic solvents or other solvents [51]. Aerogel, a very low-density and high-surface-area porous material, also has a three-dimensional structure. There is much research on aerogels as supporting layers in photothermal evaporators. Usually, aerogels are combined with other functional photothermal materials to form multi-layer photothermal evaporators for interfacial evaporation. The raw materials for preparing aerogel are abundant, environmentally friendly, and cheap, and, most importantly, their biodegradability and biocompatibility are good, and they are easy to modify and functionalize [22,51,52].

Wu et al. [53] developed a multifunctional aerogel for interfacial evaporation. Plantago stalk (PC) was selected as a cellulose matrix and crosslinked with biomass hollow carbon tubes (hct) after chemical treatment to make an aerogel with a three-dimensional porous mesh scaffold (Figure 5a). The hct crosslinked with the aerogel has excellent photothermal conversion capability while retaining the characteristics of the aerogel as a flexible solid, which is porous and lightweight. The temperature of the optimized PC@HCT aerogel rapidly reaches 46.7 °C at the gas–liquid interface at 1.0 solar irradiance, and the evaporation rate is 1.86 kg m^{-2} h^{-1}. In addition, PC@HCT aerogel has an obvious purification effect on seawater, heavy metal ion solutions, oil–water emulsions, and organic dye wastewater. It is important that the dry PC@HCT has good hardness and superelasticity in water. Furthermore, PC@HCT aerogels are green and recyclable, changing the healthy growing environment of sand and plants, thereby obtaining materials from nature and ultimately returning them to nature. Storer et al. [54] prepared a 3D photothermal aerogel composed of reduced graphene oxide (RGO) nanosheets, straw-derived cellulose fibers, and sodium alginate (SA) for solar steam generation. The photothermal aerogel shows an intense broadband light absorption of 96–97%. In the process of solar steam power generation, the 3D photothermal aerogel effectively reduces the radiation and convective energy loss while enhancing the environmental energy collection, resulting in a very high evaporation rate of 2.25 kg m^{-2} h^{-1} and an energy conversion efficiency of 88.9% under 1 solar irradiation (Figure 5b).

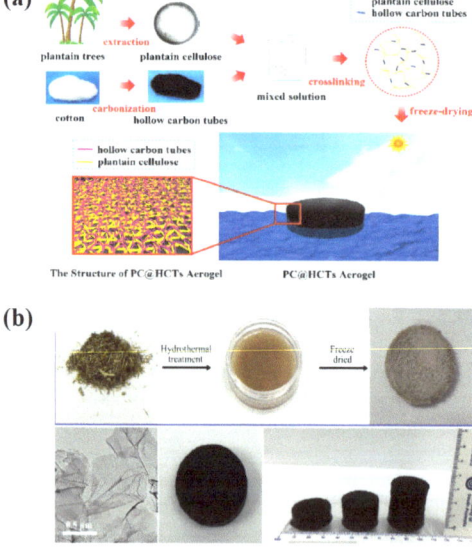

Figure 5. Schematic of the fabrication of the solar evaporator with the PC@HCTs aerogel (**a**) [53], and schematic of the fabrication of the solar evaporator with the RGO–SA–cellulose aerogel (**b**) [54].

Aerogel is not, itself, inherently able to perform photothermal conversion. However, aerogel preparation is highly scalable and can be combined with different crosslinkers to meet different functional use scenarios. The aerogel's high surface area and excellent thermal conductivity are also ideal for support layer materials in photothermal evaporators.

However, the fragile characteristics of aerogel affect its durability to a certain extent, and steps to improve its durability and mechanical properties during preparation should be considered [55,56].

2.3. Biomaterials

2.3.1. Natural Plants

Natural biomaterials have the inherent properties of hydrophilicity, cellulose components, and multi-dimensional holes and pores, making them a cost-effective source for support layers [57]. By now, a variety of plants, such as bamboo [58], pomelo peel [59], and wood [60] have been used directly or have inspired researchers to construct efficient SISG evaporators. Herein, wood has emerged as a favored candidate owing to its light weight and ability to float on the water's surface. Moreover, it exhibits exceptional durability, retaining its structural stability after long periods of soaking. Furthermore, the inherent capillary structure of the wood facilitates gradual water transport to the photothermal layer, enhancing its suitability for applications in advanced SISG technology [61]. Extensive research on wood-based photothermal evaporators has been conducted, where unmodified wood has been directly used as a support layer (Figure 6a) [62,63]. Chen et al. [64] proposed a simple, cost-efficient, and scalable brushing method to prepare an aluminophosphate-treated wood (Wood@AlP) solar steam-generation device. The wood@AlP device can float on seawater and exhibits a high solar thermal efficiency of 90.8% with a net evaporation rate of 1.423 kg m^{-2} h^{-1} under one sun illumination (Figure 6b).

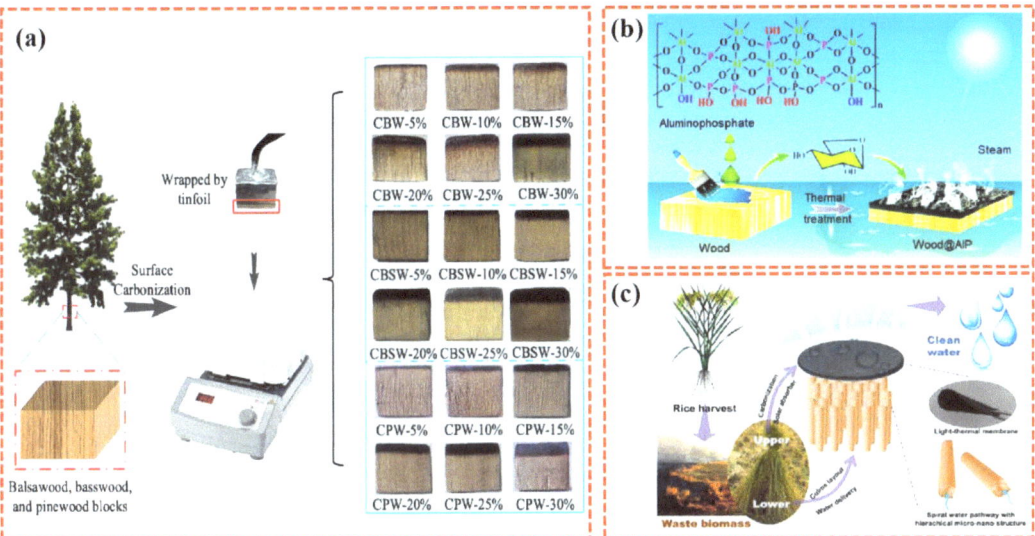

Figure 6. The fabrication of a carbonized wood-based solar evaporator through surface carbonization (**a**) [62]. Schematic image showing the fabrication of the Wood@AlP-based solar steam-generation device (**b**) [64]. Illustration of the design route of the biomass-based solar steam-generation device (**c**) [65].

It is worth noting that the prolonged exposure of natural wood to high-salinity environments can result in the blockage of its capillary structure, thereby reducing overall water transmission. In view of the negative effect of salt crystallization on the water transport in the wood support layer, a series of wood-based evaporators with good salt drainage properties were investigated [66]. Shi et al. [67] explored a wood-based composite material composed of an evaporator with high evaporation performance and high salt resistance. They successfully obtained a solar-powered wood desalination device (PPy-E-Wood) through the in situ polymerization of pyrrole monomer on pretreated elastic wood. Efficient steam generation was successfully achieved through the synergistic effect of PPy NPs' very light thermal conversion layer with a wood substrate with low thermal conductivity, micro-/nanopores, and channels. Bamboo, which has a natural porous capillary structure similar to wood, has also been selected for the support layer material in photothermal evaporators. Zhang et al. [58] developed a simple bamboo-based photothermal evaporator that uses a simple surface self-assembly technique to load polypyrrole (PPy) onto bamboo as a photothermal layer and bamboo as a support layer to make a solar evaporation device. The PPY-Bamboo solar evaporation device achieves 88% absorption of sunlight in the UV–vision-near-infrared region. The PPY-Bamboo solar evaporation device can achieve a high photothermal conversion efficiency of 76.87% under the light intensity of one sun.

In addition to biomaterials with natural porous structures, some biomaterials with hollow structures of beam tubes are also good choices for support layers. Fang et al. [65] took inspiration from rice straw and designed an evaporator that mimics the structure of rice straw as a support layer. Rice straw water transport is smooth, and the stratified channel of the stem has a spiral winding space and other multi-level structures, providing a variety of water channels, avoiding the scale, blockage, or scaling that may be caused by sand or soil particles, and maintaining its stable and anti-scaling capillary effect (Figure 6c). With the increasing innovations in natural plant-based support layer materials, diverse and unexpected naturally porous biomaterials, such as coconut shell [68], carrot [69], and loofah [70], can also be studied.

2.3.2. Carbonized Biomaterials

The carbonization pre-treatment of natural green plants can enhance the material's photothermal conversion ability. After carbonization pre-treatment, the biomaterial has the function of integrating the support layer and the photothermal layer, avoiding the need to re-design the photothermal layer. The biomass carbon prepared in this manner has a photothermal conversion ability, while retaining the supporting role of natural biological materials. Carbonized greens, such as carbonized woods and other plant-based carbon materials, are increasingly being explored for their effectiveness in interfacial water evaporation applications. These carbonized materials have shown promising properties, including high photothermal conversion efficiency due to their own dark aspect, good hydrophilicity inherited from proteins and celluloses, and scalability. Chen et al. [71] selected natural corn cob for a simple carbonization treatment (Figure 7a). Carbonized corn cob (C-corncobs) showed a rough and black surface, and maintained its natural porous structure. The surface carbonization layer acts as the light-absorbing layer. The "vesicle" structure of the central corn cob is similar to that of the foam composite material, and this special "vesicle" structure facilitates thermal management and water supply, and provides sufficient steam channels for achieving light absorption and photothermal conversion. In addition to corn cobs, Xu et al. [72] found that mushrooms can also be used for solar evaporation. The natural porous capillary fiber structure of mushrooms can provide good water transport channels and inhibit heat diffusion. The subsequent carbonization treatment of natural mushrooms can further improve their evaporation efficiency. Zeng et al. [73] used carbonized waste durian skins to develop a new type of solar evaporator with a three-dimensional photothermal structure. The carbonized durian has a macroscopic three-dimensional pyramid, and a microscopic porous and petal-like structure that contributes to ideal light capture and absorption, and enables an extremely high absorption rate of up to 99% of the solar spectrum. The abundant porous microstructure inside the carbonized durian provides a good capillary effect for adequate water supply. There are many other green plants used in the preparation of biological carbon materials, such as oranges and daikon [74–77] (Figure 7b–d).

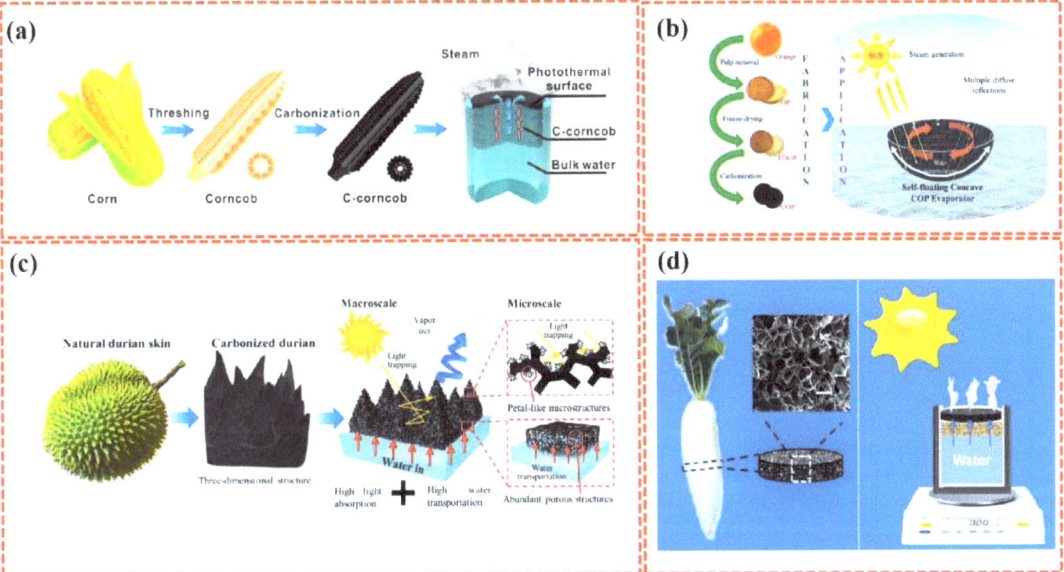

Figure 7. Schematic of the C-corncob-derived SISG device (**a**) [71]. Schematic of the 3D carbonized orange peel device (**b**) [75]. Schematic of a carbonized durian solar evaporator. Excellent solar light absorption is maintained by the light trapping of three-dimensional pyramid porous structures together with petal-like microstructures (**c**) [73]. Schematic of the carbonized daikon device (**d**) [77].

Studies are trying to improve the evaporation efficiency of carbonized biomaterials through simple modification while maintaining the advantages of easy degradation and low pollution. Therefore, modifying carbonized biomaterials further to improve their comprehensive evaporation performance or provide them with multiple functions has become a new research frontier.

2.3.3. Modified Biomaterials

Modified biological materials are generally altered to enhance their performance as either the support layer or photothermal layer in interfacial water evaporators. However, basic biochar materials can serve as a support layer with a certain photothermal transfer ability. Under carbonization at high temperatures, natural structures such as vascular bundles in plants can be destroyed, affecting the water transport capacity inside the material. Solar interface evaporation often requires materials with strong water transfer capacities. In order to solve this problem, researchers have proposed incorporating modified materials that enhance water transport into carbonized biological materials. The resulting composite materials exhibit improved water transport performance. Guo et al. [78] employed loofah to prepare a MOF-801@CL composite material, which has good hydrophilicity and excellent water evaporation stability. Firstly, the loofah was selected as the substrate biomaterial. The loofah is easy to obtain and of low cost, and degrades easily without polluting the environment. By using MOF-801 as an adsorbent, a network structure similar to CL was synthesized using the in situ growth method, and MOF-801@CL was prepared. MOF-801 has good hydrophilicity and a rich pore structure, which can not only improve the hydrophilicity of CL but also provide additional water transport channels. Under the synergistic effect of the excellent water absorption of MOF-801, and the excellent photothermal conversion performance of CL, the water evaporation rate of MOF-801@CL is 1.42 kg m^{-2} h^{-1} under one sun, which is about 1.2 times that of CL, and the solar steam conversion efficiency is 88.9%. After the addition of MOF-801, the evaporation rate of CL was significantly increased.

In addition to the good water transport capacity required for materials used for interfacial water evaporation, some researchers have also loaded modified materials that can enhance the photothermal properties of the materials into the support layer substrate. Zhang et al. [79] selected sugarcane as the substrate due to its highly developed pore structure and excellent hydrophilicity. It is well known that conventional biomass carbon, including carbonized sugarcane, typically exhibits unsatisfactory absorption in the near-infrared region. Therefore, antimony-doped tin oxide (ATO) was introduced as a modified material to enhance the photothermal properties of biomaterials. ATO was loaded into sugarcane and then carbonized to obtain a composite solar evaporator CS@ATO with a solar energy absorption rate of 99%, a solar energy evaporation rate of 1.43 kg m^{-2} h^{-1}, and an efficiency of 95.3%. In addition, the fabricated composite material was effective at treating wastewater, as experimental results showed that the concentration of heavy metal ions in wastewater was reduced by 3–4 orders of magnitude after CS@ATO treatment, with almost complete removal of dye from the wastewater.

By employing natural plants, carbonized biomaterials, and modified biological materials in the ways mentioned above, researchers aimed to create biomaterials that not only perform well in interfacial water evaporators but that also possess characteristics like biocompatibility, sustainability, and cost-effectiveness, which are advantageous for practical applications in water purification and desalination technologies.

To further clarify the comparison of the aforementioned support layer materials employed in the field of SISG evaporators, Table 1 was created to provide a brief summary of each type of material.

Table 1. The brief review of different types of support layer material in multi-layer SISG evaporators.

Support Layer Category	Typical Materials	Main Advantages
Commercial polymer foam	Polyurethane (PU) foams Polystyrene (PS) foams Polydimethylsiloxane (PDMS) foam	Light weight, self-floating, low thermal, conductivity.
Functional polymer foam	Coating Functionalized Polymer Foam Substrate Modified Functional Polymer Foam	Light weight, self-floating, high-performance, multi-functionalization.
Gels	Hydrogel Aerogel	Environmentally friendly, unique transporting water structure.
Biomaterials	Natural plants Carbonized biomaterials Modified biomaterials	Cost-effective, rich source environmentally friendly.

3. Support Layer Structure Optimization

In addition to appropriate material selection, optimizing the structural design of the support layer, which depends on various support layer materials, is crucial for improving the efficiency of the SISG evaporator. The optimization of interfacial evaporator structure addresses challenges, including heat loss and low evaporation efficiency. In addition to the traditional two-dimensional structure device, more efficient three-dimensional evaporator structures have been designed, such as conical structures [80], umbrella structures, and arch structures [81], and some emerging technologies, such as 3D printing technology, mainly based on polymers [82,83], are also being used in support layer structure design.

3.1. Cone Structure

The 3D cone structure is particularly advantageous due to its curved surface geometry, which facilitates multiple reflections and refractions of sunlight within the cone. This design enhances the probability of sunlight being captured and absorbed across a broad spectrum of wavelengths. From a macroscopic perspective, the cone structure promotes the efficient utilization of broadband light, thereby maximizing the photothermal conversion efficiency within the device [84]. Lv et al. [85] developed a 3D (3D) graded inverted conical solar evaporator consisting of a 3D copper foam skeleton cone and a graphene oxide functionalized micro-/nanostructure decorated on its surface by optimizing the foam structure of the support layer (Figure 8a). The high surface area of the skeleton structure provides sufficient space for the diffusion of steam from the boundary to the environment. The synergistic effect of the superhydrophilic skeleton and the air-planking paper wrapped around the skeleton prevents not only insufficient evaporation due to insufficient hydration but also heat loss due to the excessive evaporation of water. The rGo photothermal layer modified on its surface further improves the evaporation efficiency. Similarly, Xie et al. [86] developed a wood conical evaporator, which has multiple light reflections on the surface of the conical device, reducing diffuse reflection and promoting light absorption. When using this wood cone evaporator, the evaporation rate and efficiency reached 1.79 kg m^{-2} h^{-1}. Bu et al. [87] designed a 3D carbon fiber cotton-based conical evaporator (CFC cone) with an adjustable water supply. The 3D CFC cone evaporator was designed to effectively collect incident sunlight, with an evaporation rate of 3.27 kg m^{-2} h^{-1} and a photothermal conversion efficiency of 194.4% under one sun. On the basis of the conical structure, further micropore structure optimization was carried out, and the bionic tree structure inspired by natural trees retained the advantages of the conical structure while achieving good co-ordination of fast and high-flux water transport and salt drainage [88].

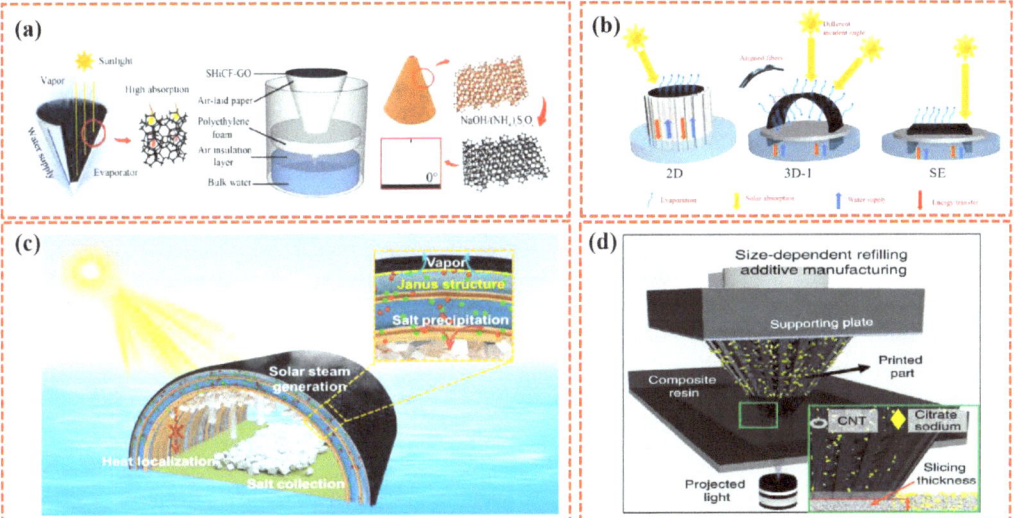

Figure 8. Schematics of the proposed light absorption principle, evaporator, fabrication process, and characterization of the 3D inverted conical solar evaporator (**a**) [85]. Some schematic diagrams of the arch structure evaporation device (**b**,**c**) [89,90]. Schematic configuration of size-dependent resin refilling induced additive manufacturing based on the continuous DLP 3D printing system (**d**) [91].

3.2. Arch Structure

Coincidentally, the simple arch structure can also significantly enhance the overall evaporation performance. Xu et al. [89] prepared a three-dimensional arched solar interface evaporator through material selection and simple structural design (Figure 8b). The photothermal layer originally used for polymerized PPY on air cushion paper combines strong capillary action and photon capture capabilities, and the simple macro-support layer structure design (3D arch structure) can greatly reduce heat conduction losses, thereby increasing the surface evaporation temperature. Chen et al. [90] demonstrated a Janus arched solar evaporator with a hydrophobic top photothermal layer for effective light absorption, solar thermal conversion, and steam evaporation with an evaporation rate of 2.82 kg m^{-2} h^{-1} in pure water. The arched design of the evaporator makes full use of the upper space and effectively reduces the evaporator footprint. The ratio of evaporation area to occupied area is 1.57:1 (Figure 8c).

3.3. Other 3D Structures

Other innovative 3D structured SISG evaporators are also being developed. For instance, Wu et al. [91] developed a pyramid evaporator using 3D printing technology (Figure 8d). Surface-distributed micropores are formed on the prepared surface, endowing the pyramid evaporator with an ultra-fast water spreading property. Due to the designed morphology of the 3D structure with asymmetric grooves and the gradient microcavity arrays, the liquid film spreads on the structure surface displays a position-related liquid film thickness and temperature gradient along the sidewall, which further leads to the thermocapillary force inside the liquid film and the ability to capture energy from the surrounding environment to enhance water evaporation and energy efficiency. This design leverages the unique geometry of a pyramid to enhance solar energy absorption and heat retention, thereby improving the efficiency of water evaporation. Zhang et al. [92] proposed a three-dimensional (3D) cup-shaped evaporator based on carbonized sorghum straw (CSS). The 3D structure of this cup-shaped evaporator incorporates features that effectively absorb heat loss from reflected light and thermal radiation. The excellent light absorption, super hydrophile properties, and good thermal insulation of the CSS give the 3D cup evaporator an excellent evaporation rate (3.27 kg m^{-2} h^{-1}) and energy efficiency (131.2%) under one sun irradiation. Three-dimensional cup evaporators also show excellent salt resistance and good acid and alkali resistance in the production of clean water from seawater and wastewater.

The trend towards 3D structure design represents the future of enhancing the evaporation rate and efficiency in interfacial evaporators. Compared to traditional two-dimensional designs, 3D solar-driven interface evaporators offer additional surface area under the same solar irradiation. This expanded surface area effectively accelerates the evaporation. Reasonable structural design and the synergistic effect of high-performance materials achieve high utilization efficiency in the evaporator. This approach not only enhances evaporation rates but also introduces innovative possibilities for advancing research in water treatment through interfacial evaporation technology [87,93–96].

4. Conclusions and Prospect

In this paper, the material selection and structural design of support layers in the field of multi-layer SISG progress were reviewed. Support layer material selection focuses on polymers and biomaterials. The polymers, which are mainly chosen as support layer materials, include foams and gels, while the biomaterials include natural plants, and carbonized and further modified biomaterials. Original unmodified materials like polymer foams and natural plants offer simplicity in preparation, and they often lack durability for practical applications, especially in harsh environments such as high-salinity conditions. Modified support layer materials have shown promise in enhancing performance to some extent, yet scaling them from laboratory settings to practical large-scale applications remains a challenge. In addition to the innovation of supporting layer material selection, recent research progress emphasizes the fact that structural optimization is also a key way to improve the efficiency of photothermal evaporators. Innovative 3D support layer structures, such as cone and arch designs, effectively regulate the light absorption area and steam flow resistance, presenting new avenues for surface structure engineering in enhancing photothermal evaporator performance.

The design of the support layer of the photothermal evaporator still faces many challenges. Compared with photothermal materials, the selection of support layer materials is relatively limited, and most studies focus on solid materials. The field of water treatment in the real world is much more complex. The support layer is not limited to the role of water transport support. The functionalization and integration of the support layer have become the focus of the subsequent design of photothermal evaporators. Broadening the selection of materials, improving the performance of the support layer, and making it commercially applicable will need to be considered in future research. It is believed that SISG technology, supported by various advanced support layers, can help drive eco-friendly, cost-effective, and sustainable development in the field of water purification.

Author Contributions: Conceptualization, H.Y. and P.W.; investigation, H.Y., L.L., Z.Z., Y.X., H.G. and B.Y.; writing—original draft preparation, H.Y. and P.W.; supervision, P.W., X.Y., K.L., Y.L., X.H. and Y.Y.; funding acquisition, P.W., Y.L. and X.H. All authors have read and agreed to the published version of the manuscript.

Funding: This research was funded by the National Natural Science Foundation of China (No. 52202083/No.52203102), Natural Science Foundation of Sichuan Province (24GJHZ0024, 2024NS-FSC0263), Sichuan Province Engineering Technology Research Center of Novel CN Polymeric Materials (CNP-C-240206) and the Research Startup Foundation of Chengdu University, China (2081920009).

Institutional Review Board Statement: Not applicable.

Data Availability Statement: Not applicable.

Conflicts of Interest: The authors declare no conflicts of interest.

References

1. Tao, P.; Ni, G.; Song, C.; Shang, W.; Wu, J.; Zhu, J.; Chen, G.; Deng, T. Solar-driven interfacial evaporation. *Nat. Energy* **2018**, *3*, 1031–1041. [CrossRef]
2. Kurihara, M. Seawater Reverse Osmosis Desalination. *Membranes* **2021**, *11*, 243. [CrossRef] [PubMed]
3. de Moraes, E.G.; Sangiacomo, L.; Stochero, N.P.; Arcaro, S.; Barbosa, L.R.; Lenzi, A.; Siligardi, C.; de Oliveira, A.N. Innovative thermal and acoustic insulation foam by using recycled ceramic shell and expandable styrofoam (EPS) wastes. *Waste Manag.* **2019**, *89*, 336–344. [CrossRef]
4. Lee, T.; Rahardianto, A.; Cohen, Y. Flexible reverse osmosis (FLERO) desalination. *Desalination* **2019**, *452*, 123–131. [CrossRef]
5. Abid, M.B.; Wahab, R.A.; Salam, M.A.; Moujdin, I.A.; Gzara, L. Desalination technologies, membrane distillation, and electrospinning, an overview. *Heliyon* **2023**, *9*, e12810. [CrossRef]
6. Alawad, S.M.; Khalifa, A.E. Analysis of water gap membrane distillation process for water desalination. *Desalination* **2019**, *470*, 114088. [CrossRef]
7. Liu, X.; Li, L.; Wang, M.; Wang, D.; Yan, H.; Li, K.; Li, Y.; Yang, Y.; You, Y.; Yang, X.; et al. In-situ polymerization of PANI nanocone array on PEN nanofibrous membranes for solar-driven interfacial evaporation. *Sep. Purif. Technol.* **2024**, *344*, 127109. [CrossRef]
8. Abdelkader, B.A.; Antar, M.A.; Khan, Z. Nanofiltration as a Pretreatment Step in Seawater Desalination: A Review. *Sci. Eng.* **2018**, *43*, 4413–4432. [CrossRef]
9. Solouki, S.; Karrabi, M.; Eftekhari, M. Application of a functionalized thin-film composite nanofiltration membrane in water desalination. *J. Mol. Liq.* **2024**, *399*, 124399. [CrossRef]
10. Sharshir, S.W.; Algazzar, A.M.; Elmaadawy, K.A.; Kandeal, A.W.; Elkadeem, M.R.; Arunkumar, T.; Zang, J.; Yang, N. New hydrogel materials for improving solar water evaporation, desalination and wastewater treatment: A review. *Desalination* **2020**, *491*, 114564. [CrossRef]
11. Wang, J.; Kong, Y.; Liu, Z.; Wang, H. Solar-driven interfacial evaporation: Design and application progress of structural evaporators and functional distillers. *Nano Energy* **2023**, *108*, 108115. [CrossRef]
12. Zhao, Q.; Yang, Y.; Pan, C.; Zhu, B.; Sha, Z.; Wei, Y.; Que, W. Integrated strategy of solar evaporator and steam collector configurations for interfacial evaporation water purification. *Sol. Energy* **2023**, *266*, 112187. [CrossRef]
13. Lu, X.; Mu, C.; Liu, Y.; Wu, L.; Tong, Z.; Huang, K. Recent advances in solar-driven interfacial evaporation coupling systems: Energy conversion, water purification, and seawater resource extraction. *Nano Energy* **2024**, *120*, 109180. [CrossRef]
14. Han, H.; Huang, K.; Meng, X. Review on solar-driven evaporator: Development and applications. *J. Ind. Eng. Chem.* **2023**, *119*, 77–89. [CrossRef]
15. Chen, C.; Kuang, Y.; Hu, L. Challenges and Opportunities for Solar Evaporation. *Joule* **2019**, *3*, 683–718. [CrossRef]
16. Zhang, Q.; Xu, W.; Wang, X. Carbon nanocomposites with high photothermal conversion efficiency. *Sci. China Mater.* **2018**, *61*, 905–914. [CrossRef]
17. Zhu, H.; Jiang, X. Development of a General Fabrication Strategy for Carbonaceous Noble Metal Nanocomposites with Photothermal Property. *Nanoscale Res. Lett.* **2020**, *15*, 17. [CrossRef]

18. Huang, J.; Fu, J.; Li, L.; Ma, J. Mg-based metallic glass nanowires with excellent photothermal effect. *Scr. Mater.* **2023**, *222*, 115036. [CrossRef]
19. Hasanzadeh, R.; Azdast, T.; Lee, P.C.; Park, C.B. A review of the state-of-the-art on thermal insulation performance of polymeric foams. *Therm. Sci. Eng. Prog.* **2023**, *41*, 101808. [CrossRef]
20. Sakhadeo, N.N.; Patro, T.U. Exploring the Multifunctional Applications of Surface-Coated Polymeric Foams—A Review. *Ind. Eng. Chem. Res.* **2022**, *61*, 5366–5387. [CrossRef]
21. Hu, X.; Yang, J.; Tu, Y.; Su, Z.; Guan, Q.; Ma, Z. Hydrogel-Based Interfacial Solar-Driven Evaporation: Essentials and Trails. *Gels* **2024**, *10*, 371. [CrossRef] [PubMed]
22. Li, J.; Liu, Q.; He, J.; Zhang, Y.; Mu, L.; Zhu, X.; Yao, Y.; Sun, C.-L.; Qu, M. Aerogel-based solar interface evaporation: Current research progress and future challenges. *Desalination* **2024**, *569*, 117068. [CrossRef]
23. Liu, X.; Tian, Y.; Caratenuto, A.; Chen, F.; Zheng, Y. Biomass-Based Materials for Sustainably Sourced Solar-Driven Interfacial Steam Generation. *Adv. Energy Mater.* **2023**, *25*, 2300778. [CrossRef]
24. Wang, S.-X.; Zhao, H.-B.; Rao, W.-H.; Huang, S.-C.; Wang, T.; Liao, W.; Wang, Y.-Z. Inherently flame-retardant rigid polyurethane foams with excellent thermal insulation and mechanical properties. *Polymer* **2018**, *153*, 616–625. [CrossRef]
25. Kirpluks, M.; Kalnbunde, D.; Benes, H.; Cabulis, U. Natural oil based highly functional polyols as feedstock for rigid polyurethane foam thermal insulation. *Ind. Crop. Prod.* **2018**, *122*, 627–636. [CrossRef]
26. An, W.; Sun, J.; Liew, K.M.; Zhu, G. Flammability and safety design of thermal insulation materials comprising PS foams and fire barrier materials. *Mater. Des.* **2016**, *99*, 500–508. [CrossRef]
27. Shi, L.; Wang, Y.; Zhang, L.; Wang, P. Rational design of a bi-layered reduced graphene oxide film on polystyrene foam for solar-driven interfacial water evaporation. *J. Mater. Chem. A* **2017**, *5*, 16212–16219. [CrossRef]
28. Nguyen, T.K.T.; Dao, Q.K.; Tanaka, D.; Nghiem, L.H.T.; Nguyen, M.V.; Nguyen, Z.H.; Pham, T.T. Flexible, affordable and environmentally sustainable solar vapor generation based on ferric tannate/bacterial cellulose composite for efficient desalination solutions. *RSC Adv.* **2021**, *11*, 31641–31649. [CrossRef] [PubMed]
29. Kiriarachchi, H.D.; Hassan, A.A.; Awad, F.S.; El-Shall, M.S. Metal-free functionalized carbonized cotton for efficient solar steam generation and wastewater treatment. *RSC Adv.* **2021**, *12*, 1043–1050. [CrossRef] [PubMed]
30. Wang, Y.; Zhao, L.; Zhang, F.; Yu, K.; Yang, C.; Jia, J.; Guo, W.; Zhao, J.; Qu, F. Synthesis of a Co-Sn Alloy-Deposited PTFE Film for Enhanced Solar-Driven Water Evaporation via a Super-Absorbent Polymer-Based "Water Pump" Design. *ACS Appl. Mater.* **2021**, *13*, 26879–26890. [CrossRef]
31. Xiao, Y.; Wang, X.; Li, C.; Peng, H.; Zhang, T.; Ye, M. A salt-rejecting solar evaporator for continuous steam generation. *J. Eeviron. Chem. Eng.* **2021**, *9*, 105010. [CrossRef]
32. Chang, C.; Liu, M.; Li, L.; Chen, G.; Pei, L.; Wang, Z.; Ji, Y. Salt-rejecting rGO-coated melamine foams for high-efficiency solar desalination. *J. Mater. Res.* **2021**, *37*, 294–303. [CrossRef]
33. Lal, S.; Batabyal, S.K. Activated carbon-cement composite coated polyurethane foam as a cost-efficient solar steam generator. *J. Clean. Prod.* **2022**, *379*, 134302. [CrossRef]
34. Chen, J.; Li, B.; Hu, G.; Aleisa, R.; Lei, S.; Yang, F.; Liu, D.; Lyu, F.; Wang, M.; Ge, X.; et al. Integrated Evaporator for Efficient Solar-Driven Interfacial Steam Generation. *Nano Lett.* **2020**, *20*, 6051–6058. [CrossRef] [PubMed]
35. Lin, B.; Yuen, A.C.Y.; Oliver, S.; Liu, J.; Yu, B.; Yang, W.; Wu, S.; Yeoh, G.H.; Wang, C.H. Dual functionalisation of polyurethane foam for unprecedented flame retardancy and antibacterial properties using layer-by-layer assembly of MXene chitosan with antibacterial metal particles. *Compos. Part B Eng.* **2022**, *244*, 110147. [CrossRef]
36. Pinto, J.; Magrì, D.; Valentini, P.; Palazon, F.; Heredia-Guerrero, J.A.; Lauciello, S.; Barroso-Solares, S.; Ceseracciu, L.; Pompa, P.P.; Athanassiou, A.; et al. Antibacterial Melamine Foams Decorated with in Situ Synthesized Silver Nanoparticles. *ACS Appl. Mater. Interfaces* **2018**, *10*, 16095–16104. [CrossRef]
37. Wang, J.; Yang, W.; He, F.; Xie, C.; Fan, J.; Wu, J.; Zhang, K. Superhydrophobic Melamine-formaldehyde Foam Prepared by In-situ Coprecipitation. *Chem. Lett.* **2018**, *47*, 414–416. [CrossRef]
38. Shi, H.-G.; Li, S.-L.; Cheng, J.-B.; Zhao, H.-B.; Wang, Y.-Z. Multifunctional Photothermal Conversion Nanocoatings Toward Highly Efficient and Safe High-Viscosity Oil Cleanup Absorption. *ACS Appl. Mater. Interfaces* **2021**, *13*, 11948–11957. [CrossRef]
39. Wang, Z.; Niu, J.; Wang, J.; Zhang, Y.; Wu, G.; Liu, X.; Liu, Q. Rational Design of Photothermal and Anti-Bacterial Foam With Macroporous Structure for Efficient Desalination of Water. *Front. Chem.* **2022**, *10*, 912489. [CrossRef] [PubMed]
40. Gnanasekaran, A.; Rajaram, K. Rational design of different interfacial evaporators for solar steam generation: Recent development, fabrication, challenges and applications. *Renew. Sust. Enegr. Rev.* **2024**, *192*, 114202. [CrossRef]
41. Ullah, F.; Othman, M.B.H.; Javed, F.; Ahmad, Z.; Akil, H.M. Classification, processing and application of hydrogels: A review. *Mater. Sci. Eng. C* **2015**, *57*, 414–433. [CrossRef]
42. Jing, X.; Liu, F.; Abdiriym, T.; Liu, X. Hydrogels as promising platforms for solar-driven water evaporators. *Chem. Eng. J.* **2024**, *479*, 147519. [CrossRef]
43. Zhao, Q.; Wen, H.; Wu, J.; Wen, X.; Xu, Z.; Duan, J. Galactomannan/graphene oxide/Fe_3O_4 hydrogel evaporator for solar water evaporation for synergistic photothermal power generation. *Desalination* **2024**, *570*, 117064. [CrossRef]
44. Wang, J.; Guo, Z.; Xiao, B.; Xiong, X.; Liu, G.; Wang, X. Reduced graphene oxide/$Cu_7·_2S_4$ composite hydrogels for highly efficient solar steam generation. *Mater. Today. Sustain.* **2022**, *18*, 100121. [CrossRef]

45. Xiong, Y.; Hu, D.; Huang, L.; Fang, Z.; Jiang, H.; Mao, Q.; Wang, H.; Tang, P.; Li, J.; Wang, G.; et al. Ultra-high strength sodium alginate/PVA/PHMB double-network hydrogels for marine antifouling. *Prog. Org. Coat.* **2024**, *187*, 108175. [CrossRef]
46. Wang, B.; Zhu, H.; Shutes, B. Multi-strategy coupling of custom hydrogel evaporators for sustained, high-efficiency clean water production and anti-pollution. *Nano Today* **2024**, *57*, 102370. [CrossRef]
47. Chu, A.; Yang, M.; Chen, J.; Zhao, J.; Fang, J.; Yang, Z.; Li, H. Biomass-enhanced Janus sponge-like hydrogel with salt resistance and high strength for efficient solar desalination. *Green. Energy Environ.* **2023**, *in press*. [CrossRef]
48. Zhao, J.; Chu, A.; Chen, J.; Qiao, P.; Fang, J.; Yang, Z.; Duan, Z.; Li, H. Spongy polyelectolyte hydrogel for efficient Solar-Driven interfacial evaporation with high salt resistance and compression resistance. *Chem. Eng. J.* **2024**, *485*, 150118. [CrossRef]
49. Peng, B.; Lyu, Q.; Gao, Y.; Li, M.; Xie, G.; Xie, Z.; Zhang, H.; Ren, J.; Zhu, J.; Zhang, L.; et al. Composite Polyelectrolyte Photothermal Hydrogel with Anti-biofouling and Antibacterial Properties for the Real-World Application of Solar Steam Generation. *ACS Appl. Mater. Interfaces* **2022**, *14*, 16546–16557. [CrossRef]
50. Peng, B.; Gao, Y.; Lyu, Q.; Xie, Z.; Li, M.; Zhang, L.; Zhu, J. Cationic Photothermal Hydrogels with Bacteria-Inhibiting Capability for Freshwater Production via Solar-Driven Steam Generation. *ACS Appl. Mater. Interfaces* **2021**, *13*, 37724–37733. [CrossRef]
51. Garg, S.; Singh, S.; Shehata, N.; Sharma, H.; Samuel, J.; Khan, N.A.; Ramamurthy, P.C.; Singh, J.; Mubashir, M.; Bokhari, A.; et al. Aerogels in wastewater treatment: A review. *J. Taiwan Inst. Chem. E* **2023**, 105299. [CrossRef]
52. Rai, N.; Chauhan, I. Multifunctional Aerogels: A comprehensive review on types, synthesis and applications of aerogels. *J. Sol-Gel. Sci. Technol.* **2023**, *105*, 324–336. [CrossRef]
53. Wu, J.; Yang, X.; Jia, X.; Yang, J.; Miao, X.; Shao, D.; Song, H.; Li, Y. Full biomass-derived multifunctional aerogel for solar-driven interfacial evaporation. *Chem. Eng. J.* **2023**, *471*, 144684. [CrossRef]
54. Storer, D.P.; Phelps, J.L.; Wu, X.; Owens, G.; Khan, N.I.; Xu, H. Graphene and Rice-Straw-Fiber-Based 3D Photothermal Aerogels for Highly Efficient Solar Evaporation. *ACS Appl. Mater. Interfaces* **2020**, *12*, 15279–15287. [CrossRef] [PubMed]
55. Xiao, J.-K.; Gong, J.-Z.; Dai, M.; Zhang, Y.-F.; Wang, S.-G.; Lin, Z.-D.; Du, F.-P.; Fu, P. Reduced graphene oxide/Ag nanoparticle aerogel for efficient solar water evaporation. *J. Alloy. Compd.* **2023**, *930*, 167404. [CrossRef]
56. Jian, H.; Wang, Y.; Li, W.; Ma, Y.; Wang, W.; Yu, D. Reduced graphene oxide aerogel with the dual-cross-linked framework for efficient solar steam evaporation. *Colloid. Surface A* **2021**, *629*, 127440. [CrossRef]
57. Fillet, R.; Nicolas, V.; Fierro, V.; Celzard, A. A review of natural materials for solar evaporation. *Sol. Energ. Mater. Sol. C* **2021**, *219*, 110814. [CrossRef]
58. Zhang, P.; Xie, M.; Jin, Y.; Jin, C.; Wang, Z. A Bamboo-Based Photothermal Conversion Device for Efficient Solar Steam Generation. *ACS Appl. Mater. Interfaces* **2022**, *4*, 2393–2400. [CrossRef]
59. Zhang, C.; Xiao, P.; Ni, F.; Yan, L.; Liu, Q.; Zhang, D.; Gu, J.; Wang, W.; Chen, T. Converting Pomelo Peel into Eco-friendly and Low-Consumption Photothermic Biomass Sponge toward Multifunctioal Solar-to-Heat Conversion. *ACS Sustain. Chem. Eng.* **2020**, *8*, 5328–5337. [CrossRef]
60. Chen, C.; Li, Y.; Song, J.; Yang, Z.; Kuang, Y.; Hitz, E.; Jia, C.; Gong, A.; Jiang, F.; Zhu, J.Y.; et al. Highly Flexible and Efficient Solar Steam Generation Device. *Adv. Mater.* **2017**, *29*, 1701756. [CrossRef]
61. Song, D.; Zheng, D.; Li, Z.; Wang, C.; Li, J.; Zhang, M. Research Advances in Wood Composites in Applications of Industrial Wastewater Purification and Solar-Driven Seawater Desalination. *Polymers* **2023**, *15*, 4712. [CrossRef] [PubMed]
62. Li, Y.; Li, Q.; Qiu, Y.; Feng, H. High-efficiency wood-based evaporators for solar-driven interfacial evaporation. *Sol. Energy* **2022**, *244*, 322–330. [CrossRef]
63. Li, W.; Li, F.; Zhang, D.; Bian, F.; Sun, Z. Porous wood-carbonized solar steam evaporator. *Wood Sci. Technol.* **2021**, *55*, 625–637. [CrossRef]
64. Chen, T.; Wu, Z.; Liu, Z.; Aladejana, J.T.; Wang, X.; Niu, M.; Wei, Q.; Xie, Y. Hierarchical Porous Aluminophosphate-Treated Wood for High-Efficiency Solar Steam Generation. *ACS Appl. Mater. Interfaces* **2020**, *12*, 19511–19518. [CrossRef]
65. Fang, Q.; Li, T.; Chen, Z.; Lin, H.; Wang, P.; Liu, F. Full Biomass-Derived Solar Stills for Robust and Stable Evaporation To Collect Clean Water from Various Water-Bearing Media. *ACS Appl. Mater. Interfaces* **2019**, *11*, 10672–10679. [CrossRef]
66. Huang, W.; Hu, G.; Tian, C.; Wang, X.; Tu, J.; Cao, Y.; Zhang, K. Nature-inspired salt resistant polypyrrole–wood for highly efficient solar steam generation. *Ssutain. Energ. Fuels* **2019**, *3*, 3000–3008. [CrossRef]
67. Shi, L.; Zhang, M.; Du, X.; Liu, B.; Li, S.; An, C. In situ polymerization of pyrrole on elastic wood for high efficiency seawater desalination and oily water purification. *J. Mater. Sci.* **2022**, *57*, 16317–16332. [CrossRef]
68. Pham, T.T.; Nguyen, T.H.; Nguyen, T.A.H.; Pham, D.D.; Nguyen, D.C.; Do, D.B.; Nguyen, H.V.; Ha, M.H.; Nguyen, Z.H. Durable, scalable and affordable iron (III) based coconut husk photothermal material for highly efficient solar steam generation. *Desalination* **2021**, *518*, 115280. [CrossRef]
69. Long, Y.; Huang, S.; Yi, H.; Chen, J.; Wu, J.; Liao, Q.; Liang, H.; Cui, H.; Ruan, S.; Zeng, Y.-J. Liu, Carrot-inspired solar thermal evaporator. *J. Mater. Chem. A* **2019**, *7*, 26911. [CrossRef]
70. Jia, X.; Liu, X.; Guan, H.; Fan, T.; Chen, Y.; Long, Y.-Z. A loofah-based photothermal biomass material with high salt-resistance for efficient solar water evaporation. *Compos. Commun.* **2023**, *37*, 101430. [CrossRef]
71. Chen, T.; Xie, H.; Qiao, X.; Hao, S.; Wu, Z.; Sun, D.; Liu, Z.; Cao, F.; Wu, B.; Fang, X. Highly Anisotropic Corncob as an Efficient Solar Steam-Generation Device with Heat Localization and Rapid Water Transportation. *ACS Appl. Mater. Interfaces* **2020**, *12*, 50397–50405. [CrossRef]

72. Xu, N.; Hu, X.; Xu, W.; Li, X.; Zhou, L.; Zhu, S.; Zhu, J. Mushrooms as Efficient Solar Steam-Generation Devices. *Adv. Mater.* **2017**, *29*, 1606762. [CrossRef]
73. Zeng, L.; Deng, D.; Zhu, L.; Wang, H.; Zhang, Z.; Yao, Y. Biomass photothermal structures with carbonized durian for efficient solar-driven water evaporation. *Energy* **2023**, *273*, 127170. [CrossRef]
74. Zhang, Q.; Yang, X.; Deng, H.; Zhang, Y.; Hu, J.; Tian, R. Carbonized sugarcane as interfacial photothermal evaporator for vapor generation. *Desalination* **2022**, *526*, 115544. [CrossRef]
75. Roy, A.; Tariq, M.Z.; La, M.; Choi, D.; Park, S.J. 3D carbonized orange peel: A self-floating solar absorber for efficient water evaporation. *Desalination* **2024**, *573*, 117191. [CrossRef]
76. Li, Z.; Wei, S.; Ge, Y.; Zhang, Z.; Li, Z. Biomass-based materials for solar-powered seawater evaporation. *Sci. Total Environ.* **2023**, *858*, 160003. [CrossRef]
77. Zhu, M.; Yu, J.; Ma, C.; Zhang, C.; Wu, D.; Zhu, H. Carbonized daikon for high efficient solar steam generation. *Sol. Energ. Mater. Sol. C* **2019**, *191*, 83–90. [CrossRef]
78. Guo, M.X.; Wu, J.B.; Zhao, H.Y.; Li, F.H.; Min, F.Q. Carbonized loofah and MOF-801 of synergistic effect for efficient solar steam generation. *Int. J. Energ. Res.* **2021**, *45*, 10599–10608. [CrossRef]
79. Zhang, W.; Zhang, L.; Li, T.; Wu, D.; Zhang, C.; Zhu, H. Efficient solar-driven interfacial water evaporation enabled wastewater remediation by carbonized sugarcane. *J. Water. Process Eng.* **2022**, *49*, 102991. [CrossRef]
80. Cao, N.; Lu, S.; Yao, R.; Liu, C.; Xiong, Q.; Qin, W.; Wu, X. A self-regenerating air-laid paper wrapped ASA 3D cone-shaped Janus evaporator for efficient and stable solar desalination. *Chem. Eng. J.* **2020**, *397*, 125522. [CrossRef]
81. Zou, M.; Zhang, Y.; Cai, Z.; Li, C.; Sun, Z.; Yu, C.; Dong, Z.; Wu, L.; Song, Y. 3D Printing a Biomimetic Bridge-Arch Solar Evaporator for Eliminating Salt Accumulation with Desalination and Agricultural Applications. *Adv. Mater.* **2021**, *33*, e2102443. [CrossRef] [PubMed]
82. Chaule, S.; Hwang, J.; Ha, S.J.; Kang, J.; Yoon, J.C.; Jang, J.H. Rational Design of a High Performance and Robust Solar Evaporator via 3D-Printing Technology. *Adv. Mater.* **2021**, *33*, 2102649. [CrossRef] [PubMed]
83. Khalil, A.; Ahmed, F.E.; Hilal, N. The emerging role of 3D printing in water desalination. *Sci. Total. Environ.* **2021**, *790*, 148238. [CrossRef] [PubMed]
84. Wang, Y.; Wang, C.; Song, X.; Huang, M.; Megarajan, S.K.; Shaukat, S.F.; Jiang, H. Improved light-harvesting and thermal management for efficient solar-driven water evaporation using 3D photothermal cones. *J. Mater. Chem. A* **2018**, *6*, 9874–9881. [CrossRef]
85. Lv, F.; Miao, J.; Hu, J.; Orejon, D. 3D Solar Evaporation Enhancement by Superhydrophilic Copper Foam Inverted Cone and Graphene Oxide Functionalization Synergistic Cooperation. *Small* **2023**, *19*, e2208137. [CrossRef] [PubMed]
86. Xie, M.; Zhang, P.; Cao, Y.; Yan, Y.; Wang, Z.; Jin, C. A three-dimensional antifungal wooden cone evaporator for highly efficient solar steam generation. *NPJ Clean Water* **2023**, *6*, 12. [CrossRef]
87. Bu, Y.; Zhou, Y.; Lei, W.; Ren, L.; Xiao, J.; Yang, H.; Xu, W.; Li, J. A bioinspired 3D solar evaporator with balanced water supply and evaporation for highly efficient photothermal steam generation. *J. Mater. Chem. A* **2022**, *10*, 2856–2866. [CrossRef]
88. Xu, Z.; Ran, X.; Zhang, Z.; Zhong, M.; Wang, D.; Li, P.; Fan, Z. Designing a solar interfacial evaporator based on tree structures for great coordination of water transport and salt rejection. *Mater. Horiz.* **2023**, *10*, 1737–1744. [CrossRef]
89. Xu, Z.; Ran, X.; Wang, D.; Zhong, M.; Zhang, Z. High efficient 3D solar interfacial evaporator: Achieved by the synergy of simple material and structure. *Desalination* **2022**, *525*, 115495. [CrossRef]
90. Chen, Y.; Hou, R.; Yang, L.; Chen, C.; Cui, J.; Zhou, T.; Zhao, Y.; Song, J.; Fan, Z.; Tang, Y.; et al. Elastic, Janus 3D evaporator with arch-shaped design for low-footprint and high-performance solar-driven zero-liquid discharge. *Desalination* **2024**, *583*, 117644. [CrossRef]
91. Wu, L.; Dong, Z.; Cai, Z.; Ganapathy, T.; Fang, N.X.; Li, C.; Yu, C.; Zhang, Y.; Song, Y. Highly efficient three-dimensional solar evaporator for high salinity desalination by localized crystallization. *Nat. Commun.* **2020**, *11*, 521. [CrossRef]
92. Zhang, Z.; Feng, Z.; Qi, H.; Chen, Y.; Chen, Y.; Deng, Q.; Wang, S. Carbonized sorghum straw derived 3D cup-shaped evaporator with enhanced evaporation rate and energy efficiency. *Sustain. Mater. Technol.* **2022**, *32*, e00414. [CrossRef]
93. Zhang, L.; Bai, B.; Hu, N.; Wang, H. Low-cost and facile fabrication of a candle soot/adsorbent cotton 3D-interfacial solar steam generation for effective water evaporation. *Sol. Energ. Mater. Sol. C* **2021**, *221*, 110876. [CrossRef]
94. Sun, S.; Shi, C.; Kuang, Y.; Li, M.; Li, S.; Chan, H.; Zhang, S.; Chen, G.; Nilghaz, A.; Cao, R.; et al. 3D-printed solar evaporator with seashell ornamentation-inspired structure for zero liquid discharge desalination. *Water Res.* **2022**, *226*, 119279. [CrossRef] [PubMed]
95. Zhang, X.; Yan, Y.; Li, N.; Yang, P.; Yang, Y.; Duan, G.; Wang, X.; Xu, Y.; Li, Y. A robust and 3D-printed solar evaporator based on naturally occurring molecules. *Sci. Bull.* **2023**, *68*, 203–213. [CrossRef] [PubMed]
96. Chen, Y.; Yang, J.; Zhang, D.; Wang, S.; Jia, X.; Li, Y.; Shao, D.; Feng, L.; Song, H.; Tang, S. A wood-inspired bimodal solar-driven evaporator for highly efficient and durable purification of high-salinity wastewater. *J. Mater. Chem. A* **2023**, *11*, 2349–2359. [CrossRef]

Disclaimer/Publisher's Note: The statements, opinions and data contained in all publications are solely those of the individual author(s) and contributor(s) and not of MDPI and/or the editor(s). MDPI and/or the editor(s) disclaim responsibility for any injury to people or property resulting from any ideas, methods, instructions or products referred to in the content.

Article

Porosity Effect of Polystyrene Membranes on Desalination Performance: A Combined Experimental and Numerical Heat and Mass Transfer Study in Direct Contact Membrane Distillation

Haneen Abdelrazeq and Majeda Khraisheh *

Department of Chemical Engineering, College of Engineering, Qatar University, Doha P.O. Box 2713, Qatar; ha082881@student.qu.edu.qa
* Correspondence: m.khraisheh@qu.edu.qa

Abstract: Membrane distillation (MD) is a thermal-based membrane operation with high potential for use in the treatment of aqueous streams. In this study, the linear relationship between the permeate flux and the bulk feed temperature for different electrospun polystyrene membranes is discussed. The dynamics of combined heat and mass transfer mechanisms across different membrane porosities of 77%, 89%, and 94%, each with different thicknesses, are examined. The main results for the effect of porosity with respect to the thermal efficiency and evaporation efficiency of the DCMD system are reported for electrospun polystyrene membranes. A 14.6% increase in thermal efficiency was noted for a 15% increase in membrane porosity. Meanwhile, a 15.6% rise in porosity resulted in a 5% increase in evaporation efficiency. A mathematical validation along with computational predictions is presented and interlinked with the maximum thermal and evaporation efficiencies for the surface membrane temperatures at the feed and temperature boundary regions. This work helps to further understand the interlinked correlations of the surface membrane temperatures at the feed and temperature boundary regions with respect to the change in membrane porosity.

Keywords: polystyrene membranes; heat transfer; mass transfer; thermal efficiency; evaporation efficiency

1. Introduction

Water desalination is key in the production of fresh water by eliminating undesirable particles from salty water. The seawater stream is mainly the feed that is heated to the point where vapor molecules transfer through the pores of the membrane and condense on the permeate side. Reverse osmosis (RO) is an economically widely used technology that is mostly used for desalinating brine at levels close to those of seawater (<45,000 mg/L) [1,2]. Unfortunately, RO remains an energy-intensive technology [3]. When it comes to energy constraints, MD processes require an energy range of 120.6–1701.8 kWh/m^3 for heating the feedwater compared to only 2.5–7.0 kWh/m^3 for the RO process. This huge energy requirement hinders the commercialization potential of advanced MD systems. From an industrial viewpoint, the desalination and treatment of high-salinity brines are inherently energy-intensive [4]. Particularly in MD processes, due to latent heat needed for the evaporation of the feed, the energy requirement significantly increases. The criteria for evaluating the energy performance of an MD system are divided into two main parts: (i) standard measures directly related to the fundamentals of the system, and (ii) developed measures based on the specificity of the employed system [5,6]. Furthermore, combinations of desirable MD features, such as modularity and scalability, have led to a number of simulation and experimental investigations focusing on innovative MD processes for the treatment of hypersaline feeds, especially at larger scales [7].

Membrane distillation (MD) is one of the most widely used thermally driven techniques resulting from a combined heat and mass transfer mechanism through a hydrophobic microporous membrane [8,9]. Generally, saline water is purified in MD, where vapor passes across the pores of the hydrophobic membranes from the hot feed side to the cold permeate side. The temperature slope occurring between the liquid and vapor interfaces at the entrance of the membrane pores results in the driving force of MD [10–12].

In direct contact membrane distillation (DCMD), the two liquid streams come into direct contact with the membrane from both sides, as illustrated in Figure 1. A complex heat transfer system occurs that is limited to the membrane thickness in DCMD. Therefore, the system's permeate impacts the mass transfer coefficients in the two counter-current streams [13–16]. The feed temperature in the system and the type of membranes used in DCMD must be optimized in such a way that condensation is prevented within the pores of the membranes. Furthermore, the type of polymer membrane material employed in DCMD systems depends mainly on the specific requirements and constraints of the application, such as the type of feed solution being treated, the desired separation performance, the operating conditions, and the cost.

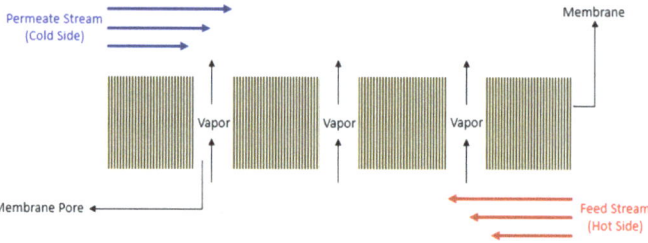

Figure 1. Illustration of direct contact membrane distillation.

To date, commercial membranes have been common contributors to the MD process as, until now, there is no commercial membrane that is specially designed for MD [17]. Nonetheless, recent review articles reported valuable summaries on water purification technologies and porous membrane materials, with reference to membrane properties in MD [18–23]. For instance, Yang et al. [24] provided an excellent overview of new polymeric membranes and compared different classes of polymeric membranes for water purification applications. The research group determined that the fouling resistance and permeability can be improved by the addition of a microporous support. The most common types of polymers used are polytetrafluoroethylene (PTFE), polyvinylidene fluoride (PVDF), polypropylene (PP), and polystyrene (PS) [25–36]. Polystyrene is abundant and can reach up to several million tons of annual production [17]. It has also been distinguished for its competitive cost compared with PVDF and PTFE [37].

Polystyrene (PS) membranes are commonly used in DCMD processes. They are known for their good thermal stability, chemical resistance, and mechanical strength, which makes them appropriate for a wide variety of MD applications [13,38]. An excellent technique by which the membrane's properties can be controlled is called electrospinning. This fabrication method produces nanofibrous membranes from polystyrene and other polymeric materials with controlled porosity and fiber diameters and high hydrophobicity, making it a very good candidate for applications in membrane distillation systems.

In DCMD, heat inside the feed and permeate solutions is transferred in the forms of sensible and latent heat, and convective heat. Therefore, it is essential to consider the mechanisms of heat conduction and heat convection within the study of heat evolution inside a DCMD module. The pores in the membrane allow the water vapor to pass through while blocking the impurities. The size and distribution of the pores can have a significant effect on the mass transfer properties of the membrane. If the pores in the membrane are too small, the mass transfer rate will be limited because the water vapor will have difficulty passing through the pores. On the other hand, if the pores are too large, the membrane may be less effective at separating impurities from the water vapor. Figure 1 illustrates a schematic diagram of a counter-current DCMD module. In addition to the pore size, the thickness also plays a key role in impacting the heat transfer mechanisms at the membrane boundary where the feed and the permeate solutions are affected by each other [39]. Porosity can affect mass transfer properties by influencing the surface area of the membrane available for mass transfer, and it can also affect the permeability of the membrane. Therefore, the selection of a membrane with the appropriate pore size, thickness, and porosity is key to optimizing the mass and heat transfer properties of the system.

Existing studies in the literature majorly focused on developing optimized models with respect to changes in process conditions and experimental parameters. Eleiwi et al. proposed a dynamic model for the DCMD process, which considers the time evolution of the heat and mass transfer mechanisms throughout the feed and the permeate solutions [39]. Kuang et al. studied the variation in the mass flow rate and concentration in DCMD using computational fluid dynamics, where the water production increased by 28.3% using a 1 mm long module [40]. Elmarghany et al. conducted a thermal investigation for a similar system, where it was shown that increasing the feed temperature negatively affected the thermal performance due to heat loss from the membrane cell [41]. This paper utilizes numerical predictions to optimize the membrane surface temperature based on controlled process parameters and to provide insights into how the porous structure of polystyrene membranes affects the efficiency of DCMD systems at a larger scale.

Several efforts have been made to address the effect of varying process conditions on the heat and mass transfer. Zhang et al. [42] presented a novel model discussing the effect of varying feed salinity (3.5 wt%, 7.5 wt%, and 20 wt%) on hollow-fiber DCMD using MATLAB. Although the proposed simulation could not precisely predict the surface temperatures at higher feed concentrations, the highest water recovery of 86.8% was achieved in seawater compared with that in desalination brine, with 72.1%, at a feed temperature of 50 °C. Similarly, de Sampaio [43] studied the modeling of DCMD consisting of shell and hollow fiber tubes for a desalination plant utilizing heat recovery. For a single DCMD unit, mass and energy conservation and thermodynamic modeling were included. The data showed good agreement with the experimental values in the literature. Additionally, Ansari et al. [44] tested commercial membranes in DCMD, and their results showed a less than 7% deviation with respect to computational modeling data. The research findings showed a 2.3-fold improvement in water flux when the Reynolds number was increased from 80 to 1600, along with a 23% increase in thermal efficiency when the membrane porosity was increased from 40% to 70%. Moreover, other research groups focused on addressing the fouling phenomenon in DCMD using a cake filtration theory to signify the reduction in flux due to inorganic fouling [45]. This led to a significant enhancement in the overall efficiency of MD and a reduction in the number of membrane washing cycles. Nonetheless, it is also important to understand the impact of porosity on heat and mass transfer in efforts to optimize the design of direct contact membrane distillation systems and improve their performance. Due to the scarcity of studies that attempt to study the relationship between the porosity effect and mass flux and its effect on the mass and heat transfer with respect to the variable operating conditions of pilot-scale MD systems, this study helps to further understand the mechanism of heat flux flow through the membrane pores and offers a numerical validation along with computational predictions that aim to find the optimum surface temperatures that are interlinked with the maximum thermal and evaporation efficiencies at the thermal boundary regions at a pilot scale. Moreover, to the best of our knowledge, the current literature has not yet explored the desalination potential of electrospun polystyrene membranes in DCMD at a pilot scale The scope of this work is limited to the optimization of membrane surface temperatures based on controlled process parameters, providing insights into how the porous structure of polystyrene membranes affects the efficiency of a pilot DCMD unit.

Moreover, the theoretical and predicative results are validated with laboratory experiments, without considering inorganic fouling.

2. Experimental

2.1. Materials

Polystyrene pellets (PS, Mw = 192,000, CAS Number: 9003-53-6) and N,N-dimethyl formamide (DMF, CAS Number: 68-12-2, 99.8% purity) were purchased from Sigma Aldrich, Saint Louis, MO, USA and used as is.

2.2. Membrane Synthesis

A custom-made electrospinning system was used for membrane fabrication in this work (Figure 1). An amount of 20 g of polystyrene (molecular weight: 192,000) was dissolved in 60 mL DMF and 40.0 mL acetone. PS and DMF were first set to stir in a beaker. This was followed by continuous stirring for acetone at room temperature for up to 24 h to ensure the polymers dissolved. Then, 0.1%, 0.5%, and 1% of PTFE powder were added to the synthesized polystyrene membranes M1, M2, M3, and M4. The spinning parameters used were as follows: volume of polymer solution, 10 mL; voltage, 14 kV; distance between the needle and collector, 15 cm; needle diameter, 20-gauge needle; flow rate, 6 mL/h; and drum RPM, 340 RPM. The fabricated membranes were then left in a vacuum oven overnight to eliminate any residuals. Afterwards, they were cold-pressed at 1 Ton for 1 min. The electrospun membranes are listed in Table 1, where sample masses of 0.1629 g, 0.1420 g, 0.1342 g, and 0.0510 g correspond to M1, M2, M3, and M4, respectively.

Table 1. Details of the used PS flat-sheet membranes.

Characteristics	M1	M2	M3	M4
BET surface area (m^2/g)	30.17	31.21	51.77	57.87
Average pore diameter (4V/A) (µm)	0.0308	0.0131	0.0225	0.0276
Porosity (%)	77	89	94	94

2.3. DCMD Pilot-Scale Investigation

A pilot unit was used for all the experimental tests using the electrospun flat-sheet polystyrene membranes. An experimental feed/permeate flow rate of 40 L/h was used in the predictions to evaluate the optimum permeate flux at fixed inlet feed and permeate temperatures of 70 °C and 20 °C, respectively. The feed spacer was the same for all experiments, and its influence on concentration polarization was not investigated. A similar pilot unit was included in our previous study for testing commercial polyethylene membranes [46].

3. Mathematical Modeling

3.1. Heat Transfer

3.1.1. Heat Transfer from the Feed Side to the Surface of the Electrospun Membrane

Convection is used to transfer heat through the feed boundary layer, and Newton's law of cooling governs this process by the following equation:

$$Q_f = h_f \left(T_{b,f} - T_{m,f} \right) \tag{1}$$

where Q_f is the convective heat flux, h_f is the boundary layer heat transfer coefficient on the membrane's feed side, and $T_{b,f}$ and $T_{m,f}$ are the average feed temperatures for the bulk and surface of the membrane from the feed side, respectively. The transfer of heat across the membrane can be categorized into two segments: the first is the transfer of heat through the membrane by conduction, which includes the polymer matrix and pores filled with gas;

the second is the transfer of heat through the membrane by the latent heat of water vapor movement.

3.1.2. Second Stage: Heat Transfer through the Membrane Layer

The conducted heat transfer across the membrane (Q_C) is added to the evaporative mass flow (Q_v) through the membrane pores to obtain the total heat flux across the membrane (Q_m).

$$Q_c = \frac{k_m}{\delta}\left(T_{m,f} - T_{m,p}\right) \quad (2)$$

$$Q_v = J_w \Delta H_v \quad (3)$$

The enthalpy of the water (ΔH_v) can be calculated using the following equation:

$$\Delta H_v = \left(\left(1.7535 * T_{m,f}\right) + 2024.3\right) \quad (4)$$

The effective thermal conductivity of the membrane (k_m) is equal to the product of the thermal conductivity of the solid membrane (k_{mem}) and the thermal conductivity of the membrane gas (k_{gas}) (air and water vapor).

$$k_m = \left(\left(\frac{\epsilon}{k_{gas}}\right) + \left(\frac{1-\epsilon}{k_{mem}}\right)\right)^{-1} \quad (5)$$

The total heat flux across the membrane (Q_m) can be described as the following:

$$Q_m = Q_c + Q_v = h_m\left(T_{m,f} - T_{m,p}\right) + J_w \Delta H_v \quad (6)$$

3.1.3. Third Stage: Heat Transfer from the Membrane Surface to the Permeate Stream

Convection is used to transmit heat across the boundary layer from the permeate-side membrane surface to the permeate bulk. The permeate heat flux, Q_p, depends on the permeate heat transfer coefficient (h_p) and temperature difference between the bulk permeate temperature ($T_{b,p}$) and the interfacial membrane temperature ($T_{m,p}$) on the permeate side. In this work, the DCMD process is assumed to be a steady-state process in order to calculate the surface temperature on both the feed and permeate sides of the membrane. The overall heat transfer fluxes of the feed, membrane, and permeate sides of the module are assumed to be under steady-state conditions ($Q_f = Q_m = Q_p$).

$$Q_p = h_p\left(T_{m,p} - T_{b,p}\right) \quad (7)$$

In the DCMD process, the vapor pressure difference arising from the temperature difference between the two surfaces of the membrane is the driving force for water vapor transfer across the membrane. The temperature difference between $T_{m,f}$ (the membrane/feed interface) and $T_{m,p}$ (the membrane/permeate interface) is the driving force for water vapor transfer through the pores of the membrane. However, one of the limitations in DCMD systems is the change in the membrane/interface temperature with respect to the bulk temperature in the process. This occurs due to heat lost from the feed stream side of the membrane surface and heat gained from the permeate stream side of the membrane surface. $T_{m,f}$ and $T_{m,p}$ are calculated using the following equations:

$$T_{m,f} = \frac{k_m\left(T_{b,p} + \frac{h_f}{h_p}T_{b,f}\right) + \delta\left(h_f T_{b,f} - J_w \Delta H_v\right)}{k_m + h_f\left(\delta + \frac{k_m}{h_p}\right)} \quad (8)$$

$$T_{m,p} = \frac{k_m\left(T_{b,f} + \frac{h_p}{h_f}T_{b,p}\right) + \delta\left(h_p T_{b,f} + J_w \Delta H_v\right)}{k_m + h_p\left(\delta + \frac{k_m}{h_f}\right)} \quad (9)$$

In this work, the DCMD process is assumed to be a steady-state process in order to calculate the surface temperature of both the feed and permeate sides of the membrane. A number of assumptions were made to assess the significance of the different heat transfer mechanisms using the pilot DCMD system: for example, the operating conditions are in a steady state; there is negligible heat loss; the membrane pores have uniform sizes; the water has constant physical properties; the water flow is laminar in the x-direction; and there is a constant total pressure of 1 atm. As such, the heat balance guarantees that the three consecutive heat transfer methods satisfy the following equation:

$$Q_f = Q_m = Q_p \quad (10)$$

After a certain period of time, the concentration polarization in the desalination process influences the transfer as a result of salt molecules building up on the membrane surface. The ratio of the solute concentration on the feed membrane surface ($C_{m,f}$) to the concentration of the feed bulk ($C_{b,f}$) is known as the concentration polarization coefficient (ϕ):

$$\phi = \frac{C_{m,f}}{C_{b,f}} \quad (11)$$

$$C_{m,f} = C_{b,f} * \exp\left(\frac{J_w}{k_s * \rho_{b,f}}\right) \quad (12)$$

where $\rho_{b,f}$ is the density of the feed flow, and k_s represents the solute mass transfer coefficient as follows:

$$k_s = Sh * \frac{D_e}{D_h} \quad (13)$$

where D_h is the hydraulic diameter of the hot channel, and Sh is the Sherwood number, which is determined using the Graetz–Leveque equation for laminar flow:

$$Sh = 1.86\left(Re * Sc * \frac{D_h}{L}\right)^{\frac{1}{3}} \quad (14)$$

In the following equations, Sc represents the Schmidt numbers, Re represents the Reynolds number, and Pr represents the Prandtl number:

$$Sc = \frac{\mu_{m,f}}{\rho_{b,f} * D_e} \quad (15)$$

$$h = \frac{Nu * k}{D_h} \quad (16)$$

$$Pr = \frac{v}{\alpha} = \frac{\mu * c_p}{k} \quad (17)$$

where k is the average thermal conductivity of the fluid on the membrane feed side, and Nu is the Nusselt number, which is determined using the equation shown below. For a flat-plate module and laminar flow ($Re < 2100$), the Nusselt number can be used for both the feed and permeate sides of the membrane using the following equation:

$$Nu = 1.86\left(\frac{RePrD_h}{L}\right)^{\frac{1}{3}} \quad (18)$$

3.2. Mass Transfer

In the following equation, J_w is the permeate mass flux, and D_e is the equivalent diffusion coefficient:

$$J_w = D_e * \Delta p_m = D_e * \left(P_{wf}^0 - P_{wp}^0 \right) \quad (19)$$

where P_{wf}^0 and P_{wp}^0 are the partial pressures of water on the feed and permeate sides of the membrane, respectively:

$$P_{wf}^0 = \exp\left(23.1964 - \frac{3816.44}{T_{mf} - 46.13} \right) \quad (20)$$

$$P_{wf}^0 = \exp\left(23.1964 - \frac{3816.44}{T_{mf} - 46.13} \right) \quad (21)$$

Considering the effect of salinity in the feed solution, the permeate flux can be represented by the following equation:

$$J_w = D_e \left(P_{wf}^0 * \gamma_{wf} * x_{wf} - P_{wp}^0 \right) \quad (22)$$

For an aqueous solution of NaCl, γ_{wf} can be expressed as the following:

$$\gamma_{wf} = 1 - (0.5 * x_{NaCl}) - \left(10 * x_{Nacl}^2 \right) \quad (23)$$

There are three different types of mechanisms that account for the movement of gases and vapor through porous media, which are the Poiseuille flow model, the molecular diffusion model, and the Knudsen model. The Knudsen flow and molecular diffusion models can be used in DCMD. The trans-membrane hydrostatic pressure is not applied since the feed and permeate solutions are retained inside the membrane module at a constant pressure (about 1.0 atm). The Poiseuille flow in this situation is insignificant. The ratio of Knudsen diffusion to molecular diffusion is used to calculate the combined influence of the molecular and Knudsen diffusions. The governing mechanism in the mass transmission is determined by this ratio. Considering the effect of salinity, the effective, Knudsen, and molecular diffusion coefficients are D_e, D_k, and D_m, respectively. The following are the mathematical expressions for D_e, D_k, and D_m:

$$D_e = \left(\frac{\alpha}{D_k} + \frac{1 - \alpha}{D_m} \right)^{-1} \quad (24)$$

$$D_k = \left(\frac{3 * \delta * \tau}{2 * \epsilon * d_{pore}} * \left(\frac{\pi * R * T_m}{8 * Mol_w} \right)^{0.5} \right)^{-1} \quad (25)$$

$$D_m = \left(\frac{R * T_m * \delta * \tau * P_{air,pore}}{Mol_w * \epsilon * PD_{w,a}} \right)^{-1} \quad (26)$$

The following expression, where $PD_{w,a}$ can be employed in the temperature range of 273–373 K, is used to compute the value for water–air, and T_m is the mean temperature across the membrane surfaces:

$$PD_{w,a} = 1.895 * 10^{-5} * T_m^{2.072} \quad (27)$$

$$P_{pore} = \frac{P_f + P_p}{2} \quad (28)$$

$$T_m = \frac{T_{mf} + T_{mp}}{2} \qquad (29)$$

The fictitious route across the membrane is frequently related to the membrane porosity ϵ, as in the Mackie–Meares equation, and τ is the membrane thickness, which is frequently constant.

$$\tau = \frac{1}{\epsilon} \qquad (30)$$

The steps followed for calculating the theoretical model in the current DCMD system are depicted in Figure 2.

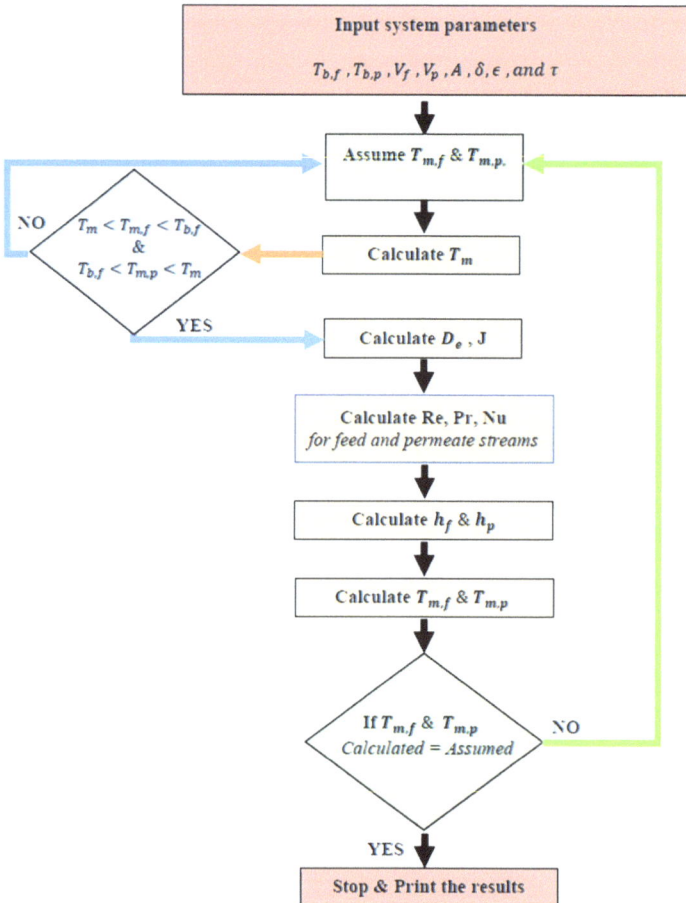

Figure 2. Flowchart for numerical optimization model in direct contact membrane distillation.

4. Results and Discussion

The accuracy of the model can be enhanced through artificial intelligence and machine learning by utilizing optimization software. Therefore, the data were simulated using Python to find the optimum conditions using the concept of iteration. Most simulation models are tested on the basis of experimental work [12,13]. Hence, DCMD runs were conducted to specifically evaluate the permeate flux under fixed experimental conditions. Hence, to further investigate the effect of varying system conditions on the membrane performance, both theoretical and predictive models were tested on the basis of experimental data. The accuracy of the theoretical mode was enhanced through the utilization of optimization methods in Python. The data were simulated using Python to find the optimum conditions using the concept of iteration.

4.1. Permeate Flux

As shown in Table 2, the increase in the Teflon percentage from 0.1% to 0.5% led to a slight reduction in the experimental permeate flux by 1.8%. This decrease in the permeate flux is attributed to the increase in membrane thickness from 190 µm to 199 µm. Furthermore, at a fixed porosity of 94%, the flux was enhanced by 2.7% and 0.44% with the addition of 1% for polystyrene membranes with 157 µm and 131 µm thicknesses, respectively. It can be observed that the optimum flux was achieved by the membrane with 14.05 LMH. This value is relatively higher than that of other membranes in DCMD, as listed in Table 3. This is attributed to the presence of PTFE beads that acted as a template for the formation of pores in the membrane, leading to an increase in its porosity. The mechanism of MD involves the application of a temperature gradient across a porous membrane, which creates a difference in vapor pressure on either side of the membrane. Water molecules in the liquid phase evaporate on the warm side of the membrane and diffuse through the pores to the cold side, where they condense into a liquid phase. The process is driven by the difference in vapor pressure across the membrane. The effectiveness of MD systems largely depends on several factors such as the pore size and thickness of the membrane, as well as the temperature and concentration of the saline feed across the membrane. A smaller pore size can reduce the rate of water vapor transport due to the longer diffusion path, which may result in a higher energy consumption to achieve the desired level of water recovery. On the other hand, larger pore sizes can increase the rate of transport and potentially lead to better energy efficiency, but may also increase the risk of wettability.

Table 2. Experimental, theoretical, and predicted permeate fluxes of polystyrene membranes under controlled parameters of $T_{b,f}$ = 60 °C and $T_{b,p}$ = 20 °C.

Porosity	Thickness (µm)	J_{exp} (LMH)	$J_{theoretical}$ (LMH)	Error (%)	$J_{predicted}$ (LMH)	Error (%)
0.77	190	13.68	12.9032	6	15.24	11
0.89	199	13.43	11.5563	14	15.15	13
0.94	157	14.05	12.1535	13	15.26	9
0.94	131	13.74	13.1158	5	14.84	8

Alternatively, a thinner membrane can enhance the heat transfer rate across the membrane, leading to enhanced energy efficiency. However, it may also increase the risk of membrane fouling, which can decrease the overall efficiency of the process. In contrast, a thicker membrane can provide better mechanical strength and durability but may result in lower energy efficiency due to the reduced heat transfer. Therefore, selecting the optimal pore size and membrane thickness requires balancing the trade-offs between membrane durability, energy consumption, and the potential occurrence of membrane wettability in order to achieve an improved desalination performance. Similar to previous studies, an optimized numerical model was used to evaluate the experimental values of the membrane/liquid interface temperatures, the thermal efficiency of the system, and the evaporation efficiency [39,47]. The predicted flux resulted in a similar trend to the experimental values. Nonetheless, the theoretical flux showed a comparatively high error of 13% in relation to that of the experimental flux. This is because the applied theoretical model investigated a wider range of applicability, leading to a higher probability of error in the obtained flux, as presented in Figure 3. For this reason, optimization was performed using Python in order to account for the variation in the surface membrane temperatures on both the feed and permeate sides to accurately predict the flux.

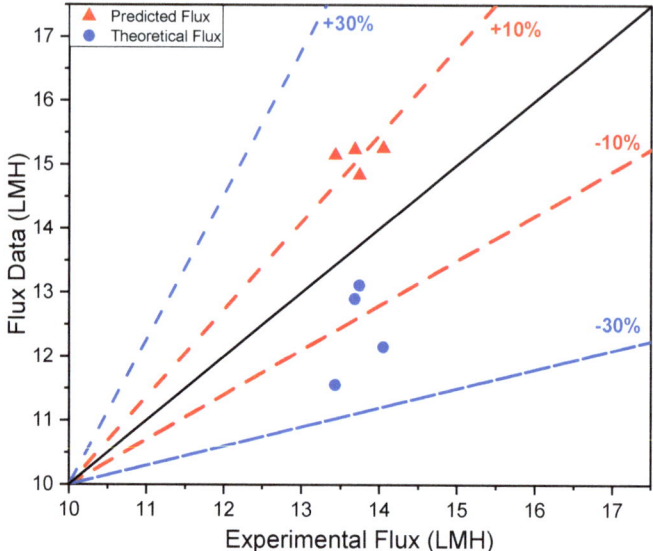

Figure 3. Numerical validation of predicted and theoretical flux with respect to experimental permeate flux for polystyrene membranes in pilot scale DCMD.

In the results depicted in Figure 4, it can be observed that there is a linear relationship between the permeate flux and the bulk feed temperature, and that, by looking at the 89% porosity, as the bulk feed temperature increased from 60 °C to 70 °C and from 70 °C to 80 °C, the permeate flux increased from almost 11.9 to 14 LMH, and from 14 to 16.2 LMH, respectively.

Additionally, the results show that there is no direct relationship between the increase in flux and the increase in membrane porosity, as, for example, the lowest porosity does not correspond to the lowest permeate flux, which means that the increase in flux is not proportional to the increase in porosity. The same conclusion can be applied to the relation between the permeate flux and membrane thickness. It is worth mentioning that the lowest experimental flux was observed at the largest thickness of 199 μm, with a porosity of 89%. However, the optimum flux was achieved at the highest porosity, with a slightly lower thickness of 157 μm.

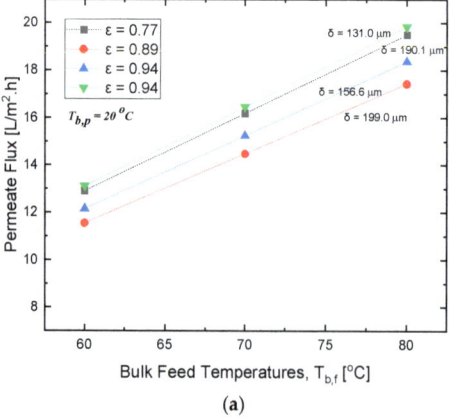

(a)

Figure 4. Cont.

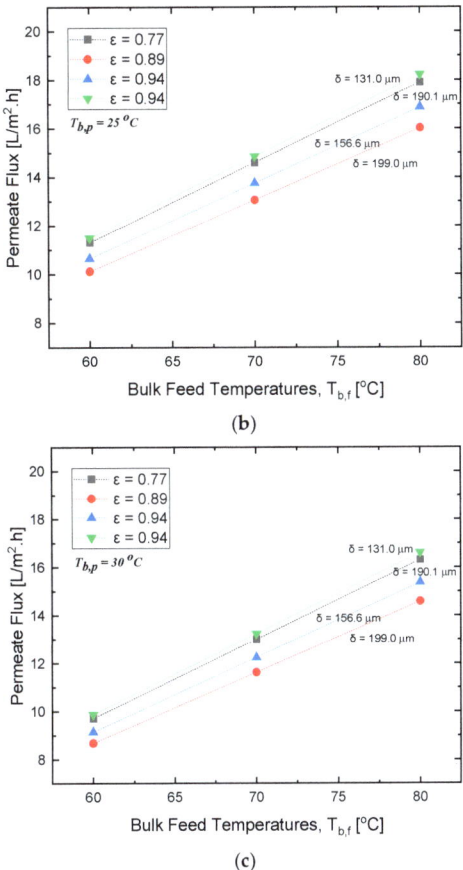

Figure 4. Effect of varying membrane porosity with respect to change in the bulk feed temperature and membrane thickness at a $T_{b,p}$ of (**a**) 20 °C, (**b**) 25 °C, and (**c**) 30 °C.

Table 3. Comparison of the flux predicted in this work with various experimental performances of different membranes in the literature.

Membrane	Feed Temperature (°C)	Feed Concentration (g/L)	Feed Flow Rate (L/min)	Experimental Flux (LMH)	Ref.
PVDF	50	35	0.6	21	[48]
	80	0.45	6	51.5	[49]
	40–90	4.65	0.14–100	55–72	[50]
PTFE	60	Seawater	4.5	45.5	[51]
	38	Various	11–22	2–5	[52]
	60	Synthetic brine	0.03	4.85–15.95	[13]
PTFE-PP	60	30	0.04	12.2	[53]
PVDF-PTFE	60	20	0.5	19	[54]
PP	40–60	-	0.5–1.7	5–25	[55]
PE	80	3.5	1.5	123	[56]
	70	Synthetic brine	1.2	122.2	[46]
PS	60	Synthetic brine	0.03	2.9–11.68	[13]
PS-PS	65	7	0.05	8.1	[57]
PS-AC	65	7	0.05	6.3	[57]

Table 3. Cont.

Membrane	Feed Temperature (°C)	Feed Concentration (g/L)	Feed Flow Rate (L/min)	Experimental Flux (LMH)	Ref.
PS-PTFE	60–80	Synthetic brine	1.5	13.68–14.05	Present work
PS-PTFE	60–80	15	1.5	Predicted flux 14.84–15.26	Present work

4.2. Effect of Porosity on Thermal Efficiency

At a low membrane porosity of 77%, the membrane showed a minimum thermal efficiency at all bulk feed temperatures. With a 15% increase in membrane porosity, a higher amount of water vapor was transported across the membrane, leading to more heat being exchanged between the two sides of the polystyrene membrane. This resulted in a 14.6% increase in thermal efficiency. Similarly, as depicted in Figure 5, the maximum thermal efficiency of 63% was reached when the membrane porosity was increased by 22%.

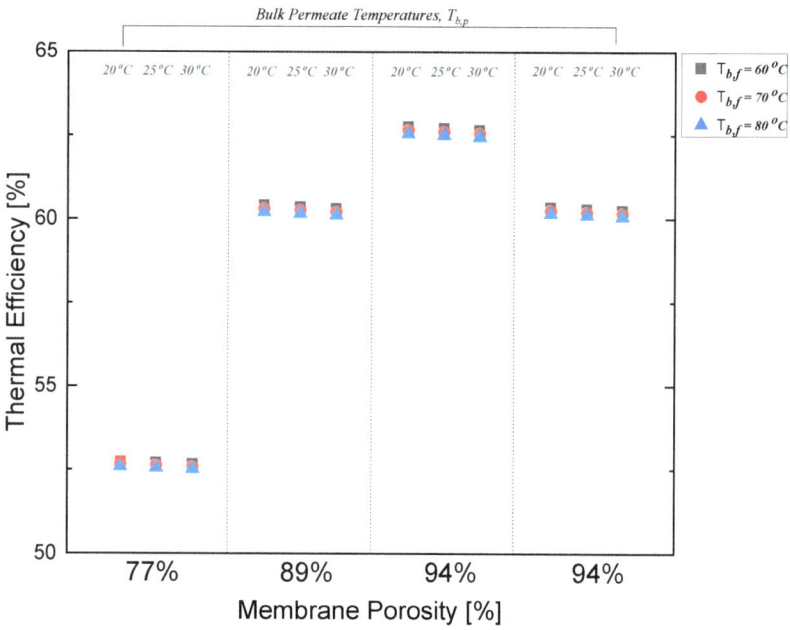

Figure 5. Effect of bulk temperatures on thermal efficiency in DCMD.

Interestingly, at an equal membrane porosity of 94%, increasing the membrane thickness by 14.5% resulted in a thermal efficiency drop by 4%. MD membranes with higher thicknesses have higher thermal resistance, which hinders the heat transfer between both the hot and cold streams. This results in a lower temperature driving force across the membrane, which decreases the rate of mass transfer and reduces the thermal efficiency of the MD process [58]. Moreover, the thermal efficiency of polystyrene membranes in DCMD is majorly affected by membrane fouling. As previously demonstrated, the accumulation of salt particles takes place on the membrane surface, coming from the synthetic feed [13].

4.3. Effect of Porosity on Evaporation Efficiency

The membrane porosity was shown to have a significant effect on the evaporation efficiency with respect to changes in the bulk permeate and feed temperatures. A rise in porosity from 77% to 89% resulted in a 5% increase in evaporation efficiency. Higher porosities allow for more efficient evaporation since a larger surface area becomes available for water vapor to pass through. However, as the porosity continued to increase up to 94%, the ability of the membrane to properly reject the dissolved solutes, coming from the feed solution, decreased. This negatively impacted the evaporation efficiency, decreasing it substantially by 5%.

As per the results depicted in Figure 6, as the porosity of the polystyrene membrane increased beyond 89%, the increased surface area for evaporation became offset by the decrease in the solute rejection, resulting in a decrease in evaporation efficiency. In Figure 4a–c, the relationship between membrane porosity, thickness, and bulk permeate temperature is illustrated. Taking the 60 °C bulk feed temperature as an example, the maximum permeate flux can be observed at a porosity of 94% with a thickness of 133 µm; the second highest permeate can be observed at a porosity of 77% with a thickness of 190 µm; the third highest permeate flux can be observed at a porosity of 94% with a thickness of 156.6 µm; and the lowest permeate flux can be observed at a porosity of 89% with a thickness of 199 µm.

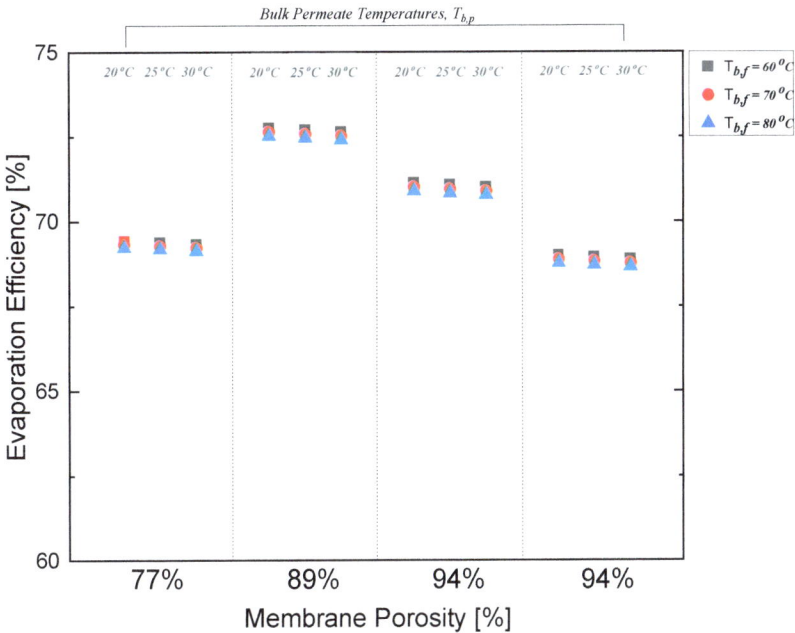

Figure 6. Effect of bulk temperatures on evaporation efficiency in DCMD.

With an increase in the bulk feed temperatures, the permeate flux increased by 51% due to the increasing vaporization of the synthetic brine at higher temperatures [38]. This is in accordance with Antoine's equation, where the vapor pressure exponentially rises with an increase in feed temperature, resulting in a rise in the permeate flux and enhancement of the overall MD process efficiency [59].

Ni et al. investigated the effect of membrane characteristics of different membrane materials [60]. Their study showed that with a decrease in membrane thickness, the permeate flux could be enhanced until a certain limit is reached. This limit is the threshold where the permeate flux is no longer improved. Theoretically, based on the literature, a reduction in thickness results in a continuous increase in the permeate flux. However, this is not the case in experimental investigations. Once the threshold is reached, the efficiency in membrane separation starts to decrease. This trend is depicted in Figure 7. However, Park and Lee investigated the energy efficiency in a pilot-scale DCMD system for hollow-fiber modules [61]. Their study showed that the thermal efficiencies of different MD modules cannot be directly compared in terms of the flux.

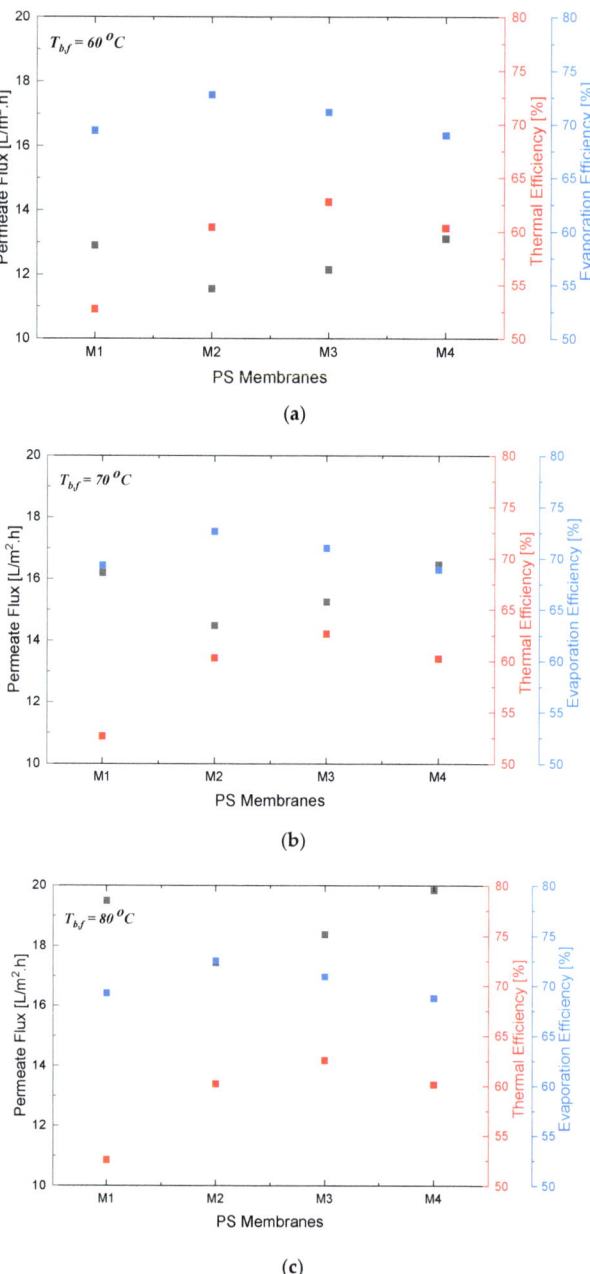

Figure 7. Permeate flux vs. both thermal efficiency and evaporation efficiency of polystyrene membranes at fixed $T_{b,p} = 20\ °C$ and changing $T_{b,f}$ from 60 °C to 80 °C in (**a**–**c**), respectively.

Figure 8 demonstrates the relationship between the permeate flux and membrane porosity by taking into consideration the change in the bulk permeate and bulk feed temperatures. At a constant porosity of 77% and a constant bulk permeate temperature of 20 °C, it can be observed that the permeate flux increased as the bulk feed temperature increased from 60 °C to 80 °C. The same trend can be observed for the bulk permeate temperatures of 25 °C and 30 °C, and for the rest of porosities as well. At a constant porosity and bulk feed temperature, a decreasing trend with respect to the permeate flux can be observed. Looking at the 77%-porosity membrane, the permeate flux decreased from 30 to 9 LMH when the bulk permeate temperature increased from 20 °C to 30 °C.

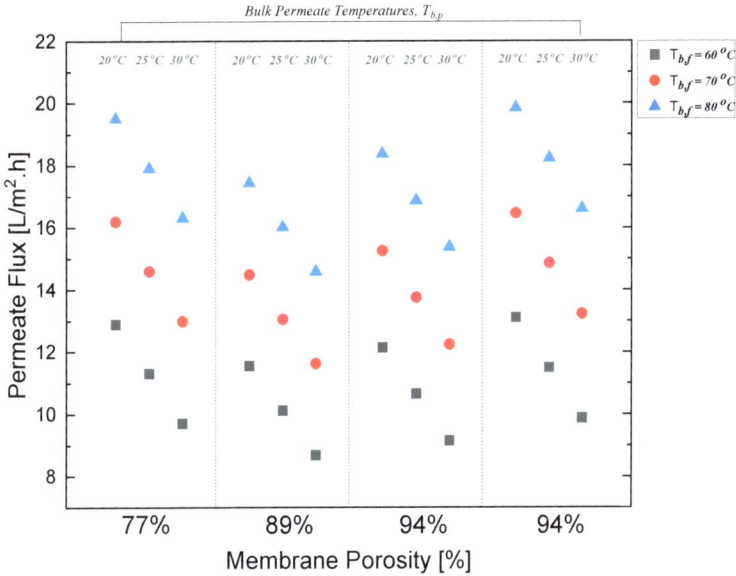

Figure 8. Effect of change in membrane porosity on permeate flux in DCMD.

At a constant bulk feed and bulk permeate temperature, and with the increase in membrane porosity from 77% to 94%, a non-linear relationship can be observed. The permeate flux first decreased from 12.9 to 11.8 when the porosity increased from 77% to 89%; then, the permeate flux increased from 11.9 to 12.1, and again to 13.1, when the porosity increased from 89% to 94%; afterwards, it stayed constant at 94%.

4.4. Effect of Porosity on Mass Transfer Coefficient

On the other hand, the design and manufacturing of various separation tools are usually quantified by the mass transfer coefficient. The mass transfer coefficient in any MD configuration is significantly dependable upon the membrane's temperature and characteristics [62]. In Figure 9a, it can be seen that at a constant permeate flux and pore diameter, the mass transfer coefficient values decreased with the increase in the mean average temperature. The maximum mass transfer coefficient value occurred at the highest permeate flux of 94% with a pore diameter of 0.0276 µm, and the lowest mass transfer coefficient occurred at a permeate flux of 89% with a pore diameter of 0.0131 µm. Thus, it can be observed that there is no direct relationship between the increase in the permeate flux, pore size diameter, and variation in the mass transfer coefficient. The same conclusion can be drawn from Figure 9b,c.

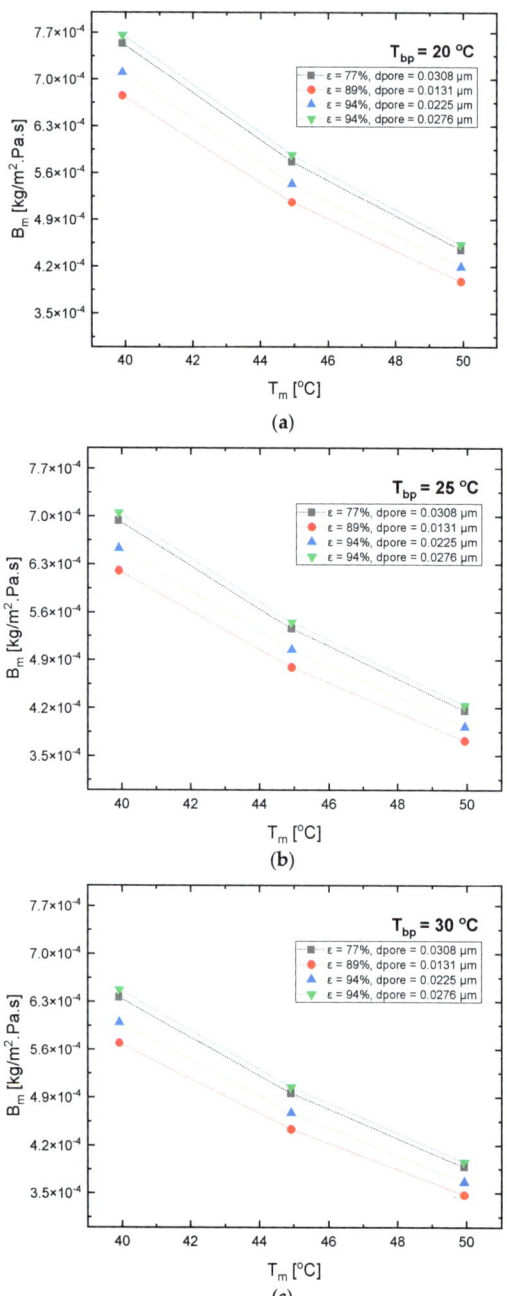

Figure 9. Mass transfer coefficient with respect to change in bulk feed and permeate temperature.

As shown in Figure 9b,c, the bulk permeate temperature was increased from 20 °C to 25 °C and 30 °C. Looking at the membrane with a 94% permeate flux and a pore diameter of 0.0276 m at a 40 °C mean average temperature, as the temperature increased from 20 °C to 25 °C and from 25 °C to 30 °C, the mass transfer coefficient decreased from 7.7×10^{-4} pa·s to 7.0×10^{-4} pa·s, and from 7.0×10^{-4} pa·s to 6.4×10^{-4} pa·s. This is due to the fact that an increase in temperature can cause the fluid in the pores to become more viscous, which can lower the mass transfer coefficient. This occurs since the fluid becomes more resistant to flow, and also because diffusion through the pores becomes more difficult.

5. Conclusions

This study demonstrates the effect of the porous structure of polystyrene membranes on the overall efficiency of a pilot-scale DCMD system. Such insights are difficult to obtain through conventional bench-scale DCMD setups. Supported by the experimental findings, an optimized iterative method was used to minimize the error between the initial estimates of the surface membrane temperature and the actual values, allowing them to predict the temperature accurately in each experiment. The prediction model used in this study was effective in predicting the permeate flux, as the results showed good agreement between the experimental results and the optimization model, with an error between 8 and 13%. The theoretical modeling data showed that the higher the temperature difference between the feed and permeate sides, the greater the vapor pressure difference, resulting in an increased permeate flux, until a threshold is reached, at which point the flux stops improving. The lowest experimental flux was observed at the largest thickness of 199 μm with a porosity of 89%. Additionally, for the 94% membrane porosity, increasing the membrane thickness by 14.5% resulted in a thermal efficiency drop by 4%. The results show that at porosities beyond 89%, the ability of the membrane to efficiently reject the dissolved solutes was reduced. This negatively impacted the evaporation efficiency, decreasing it substantially by 5%. The results obtained in this work can contribute to creating a good basis for future studies on the scalability of PS membranes for potential membrane-based desalination technologies in the industry. Our findings are in line with recent innovative wastewater technologies and are largely accountable for the optimization of industrial MD processes during the treatment of industrial wastewater. Yet, further investigations are still required to understand how heat loss minimization should be carried out experimentally for consideration in larger-scale applications.

Author Contributions: Conceptualization, H.A. and M.K.; methodology, H.A. and M.K.; software, H.A.; validation, H.A.; formal analysis, H.A. and M.K.; investigation, H.A.; writing—original draft preparation, H.A.; writing—review and editing, H.A. and M.K.; visualization, H.A.; supervision, M.K. All authors have read and agreed to the published version of the manuscript.

Funding: This publication was funded by the grant [IRCC-2020-016] from Qatar University.

Institutional Review Board Statement: Not applicable.

Data Availability Statement: The data presented in this study are available on request from the corresponding author.

Acknowledgments: This work was made possible by the Award [GSRA5-1-0524-18068] from the Qatar National Research Fund (a member of the Qatar Foundation). The contents herein are solely the responsibility of the authors.

Conflicts of Interest: The authors declare no conflict of interest.

Nomenclature

A	Cross-sectional area [m^2]
d_p	Pore size [μm]
d_h	Hydraulic diameter [m]
D	Diffusion coefficient [m^2/s]
h	Heat transfer coefficient [W/m^{2}K]
H_v	Heat of vaporization [kJ/kg]

J_w	Permeate flux [LMH]
K	Thermal conductivity [W/mK]
B_m	Mass transfer coefficient [kg/m^{2}sPa]
K_m	Membrane thermal conductivity [W/mK]
K_{gas}	Thermal conductivity of gas filling the pores [W/mK]
K_{mem}	Thermal conductivity of membrane material [W/mK]
Kn	Knudsen number [dimensionless number]
M_w	Molecular weight [g/mol]
Nu	Nusselt number [dimensionless number]
P	Total pressure [Pa]
P_m	Mean pressure [Pa]
Pr	Prandtl number [dimensionless number]
Q_s	Sensible heat transfer [W/m^2]
Q_v	Latent heat transfer [W/m^2]
Q_c	Conduction heat transfer [W/m^2]
R	Gas constant [J/Kmol]
Re	Reynolds number [dimensionless number]
Sc	Schmidt number [dimensionless number]
Sh	Sherwood number [dimensionless number]
T	Absolute temperature [K]

Subscripts and Superscripts:

f	Feed
p	Permeate
b	Bulk
m	Membrane
bf	Bulk feed
bp	Bulk permeate
mf	Feed side of membrane
mp	Permeate side of membrane

Greek Letters:

δ	Membrane thickness [μm]
ε	Porosity [%]
τ	Tortuosity [dimensionless number]
λ	Mean free path [m]
μ	Dynamic viscosity [Ns/m^2]
V	Kinematic viscosity [m^2/s]
ρ	Density [kg/m^3]

References

1. Kim, B.; Kwak, R.; Kwon, H.J.; Pham, V.S.; Kim, M.; Al-Anzi, B.; Lim, G.; Han, J. Purification of high salinity brine by multi-stage ion concentration polarization desalination. *Sci. Rep.* **2016**, *6*, 31850. [CrossRef]
2. Cohen-Tanugi, D.; McGovern, R.K.; Dave, S.H.; Lienhard, J.H.; Grossman, J.C. Quantifying the potential of ultra-permeable membranes for water desalination. *Energy Environ. Sci.* **2014**, *7*, 1134–1141. [CrossRef]
3. Ruiz-García, A.; Nuez, I. Simulation-based assessment of safe operating windows and optimization in full-scale seawater reverse osmosis systems. *Desalination* **2022**, *533*, 115768. [CrossRef]
4. Deshmukh, A.; Boo, C.; Karanikola, V.; Lin, S.; Straub, A.P.; Tong, T.; Warsinger, D.M.; Elimelech, M. Membrane distillation at the water-energy nexus: Limits, opportunities, and challenges. *Energy Environ. Sci.* **2018**, *11*, 1177–1196. [CrossRef]
5. Miladi, R.; Frikha, N.; Kheiri, A.; Gabsi, S. Energetic performance analysis of seawater desalination with a solar membrane distillation. *Energy Convers. Manag.* **2019**, *185*, 143–154. [CrossRef]
6. Dévora-Isiordia, G.E.; Cásares-De la Torre, C.A.; Morales-Mendívil, D.P.; Montoya-Pizeno, R.; Velázquez-Limón, N.; Aguilar-Jiménez, J.A.; Ríos-Arriola, J. Evaluation of Concentration Polarization Due to the Effect of Feed Water Temperature Change on Reverse Osmosis Membranes. *Membranes* **2022**, *13*, 3. [CrossRef]
7. Shamlou, E.; Vidic, R.; Khanna, V. Optimization-based technoeconomic comparison of multi-stage membrane distillation configurations for hypersaline produced water desalination. *Desalination* **2022**, *543*, 116098. [CrossRef]
8. Singh, D.; Sirkar, K.K. Desalination of brine and produced water by direct contact membrane distillation at high temperatures and pressures. *J. Memb. Sci.* **2012**, *389*, 380–388. [CrossRef]

9. De Sousa Silva, R.; Ramlow, H.; Cavalcanti, C.D.Á.K.; Valle, R.d.C.S.C.; Machado, R.A.F.; Marangoni, C. Steady state evaluation with different operating times in the direct contact membrane distillation process applied to water recovery from dyeing wastewater. *Sep. Purif. Technol.* **2020**, *230*, 115892. [CrossRef]
10. Rashid, K.T.; Rahman, S.B.A.; Alsalhy, Q.F. Optimum Operating Parameters for Hollow Fiber Membranes in Direct Contact Membrane Distillation. *Arab. J. Sci. Eng.* **2016**, *41*, 2647–2658. [CrossRef]
11. Gryta, M. Influence of polypropylene membrane surface porosity on the performance of membrane distillation process. *J. Memb. Sci.* **2007**, *287*, 67–78. [CrossRef]
12. Duong, H.C.; Duke, M.; Gray, S.; Cooper, P.; Nghiem, L.D. Membrane scaling and prevention techniques during seawater desalination by air gap membrane distillation. *Desalination* **2016**, *397*, 92–100. [CrossRef]
13. Abdelrazeq, H.; Khraisheh, M.; Al Momani, F.; McLeskey, J.T.; Hassan, M.K.; Gad-el-Hak, M.; Tafreshi, H.V. Performance of electrospun polystyrene membranes in synthetic produced industrial water using direct-contact membrane distillation. *Desalination* **2020**, *493*, 114663. [CrossRef]
14. Boubakri, A.; Hafiane, A.; Bouguecha, S.A.T. Application of response surface methodology for modeling and optimization of membrane distillation desalination process. *J. Ind. Eng. Chem.* **2014**, *20*, 3163–3169. [CrossRef]
15. Rattner, A.S.; Nagavarapu, A.K.; Garimella, S.; Fuller, T.F. Modeling of a flat plate membrane-distillation system for liquid desiccant regeneration in air-conditioning applications. *Int. J. Heat Mass Transf.* **2011**, *54*, 3650–3660. [CrossRef]
16. Abdelrazeq, H.; Khraisheh, M.; Ashraf, H.M.; Ebrahimi, P.; Kunju, A. Sustainable innovation in membrane technologies for produced water treatment: Challenges and limitations. *Sustainability* **2021**, *13*, 6759. [CrossRef]
17. Sallakh Niknejad, A.; Bazgir, S.; Kargari, A. Novel Triple-Layer HIPS/SBR/PP Nanofibrous Membranes for Robust DCMD Desalination. *Ind. Eng. Chem. Res.* **2021**, *60*, 2911–2920. [CrossRef]
18. Wang, Z.; Wu, A.; Ciacchi, L.C.; Wei, G. Recent Advances in Nanoporous Membranes for Water Purification. *Nanomaterials* **2018**, *8*, 65. [CrossRef]
19. Dervin, S.; Dionysiou, D.D.; Pillai, S.C. 2D nanostructures for water purification: Graphene and beyond. *Nanoscale* **2016**, *8*, 15115–15131. [CrossRef]
20. Lee, A.; Elam, J.W.; Darling, S.B. Membrane materials for water purification: Design, development, and application. *Environ. Sci. Water Res. Technol.* **2016**, *2*, 17–42. [CrossRef]
21. Werber, J.R.; Osuji, C.O.; Elimelech, M. Materials for next-generation desalination and water purification membranes. *Nat. Rev. Mater.* **2016**, *1*, 16018. [CrossRef]
22. Ahmed, F.E.; Lalia, B.S.; Hashaikeh, R. A review on electrospinning for membrane fabrication: Challenges and applications. *Desalination* **2014**, *356*, 15–30. [CrossRef]
23. Ullah, R.; Khraisheh, M.; Esteves, R.J.; McLeskey, J.T.; AlGhouti, M.; Gad-el-Hak, M.; Vahedi Tafreshi, H. Energy efficiency of direct contact membrane distillation. *Desalination* **2018**, *433*, 56–67. [CrossRef]
24. Yang, Z.; Zhou, Y.; Feng, Z.; Rui, X.; Zhang, T.; Zhang, Z. A review on reverse osmosis and nanofiltration membranes for water purification. *Polymers* **2019**, *11*, 1252. [CrossRef] [PubMed]
25. Cho, H.; Choi, Y.; Lee, S. Effect of pretreatment and operating conditions on the performance of membrane distillation for the treatment of shale gas wastewater. *Desalination* **2018**, *437*, 195–209. [CrossRef]
26. Khayet, M.; Wang, R. Mixed Matrix Polytetrafluoroethylene/Polysulfone Electrospun Nanofibrous Membranes for Water Desalination by Membrane Distillation. *ACS Appl. Mater. Interfaces* **2018**, *10*, 24275–24287. [CrossRef]
27. Wang, P.; Chung, T.S. Recent advances in membrane distillation processes: Membrane development, configuration design and application exploring. *J. Memb. Sci.* **2015**, *474*, 39–56. [CrossRef]
28. Cong, H.; Chuai, D.; Chul, Y.; Kyong, H.; Duc, L. A novel electrospun, hydrophobic, and elastomeric styrene-butadiene-styrene membrane for membrane distillation applications. *J. Memb. Sci.* **2018**, *549*, 420–427. [CrossRef]
29. Wang, K.; Hou, D.; Wang, J.; Wang, Z.; Tian, B.; Liang, P. Applied Surface Science Hydrophilic surface coating on hydrophobic PTFE membrane for robust anti-oil-fouling membrane distillation. *Appl. Surf. Sci.* **2018**, *450*, 57–65. [CrossRef]
30. Liao, Y.; Wang, R.; Tian, M.; Qiu, C.; Fane, A.G. Fabrication of polyvinylidene fluoride (PVDF) nanofiber membranes by electro-spinning for direct contact membrane distillation. *J. Memb. Sci.* **2013**, *425–426*, 30–39. [CrossRef]
31. Lee, E.J.; An, A.K.; Hadi, P.; Lee, S.; Woo, Y.C.; Shon, H.K. Advanced multi-nozzle electrospun functionalized titanium dioxide/polyvinylidene fluoride-co-hexafluoropropylene (TiO$_2$/PVDF-HFP) composite membranes for direct contact membrane distillation. *J. Memb. Sci.* **2017**, *524*, 712–720. [CrossRef]
32. Lalia, B.S.; Guillen-Burrieza, E.; Arafat, H.A.; Hashaikeh, R. Fabrication and characterization of polyvinylidenefluoride-co-hexafluoropropylene (PVDF-HFP) electrospun membranes for direct contact membrane distillation. *J. Memb. Sci.* **2013**, *428*, 104–115. [CrossRef]
33. Lalia, B.S.; Guillen, E.; Arafat, H.A.; Hashaikeh, R. Nanocrystalline cellulose reinforced PVDF-HFP membranes for membrane distillation application. *Desalination* **2014**, *332*, 134–141. [CrossRef]
34. Essalhi, M.; Khayet, M. Self-sustained webs of polyvinylidene fluoride electrospun nanofibers at different electrospinning times: 1. Desalination by direct contact membrane distillation. *J. Memb. Sci.* **2013**, *433*, 167–179. [CrossRef]
35. Lee, J.; Kim, Y.; Kim, W.; Francis, L.; Amy, G.; Ghaffour, N. Performance modeling of direct contact membrane distillation (DCMD) seawater desalination process using a commercial composite membrane. *J. Memb. Sci.* **2015**, *478*, 85–95. [CrossRef]

36. Ke, H.; Feldman, E.; Guzman, P.; Cole, J.; Wei, Q.; Chu, B.; Alkhudhiri, A.; Alrasheed, R.; Hsiao, B.S. Electrospun polystyrene nanofibrous membranes for direct contact membrane distillation. *J. Memb. Sci.* **2016**, *515*, 86–97. [CrossRef]
37. Tsyurupa, M.P.; Davankov, V.A. Porous structure of hypercrosslinked polystyrene: State-of-the-art mini-review. *React. Funct. Polym.* **2006**, *66*, 768–779. [CrossRef]
38. Elrasheedy, A.; Rabie, M.; El-Shazly, A.; Bassyouni, M.; Abdel-Hamid, S.M.S.; El Kady, M.F. Numerical investigation of fabricated mwcnts/polystyrene nanofibrous membrane for dcmd. *Polymers* **2021**, *13*, 160. [CrossRef]
39. Eleiwi, F.; Ghaffour, N.; Alsaadi, A.S.; Francis, L.; Laleg-Kirati, T.M. Dynamic modeling and experimental validation for direct contact membrane distillation (DCMD) process. *Desalination* **2016**, *384*, 1–11. [CrossRef]
40. Kuang, Z.; Long, R.; Liu, Z.; Liu, W. Analysis of temperature and concentration polarizations for performance improvement in direct contact membrane distillation. *Int. J. Heat Mass Transf.* **2019**, *145*, 118724. [CrossRef]
41. Elmarghany, M.R.; El-Shazly, A.H.; Salem, M.S.; Sabry, M.N.; Nady, N. Thermal analysis evaluation of direct contact membrane distillation system. *Case Stud. Therm. Eng.* **2019**, *13*, 100377. [CrossRef]
42. Zhang, X.; Koirala, R.; Date, A.; Jegatheesan, V. Modelling and simulation of flux prediction and salinity variation in direct contact membrane distillation for seawater desalination and brine treatment. *Desalination* **2022**, *540*, 116021. [CrossRef]
43. de Sampaio, P.A.B. Computational model and simulation of DCMD desalination systems with heat recovery. *Desalination* **2022**, *533*, 115769. [CrossRef]
44. Ansari, A.; Kavousi, S.; Helfer, F.; Millar, G.; Thiel, D.V. An improved modelling approach for the comprehensive study of direct contact membrane distillation. *Membranes* **2021**, *11*, 308. [CrossRef] [PubMed]
45. Suárez, F.; Del Río, M.B.; Aravena, J.E. Water Flux Prediction in Direct Contact Membrane Distillation Subject to Inorganic Fouling. *Membranes* **2022**, *12*, 157. [CrossRef]
46. Abdelrazeq, H.; Khraisheh, M.; Hassan, M.K. Long-Term Treatment of Highly Saline Brine in a Direct Contact Membrane Distillation (DCMD) Pilot Unit Using Polyethylene Membranes. *Membranes* **2022**, *12*, 424. [CrossRef]
47. Qtaishat, M.; Matsuura, T.; Kruczek, B.; Khayet, M. Heat and mass transfer analysis in direct contact membrane distillation. *Desalination* **2008**, *219*, 272–292. [CrossRef]
48. Phattaranawik, J.; Jiraratananon, R.; Fane, A.G. Effect of pore size distribution and air flux on mass transport in direct contact membrane distillation. *J. Memb. Sci.* **2003**, *215*, 75–85. [CrossRef]
49. Meng, S.; Mansouri, J.; Ye, Y.; Chen, V. Effect of templating agents on the properties and membrane distillation performance of TiO$_2$-coated PVDF membranes. *J. Memb. Sci.* **2014**, *450*, 48–59. [CrossRef]
50. Kim, Y.D.; Thu, K.; Ghaffour, N.; Choon Ng, K. Performance investigation of a solar-assisted direct contact membrane distillation system. *J. Memb. Sci.* **2013**, *427*, 345–364. [CrossRef]
51. Adnan, S.; Hoang, M.; Wang, H.; Xie, Z. Commercial PTFE membranes for membrane distillation application: Effect of microstructure and support material. *Desalination* **2012**, *284*, 297–308. [CrossRef]
52. Lee, J.G.; Kim, W.S.; Choi, J.S.; Ghaffour, N.; Kim, Y.D. Dynamic solar-powered multi-stage direct contact membrane distillation system: Concept design, modeling and simulation. *Desalination* **2018**, *435*, 278–292. [CrossRef]
53. Floros, I.N.; Kouvelos, E.P.; Pilatos, G.I.; Hadjigeorgiou, E.P.; Gotzias, A.D.; Favvas, E.P.; Sapalidis, A.A. Enhancement of flux performance in PTFE membranes for direct contact membrane distillation. *Polymers* **2020**, *12*, 345. [CrossRef] [PubMed]
54. Li, J.; Ren, L.F.; Shao, J.; Tu, Y.; Ma, Z.; Lin, Y.; He, Y. Fabrication of triple layer composite membrane and its application in membrane distillation (MD): Effect of hydrophobic-hydrophilic membrane structure on MD performance. *Sep. Purif. Technol.* **2020**, *234*, 116087. [CrossRef]
55. Swaminathan, J.; Chung, H.W.; Warsinger, D.M.; AlMarzooqi, F.A.; Arafat, H.A.; Lienhard, J.H. Energy efficiency of permeate gap and novel conductive gap membrane distillation. *J. Memb. Sci.* **2016**, *502*, 171–178. [CrossRef]
56. Zuo, J.; Bonyadi, S.; Chung, T.S. Exploring the potential of commercial polyethylene membranes for desalination by membrane distillation. *J. Memb. Sci.* **2016**, *497*, 239–247. [CrossRef]
57. Esteves, R.J.A.; Gornick, V.; Alqurwani, D.S.; Koenig-Lovejoy, J.; Abdelrazeq, H.; Khraisheh, M.; Forzano, A.V.; Gad-el-Hak, M.; Tafreshi, H.V.; McLeskey, J.T. Activated carbon-doped polystyrene fibers for direct contact membrane desalination. *Emergent Mater.* **2020**, *3*, 807–814. [CrossRef]
58. Gryta, M. Effectiveness of water desalination by membrane distillation process. *Membranes* **2012**, *2*, 415–429. [CrossRef]
59. Lawal, D.U.; Khalifa, A.E. Flux Prediction in Direct Contact Membrane Distillation. *Int. J. Mater. Mech. Manuf.* **2014**, *2*, 302–308. [CrossRef]
60. Ni, W.; Li, Y.; Zhao, J.; Zhang, G.; Du, X.; Dong, Y. Simulation study on direct contact membrane distillation modules for high-concentration nacl solution. *Membranes* **2020**, *10*, 179. [CrossRef]
61. Park, Y.; Lee, S. Analysis of thermal energy efficiency for hollow fiber membranes in direct contact membrane distillation. *Environ. Eng. Res.* **2019**, *24*, 347–353. [CrossRef]
62. Olatunji, S.O.; Camacho, L.M. Heat and mass transport in modeling membrane distillation configurations: A review. *Front. Energy Res.* **2018**, *6*, 130. [CrossRef]

Disclaimer/Publisher's Note: The statements, opinions and data contained in all publications are solely those of the individual author(s) and contributor(s) and not of MDPI and/or the editor(s). MDPI and/or the editor(s) disclaim responsibility for any injury to people or property resulting from any ideas, methods, instructions or products referred to in the content.

Article

Vapor Pressure and Evaporation Studies of Saline Solutions on Natural and Synthetic Fabrics for Industrial Water Treatment

Alexis López-Borrell [1,*], Jaime Lora-García [1], Salvador C. Cardona [1], María-Fernanda López-Pérez [1] and Vicent Fombuena [2]

1. Instituto de Seguridad Industrial, Radiofísica y Medioambiental (ISIRYM), Universitat Politècnica de València (UPV), Plaza Ferrándiz y Carbonell, s/n, 03801 Alcoy, Spain; jlora@iqn.upv.es (J.L.-G.); scardona@iqn.upv.es (S.C.C.); malope1@iqn.upv.es (M.-F.L.-P.)
2. Technological Institute of Materials (ITM), Universitat Politècnica de València (UPV), Plaza Ferrándiz y Carbonell 1, 03801 Alcoy, Spain; vifombor@upv.es
* Correspondence: allobor1@epsa.upv.es; Tel.: +34-(96)-652-8460

Abstract: In the present paper, we have conducted a comprehensive analysis of vapor pressures of both saturated and unsaturated solutions, alongside a study of evaporation using synthetic and natural fabrics for industrial applications in brackish water treatment under zero liquid discharge (ZLD) philosophy. By determining the vapor pressures of saturated solutions, we obtained results consistent with those of other researchers, extending the range of tested temperatures from 1 to 50 °C and successfully fitting the parameters of an Antoine-type equation. Similarly, positive results were achieved for unsaturated solutions, where various parameters of different equations accounting for the salt concentration were estimated, simplifying the fitting procedure. Natural evaporation tests from water surfaces using saturated solutions revealed that salts with higher associated vapor pressures exhibit higher evaporation rates. On the other hand, hydrated salts retain water in their structure and are significantly affected by ambient humidity. Evaporation studies on natural and synthetic fabrics with saturated NaCl and $CuSO_4 \cdot 5H_2O$ solutions showed distinct behaviors. NaCl increased both the evaporation rate and salt deposition with each cycle. In contrast, $CuSO_4 \cdot 5H_2O$ reduced the absorption capacity by blocking the fabric's structure, decreasing the evaporation efficiency over successive cycles.

Keywords: vapor pressure; evaporation; natural fibers; absorption; relative humidity

1. Introduction

Demand for water resources has become a significant issue in many parts of the world, even in regions where it was not previously a concern. The decline in water availability is primarily due to the misuse and mismanagement of water resources and water pollution. These challenges are further exacerbated by population growth, increasing demand for this essential resource and leading to overexploitation [1].

Much of the research over the years has focused on investigating water deterioration, the main pollutants, and their origin. A large part of water pollution is a consequence of industrial activity and the pollution from urban areas. Water pollution due to industrial activity is diverse, and it ranges from mining activity with heavy metal pollution, pesticide pollution and fertilizer leaks from the agri-food industry to companies that generate untreated excess water discharges [2–4]. On the other hand, a significant source of pollution is seen in urban areas due to the large amount of water consumed per inhabitant. It has been observed that water quality has been worsening over time, as it is not treated correctly, causing an impact on human health [5–8].

Conventional treatments mainly used for wastewater management involve physicochemical processes such as aerobic and anaerobic biological processes used in wastewater treatment plants, chemical coagulation-flocculation processes and the combination of physical techniques such as membrane separation treatment. More advanced techniques based on membrane separation include microfiltration (MF), ultrafiltration (UF), nanofiltration (NF) or reverse osmosis (RO) in conjunction with the other conventional processes for higher separation efficiency or to ensure the quality of the wastewater to be treated [9,10]. On the other hand, one of the working philosophies for wastewater treatment lies in Zero Liquid Discharge (ZLD) [11]. This method does not recover water from its treatment but concentrates the waste in the wastewater. Afterwards, it can be treated to either form a highly concentrated liquid or to leave a solid residue by removing the remaining liquid [12–14]. Also, in this case, depending on the type of industry operating under ZLD, it may be beneficial to revalue the waste and create a circular economy within the industry.

The ZLD methodology has been used for many years to reduce the volume of waste and concentrate it to very high levels. One of the techniques generally used is evaporation processes, either by forced convection, which requires high energy consumption, or by natural evaporation, which requires typically high evaporation surfaces such as salt ponds [15]. Due to the two main disadvantages of evaporation techniques—high energy consumption and the need for large working surfaces—a viable alternative that allows evaporation processes to operate with lower energy consumption and smaller working surfaces must be found. An alternative approach is to use parallel sheets of material immersed in the problem liquid and then exposed to air for drying. This method increases the available evaporation surface area by arranging multiple parallel sheets in a treatment bath [16].

In recent years, the production of natural fabrics has increased, as they are seen as an alternative to conventional synthetic fabrics such as glass fiber, carbon fiber or aramid to manufacture composite materials. These natural fabrics can compete with synthetics due to their low price and acceptable mechanical properties. Some fabrics or natural fibers most commonly used and with a high annual production include jute, bamboo or flax, among many others [17–19]. Bamboo fibers have a yearly production of 30 Mt, Jute fibers 2.3 Mt, and Flax fibers 0.83 Mt [20–22]. These natural fibers exhibit significant annual production rates, with even more growth expected in the coming years.

Natural and synthetic fabrics have versatile applications beyond composite materials. Their structure enables water and contaminants to be retained within, with liquid absorption capacities ranging from 100 to 850% of their initial weight [23]. The water absorption property within their structure facilitates their alignment in parallel within a treatment bath for evaporation processes. This reduces the energy consumption and required surface area and offers scalability due to its ease of implementation. An example of the versatility that natural fabrics can offer for these objectives is the article authored by the researchers, in which a comprehensive characterization of the fabrics is conducted from a physicochemical, mechanical, and absorbent capacity perspective [23]. However, the evaporation capacity when applying a technology based on the ZLD philosophy does not solely depend on the type of fabric. For instance, it is very important to determine the influence of the types of salts present in brackish water and to study the vapor pressure, as recent studies have shown [24].

To enhance the natural evaporation rate, the idea of combining absorbent surfaces, such as natural and synthetic fabrics, to increase the available evaporation surface by leveraging wind currents is explored. This technique, known as Wind-Aided Intensified eVaporation (WAIV), does not recover water; instead, it facilitates its evaporation into the environment. However, it can be employed to diminish waste volume or recover mineral salts in the water, which precipitate on the absorbent fabrics during the process [25,26].

This study investigates the behavior of various salts in a saturated state and at different concentrations to determine their vapor pressure across a wide range of temperatures. Additionally, it aims to assess the evaporation rate of these solutions on different surfaces, including natural and synthetic fabrics, along various operational cycles.

2. Materials and Methods

2.1. Materials

2.1.1. Fabrics

Six fabrics pre-selected from previous research were employed to carry out the tests during this study [23]. This selection encompasses both natural and synthetic fabrics. On the one hand, natural fabrics such as jute (Jut), bamboo (Bam), linen composed of 50% polylactic acid (LPLA) and, finally, a non-woven fabric composed of 70% palm prunings, 20% lyocell and 10% PLA (WL-T), were included. The WL-T fabric was supplied by AITEX (Alcoy, Spain). Additionally, two synthetic fabrics were employed: a non-woven polyester fabric (PES) and a fabric with an aramid taffeta structure (Ara). Both of these fabrics were supplied by Castro Composites (Pontevedra, Spain).

2.1.2. Salts

For the experiments, salts include sodium nitrate ($NaNO_3$) supplied by Scharlau Chemie S.A. (Barcelona, Spain), and calcium hydroxide ($Ca(OH)_2$), potassium nitrate (KNO_3), sodium hydrogen carbonate ($NaHCO_3$), magnesium sulphate ($MgSO_4 \cdot 7H_2O$), and copper sulphate ($CuSO_4 \cdot 5H_2O$) provided by PanReac Applichem, ITW Reagents (Barcelona, Spain). Additionally, sodium chloride (NaCl) delivered by Carlo Erba Reagents S.A. (Barcelona, Spain) was tested. All these salts were used for vapor pressure determination tests and the natural evaporation of saturated solutions using natural and synthetic fabrics.

These salts were selected because they were used in previous research to synthesize brackish water [23]. Additionally, NaCl was chosen for further study because it is the most abundant salt in seawater and may present an alternative treatment pathway.

2.2. Experimental Methodology

2.2.1. Natural Evaporation of Saturated Salts

Saturated solutions of the salts mentioned above were prepared for the evaporation studies of concentrated salts. Natural evaporation tests were conducted for each salt using three beakers, each with a different evaporation surface, to assess its effect on the evaporation rates of water. The evaporation surfaces utilized are outlined in Table 1.

Table 1. Evaporation surfaces used in the natural evaporation of saturated solutions.

Type of Vessel Used	Diameter (mm)	Evaporation Surface (mm^2)
Small (S_1)	13.4 ± 0.3	141.2 ± 5.7
Medium (S_2)	46.3 ± 0.3	1683.7 ± 25.2
Large (S_3)	66.3 ± 0.3	3452.4 ± 36.1

The tests involved placing a quantity of the saturated solution in different beakers and recording the initial mass of the solution using an analytical balance, Nahita Blue Series 5134. The analytical balance has a maximum capacity of 220 g, with a resolution of 10 mg and a weighing accuracy of ±1 mg. Mass measurements were taken every 24 h for each sample to record the water loss due to natural evaporation. These tests continued until the mass between two consecutive 24-h measurements remained unchanged.

2.2.2. Vapor Pressure Experiments for Saturated Salts

The vapor pressure was determined for the seven salts mentioned above in their saturated form. Saturated solutions were prepared in a volume of 75 mL and placed in a hermetically sealed jacketed reactor to control the temperature of the liquid and air using a Lauda Ecoline RE104 Recirculating Chiller, supplied by Lauda DR. R. Wobser GMBH & CO. KG (Lauda-Königshofen, Germany). These vapor pressure tests were conducted over a temperature range of 1 to 50 °C, in increments of 5 °C. The experimental set-up is illustrated in Figure 1.

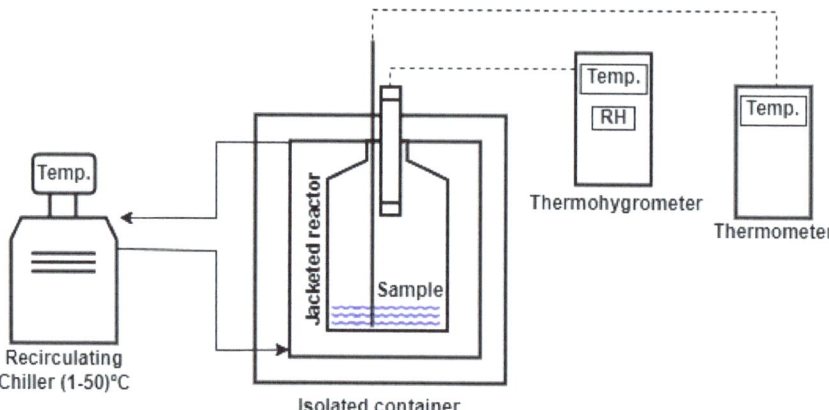

Figure 1. Experimental set-up for the determination of vapor pressures.

The relative humidity (*RH*) measurement was used to determine the vapor pressure. For *RH* measurements, a Testo 635-2 thermohygrometer was used (Instrumentos Testo S.A., Barcelona, Spain), with connectable temperature and humidity probes inserted into the hermetically sealed container and placed in the upper gas chamber. This probe has a temperature measurement range of −20 to 70 °C with an accuracy of ±0.3 °C. Additionally, the humidity measurement range is 0 to 100% with an accuracy of ±2%. The temperatures were adjusted using the recirculation chiller. The *RH* was measured 30 min after reaching the temperature setpoint to ensure equilibrium with the air. An additional flexible Pt-100 temperature probe (Instrumentos Testo S.A., Barcelona, Spain) was used to check the temperature of the liquid phase. This flexible temperature probe has a measurement range of −80 to 300 °C with an accuracy of ±0.05 °C.

For the determination of the vapor pressure, Equation (1), which relates to relative humidity, is used:

$$RH\ (\%) = \frac{P_v}{P_{v,sat}} \cdot 100 \qquad (1)$$

where *RH* is the relative humidity measured in %, P_v is the vapor pressure of the solution in Pa and $P_{v,sat}$ is the saturation vapor pressure of pure water in Pa. The variable $P_{v,sat}$ (Pa) is obtained from Equation (2) based on the IAPWS datasheets, but improving the fitting errors [27]:

$$P_{v,sat} = \frac{\exp\left(34.494 - \frac{4924.99}{T+237.1}\right)}{(T+105)^{1.57}} \qquad (2)$$

where *T* is the temperature, measured in °C. Note that this expression is valid only for temperatures above 0 °C.

2.2.3. Vapor Pressure Experiments for Unsaturated Salts

NaCl and $CuSO_4 \cdot 5H_2O$ salts were used in different molal concentrations in the temperature range from 1 to 50 °C, in 5 °C increments, for the vapor pressure tests of unsaturated solutions. The molalities chosen for the tests were 0.08, 0.17, 0.60, 1.23, 2.71, 3.68 and 5.75 mol·kg^{-1}. The vapor pressure was determined in the same way as for the tests with saturated salts using the Testo 635-2 thermohygrometer to measure the relative humidity of the air once equilibrium had been reached, according to the experimental set-up in Figure 1.

2.2.4. Evaporation on Natural and Synthetic Fabrics

Saturated solutions of NaCl and $CuSO_4 \cdot 5H_2O$ salts were used for evaporation tests on natural and synthetic fabrics. These experimental tests were conducted under laboratory conditions, with the room maintained at a temperature of 20 ± 2 °C and a relative humidity of 40 ± 5%.

The fabrics used for the evaporation tests had a surface area of 50 mm × 75 mm. To control the evaporation surface, the wet area will be confined to 50 mm × 60 mm, leaving the remaining surface available for handling the fabric. Two samples of each fabric with this surface area were prepared, and the effects of evaporation during wet and dry operation cycles were studied for each salt. To wet the samples they were inserted obliquely into the solution and kept immersed for (60 ± 1) s, then removed from the solution and allowed to drain the excess liquid for (120 ± 3) s, as shown in the first step of Figure 2. After this time, the samples were placed on a support and measurements of water weight loss were taken using an analytical balance (Nahita blue series 5134), as shown in the second step of Figure 2. The test continued until the fabric was completely dry.

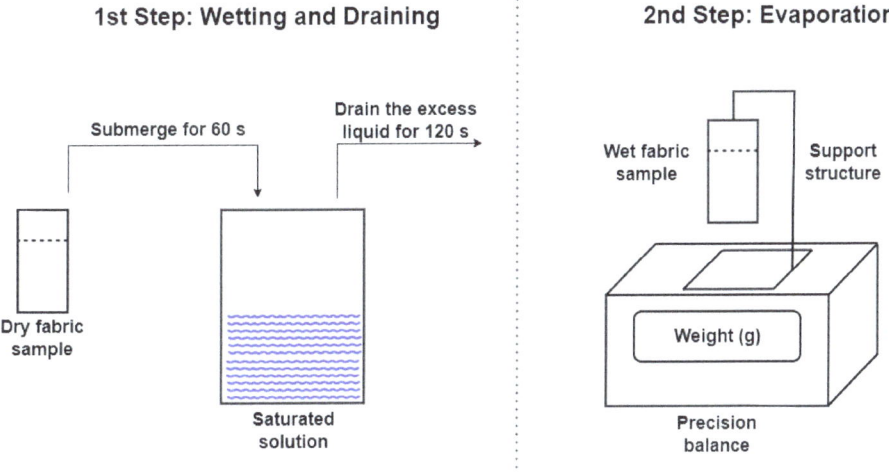

Figure 2. Experimental set-up for the evaporation experiments.

Multiple evaporation cycles were performed to study the evaporation of the saturated solutions on the fabrics. Each fabric underwent 15 wetting and drying cycles for each solution studied. The evaporation rates for cycles 1 (1C), 2 (2C), 5 (5C), and 15 (15C) were compared to observe the effect of salt precipitation on the fabrics. This approach aims to examine the impact of salt accumulation on evaporation rates and also the amount of salt deposited on the fabrics.

2.3. Parameter Fitting for Different Expressions of the Vapor Pressure of Saturated and Unsaturated Solutions

2.3.1. Fitting the Parameters of the Antoine Equation to the Experimental Vapor Pressures of Saturated Solutions

The Antoine equation has been used [28], with its parameters A, B, and C, as shown in Equation (3), fitted to the experimental vapor pressures measured for the seven saturated salt solutions at different temperatures:

$$\ln(P_v) = A - \frac{B}{C + T} \qquad (3)$$

where P_v is the vapor pressure of the saturated salt solutions in kPa, and T is the temperature in K.

2.3.2. Fitting the Parameters to the Experimental Vapor Pressures of Unsaturated Solutions

A literature review of the various expressions used to determine the vapor pressure of unsaturated solutions reveals that most of them are based on polynomial equations. These equations consider both the solution's concentration and its temperature dependence.

In this research, three fitting equations have been selected for testing. The first one is the Antoine equation (Equation (3)), where parameter estimations have been performed separately for each concentration, on the one hand, and the average, on the other hand. These fittings aim to determine if a simple expression can accurately predict the vapor pressures. The second and third equations are based on polynomial equations.

The equation described by Patil et al. [29] has three fitting parameters, as shown in Equation (4).

$$\log(P_v) = A(m) + \frac{B(m)}{T} + \frac{C(m)}{T^2} \quad (4)$$

where P_v is the vapor pressure in kPa, T is the temperature expressed in K and A, B and C are the fitting parameters that depend on the molality of the solution. This relationship is formulated through Equations (5), (6) and (7), respectively:

$$A(m) = A_0 + A_1 \cdot m + A_2 \cdot m^2 + A_3 \cdot m^3 \quad (5)$$

$$B(m) = B_0 + B_1 \cdot m + B_2 \cdot m^2 + B_3 \cdot m^3 \quad (6)$$

$$C(m) = C_0 + C_1 \cdot m + C_2 \cdot m^2 + C_3 \cdot m^3 \quad (7)$$

where m is the molality of the solution expressed in mol·kg^{-1} and A_i, B_i and C_i represent the respective fitting parameters. To simplify the optimization procedure, the objective is to minimize the total number of fitting parameters while ensuring that the final equation accurately represents the observed reality.

On the other hand, the following expression to be tested to explain the behaviour of unsaturated solutions at different concentrations is the one described by Shibue [30], as shown in Equation (8).

$$\ln(P_v) = \ln(P_c) + g(T) + h(x) \quad (8)$$

where P_v is the vapor pressure expressed in MPa, P_c is the critical pressure expressed in MPa, $g(T)$ is a temperature-dependent expression expressed in K and, finally, $h(x)$ is a mole fraction-dependent expression. P_c is obtained from Equation (9):

$$P_c = 22.064 + q_5 \cdot x + q_6 \cdot x^2 + q_7 \cdot x^3 + q_8 \cdot x^4 + q_9 \cdot x^5 + q_{10} \cdot x^6 \quad (9)$$

where x is the experimental mole fraction of the salt used and the parameters q_i are fitted values for a particular salt. In this work, the values of q_i reported by Shibue for the H_2O + NaCl mixture are assumed, as he stated these are well-studied and validated P_c values (Table 2).

Table 2. Fitted values for the P_c expression determined by Shibue [30].

Parameter	Value
q_5	9.00404×10^2
q_6	-2.92542×10^4
q_7	1.39806×10^6
q_8	-2.80756×10^7
q_9	2.41637×10^8
q_{10}	-7.18726×10^8

The function $g(T)$ is obtained from the Equation (10). The author found that with this simplified version, the fits were accurate. The critical pressure of pure water is considered at 647.096 K, so the expression only depends on the experimental temperature values.

$$g(T) = \frac{647.096}{T} \cdot \left[-7.85951783 \cdot \left\{1 - \left(\frac{T}{647.096}\right)\right\} + 1.184408256 \cdot \left\{1 - \left(\frac{T}{647.096}\right)\right\}^{1.5}\right]$$

$$+ \frac{647.096}{T} \cdot \left[-11.7866497 \cdot \left\{1 - \left(\frac{T}{647.096}\right)\right\}^3 + 22.6807411 \cdot \left\{1 - \left(\frac{T}{647.096}\right)\right\}^{3.5}\right] \quad (10)$$

$$+ \frac{647.096}{T} \cdot \left[-15.9618719 \cdot \left\{1 - \left(\frac{T}{647.096}\right)\right\}^4 + 1.80122502 \cdot \left\{1 - \left(\frac{T}{647.096}\right)\right\}^{7.5}\right]$$

Finally, the function $h(x)$ depends on the mole fraction (concentration) of the salt used and is obtained from Equation (11) for the dilute region, with a mole fraction ranging from 0 to 0.024. For the concentrated region, it is derived from Equation (12), for a mole fraction ranging from 0.024 to 0.117.

$$h(x) = \frac{a_2 \cdot x}{x + a_1^2} + a_3 \cdot x^2 \quad (11)$$

$$h(x) = \left[\frac{a_1^2 \cdot a_2}{(u+a_1^2)^2} + 2 \cdot a_3 \cdot u\right] \cdot (x - u) + b_1 \cdot (x - u)^2$$
$$+ b_2 \cdot (x - u) \cdot (x^2 - u^2) + b_2 \cdot (x - u) \cdot (x^2 - u^2) + \frac{a_2 \cdot u}{u + a_1^2} + a_3 \cdot u^2 \quad (12)$$

where x is the mole fraction, u is the value of the mole fraction where the change from the dilute to the concentrated region takes place and a_i and b_i are the fitting parameters for both expressions. Equation (11) is fitted first and then extended to Equation (12).

All parameter settings of the different equations were fitted using Matlab 2021a and the lsqnonlin function. A residual value, calculated as the sum of the squared differences between the experimental and theoretical data, was obtained to assess the quality of the fitting parameters. Additionally, corresponding confidence intervals were provided.

3. Results and Discussion

3.1. Vapor Pressure of Saturated Salt Solutions

The results of the determination of vapor pressures for saturated salt solutions are shown in Figure 3, and the values of the fitted parameters of Equation (3) can be seen in Table 3.

Table 3. Values for the fitted parameters of Antoine's equation.

Saturated salt	A	B	C	Residual
$Ca(OH)_2$	$14.5 \pm 8.40 \times 10^{-1}$	$2.91 \times 10^3 \pm 3.80 \times 10^2$	-81.0 ± 14.8	2.00×10^{-3}
$NaHCO_3$	15.3 ± 1.66	$3.30 \times 10^3 \pm 8.07 \times 10^2$	-65.5 ± 29.7	5.00×10^{-3}
NaCl	$16.2 \pm 9.40 \times 10^{-1}$	$3.84 \times 10^3 \pm 4.91 \times 10^2$	-46.9 ± 16.7	1.00×10^{-3}
$MgSO_4 \cdot 7H_2O$	20.1 ± 5.02	$5.80 \times 10^3 \pm 3.16 \times 10^3$	6.53 ± 85.9	1.50×10^{-3}
$NaNO_3$	18.6 ± 1.76	$5.21 \times 10^3 \pm 1.07 \times 10^3$	-3.43 ± 31.4	1.00×10^{-3}
KNO_3	18.3 ± 2.17	$4.90 \times 10^3 \pm 1.28 \times 10^3$	-13.1 ± 38.5	4.00×10^{-3}
$CuSO_4 \cdot 5H_2O$	$14.5 \pm 9.30 \times 10^{-1}$	$2.86 \times 10^3 \pm 4.17 \times 10^2$	-84.9 ± 16.3	2.00×10^{-3}

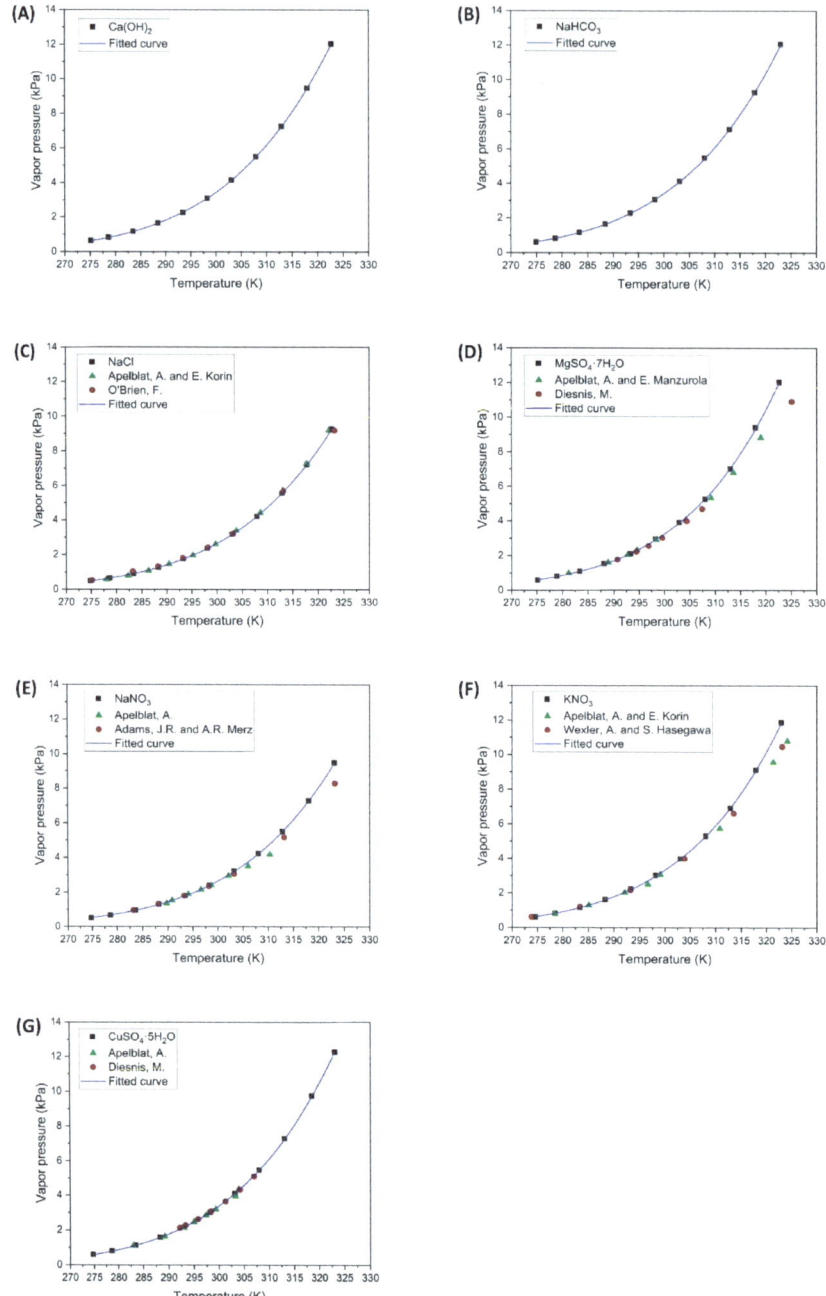

Figure 3. Vapor pressures of saturated salt solutions. (■ Experimental data; ▲ Apelblat, et al. [31–35]; ● O'Brien, F. [36], Diesnis, M. [37], Adams, J.R. and A.R. Merz [38], Wexler, A. and S. Hasegawa [39]; ⎯ Antoine fitting curve).

As shown in Figure 3, the experimental results obtained in this study align closely with those reported in the literature by other researchers who have determined the vapor pressures of saturated salt solutions.

The authors have observed that the application of Equation (2) results in values very similar to those obtained in the previous literature, as cited in the following references. These highly accurate results have been obtained with the salts $Ca(OH)_2$ and $NaHCO_3$. On the other hand, some salts do not show as much similarity with previously published results. These errors, ranging from 3% to 8%, are observed in the samples of $MgSO_4 \cdot 7H_2O$, $NaNO_3$, and KNO_3 salts in Figure 3D–F, respectively. These differences in vapor pressure measurements become more pronounced at temperatures above 35 °C. In this study, the vapor pressures determined are slightly higher than those reported by other authors. These discrepancies could be attributed to variations in the experimental equipment used or the determination of $P_{v,sat}$ as shown in Equation (2). Despite these deviations, the values determined are considered accurate.

The vapor pressures of $Ca(OH)_2$ and $NaHCO_3$ solutions have been poorly studied. The saturated $NaHCO_3$ solution exhibits high solubility within the temperature range investigated in this study, as noted by Ozcan and Miller [40]. Leon-Hidalgo et al. [41] determined the vapor pressure of a saturated sodium bicarbonate solution using an indirect hygrometer measurement at 25 °C, obtaining a value of 3.18 kPa. The present work has recorded a vapor pressure value of 3.08 kPa at 25.1 °C, closely matching Leon-Hidalgo's results. Knuutila et al. [42] investigated the sodium bicarbonate solution at the equilibrium point where sodium carbonate (Na_2CO_3) forms without reaching saturation. The values across the different concentration ranges are similar to those obtained for the saturated $NaHCO_3$ solution.

Finally, the $Ca(OH)_2$ solution has been scarcely studied due to its low solubility in water within the working temperature range. Duchesne and Reardon [43] compiled data from other authors indicating that the solubility of the Ca^{2+} ion was around 0.02 m, thus highlighting its low solubility level. Other researchers have increased the solubility of this salt by using bases, such as NaOH, at high concentrations and 25 °C, as investigated by Pallagi et al. [44]. Another study by Konno et al. [45] explored the solubility of this salt with varying concentrations of NaOH and temperatures of 25, 50, and 75 °C. The present study observed that due to this low solubility, the saturated $Ca(OH)_2$ solution exhibited relative humidity values above 95% at a temperature of 15 °C, indicating that its vapor pressure closely approximates that of pure water as the temperature increases.

The parameter fitting of Equation (3), corresponding to Antoine's equation, has demonstrated a good overlap with the experimental data, resulting in a low fitting error for all the salt solutions tested. Leon-Hidalgo et al., [41] conducted studies with both saturated and unsaturated saline solutions, fitting the saturated solutions to an Antoine-type expression. The parameters that were fitted in their study are similar to those obtained in the present research.

3.2. Vapor Pressure of Unsatured Solutions

3.2.1. NaCl Solutions

The vapor pressures of NaCl solutions at different molalities have been experimentally determined and compared with values reported by other authors. The results for the unsaturated NaCl solutions are presented in Figure 4.

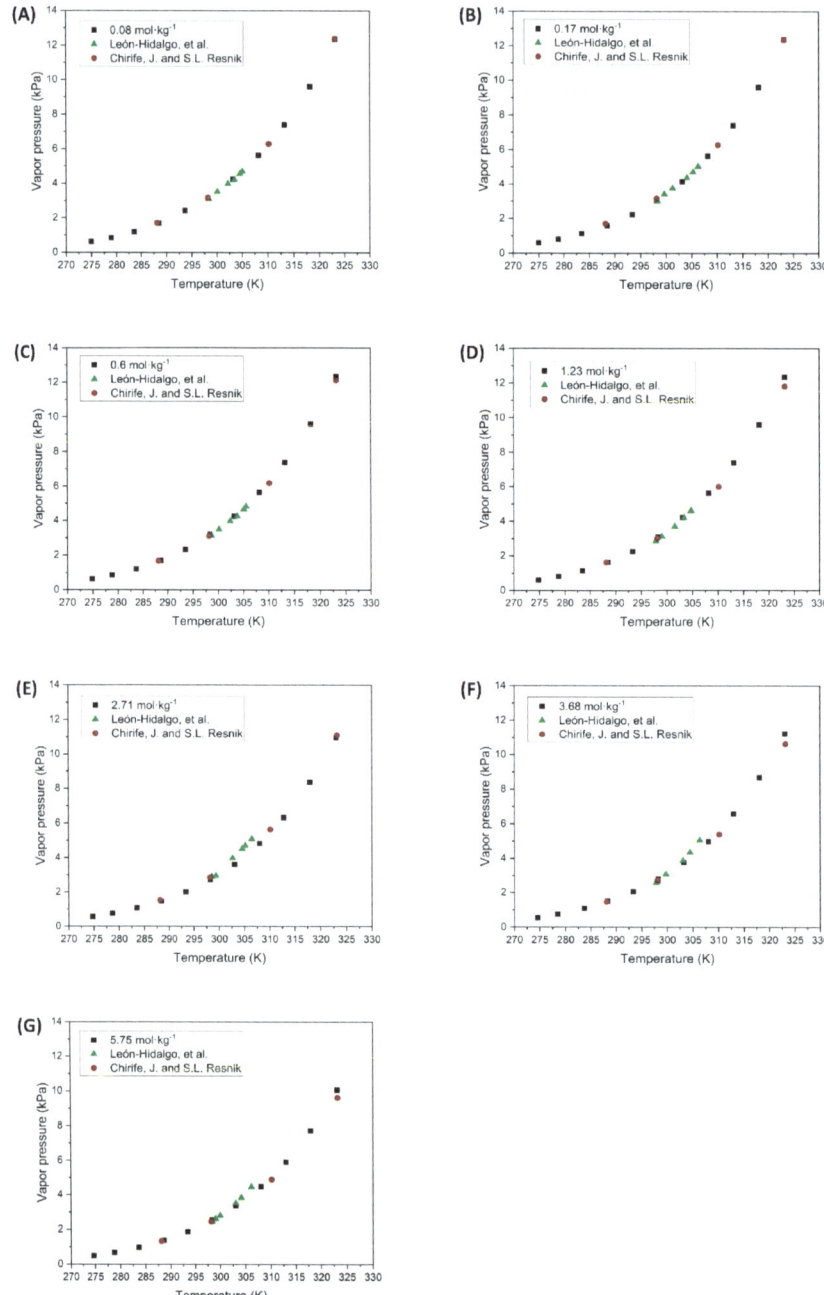

Figure 4. Vapor pressure of unsaturated NaCl solutions at different molalities. (■ Experimental data; ▲ León-Hidalgo, et al. [41]; ● Chirife, J. and S.L. Resnik [46]).

Based on the results obtained from determining the vapor pressures of unsaturated NaCl solutions, it can be concluded that these results align with those reported by other authors, as shown in Figure 4. The small deviations observed occur as the concentration increases from a molality of 2.71 mol·kg^{-1} to 5.75 mol·kg^{-1}, when comparing the results with those reported by Leon-Hidalgo et al. [41]. The main differences found at intermediate temperatures may be attributed to measurement errors due to the instrumentation used by the various authors. At elevated temperatures, the primary discrepancy may arise from the calculation of $P_{v,sat}$ as previously mentioned. Despite these slight differences, the values obtained in the present investigation are considered to be accurate.

The parameters of the Antoine-type Equation (3) were fitted for each concentration separately as well as for all the concentrations together. This was done to investigate if there is any relationship between the parameters and the salt concentration that could explain the behavior of the vapor pressures without complicating the model. The values of the fitted parameters are shown in Table 4, and the representation of the fit is illustrated in Figure 5.

Table 4. The fitted parameters of Antoine's Equation (3) for each molality separately and also for all molalities together with the NaCl solution.

Molality (mol·kg^{-1})	Parameters			
	A	B	C	Residual
0.08	13.9 ± 1.22	$2.67 \times 10^3 \pm 5.33 \times 10^2$	-89.4 ± 21.9	5.00×10^{-3}
0.17	12.9 ± 1.24	$2.19 \times 10^3 \pm 4.85 \times 10^2$	$-1.13 \times 10^2 \pm 21.7$	7.00×10^{-3}
0.60	$14.1 \pm 8.70 \times 10^{-1}$	$2.76 \times 10^3 \pm 3.86 \times 10^2$	-86.2 ± 15.6	2.00×10^{-3}
1.23	$13.2 \pm 9.60 \times 10^{-1}$	$2.32 \times 10^3 \pm 3.90 \times 10^2$	$-1.06 \times 10^2 \pm 17.1$	4.00×10^{-3}
2.71	19.2 ± 1.17	$5.43 \times 10^3 \pm 7.23 \times 10^2$	$-3.10 \times 10^{-1} \pm 20.5$	1.00×10^{-3}
3.68	15.9 ± 1.43	$3.60 \times 10^3 \pm 7.21 \times 10^2$	-56.9 ± 25.2	3.00×10^{-3}
5.75	16.0 ± 1.27	$3.70 \times 10^3 \pm 6.48 \times 10^2$	-52.9 ± 22.4	2.00×10^{-3}
All Exp. data	15.0 ± 8.69	$3.13 \times 10^3 \pm 4.09 \times 10^3$	$-73.1 \pm 1.54 \times 10^2$	12.9

Observing the results obtained from the parameter fitting of Equation (3), Figure 5 and Table 4, it is clear that fitting each concentration separately provides a good representation of the experimental points. However, the fit of all concentrations together exhibits significant deviations. The fitted curve overestimates the experimental data for concentrations greater than 2.71 mol·kg^{-1}. This discrepancy is also reflected in Table 4, where the residual error for the average fit is notably higher than that of each separate fit.

The fittings made for each concentration separately have resulted in distinct values for A, B, and C for each case. These fittings closely match the experimental points and exhibit a very low residual error, indicating that they can accurately represent the pattern of the experimental data. However, the drawback of this type of fit is that the Antoine equation does not explicitly include the concentration value, making the fit accurate only for each concentration individually. Therefore, a different model must be used to obtain an expression for determining vapor pressure at any concentration. Consequently, Patil's model, as shown in Equation (4), is proposed for this purpose.

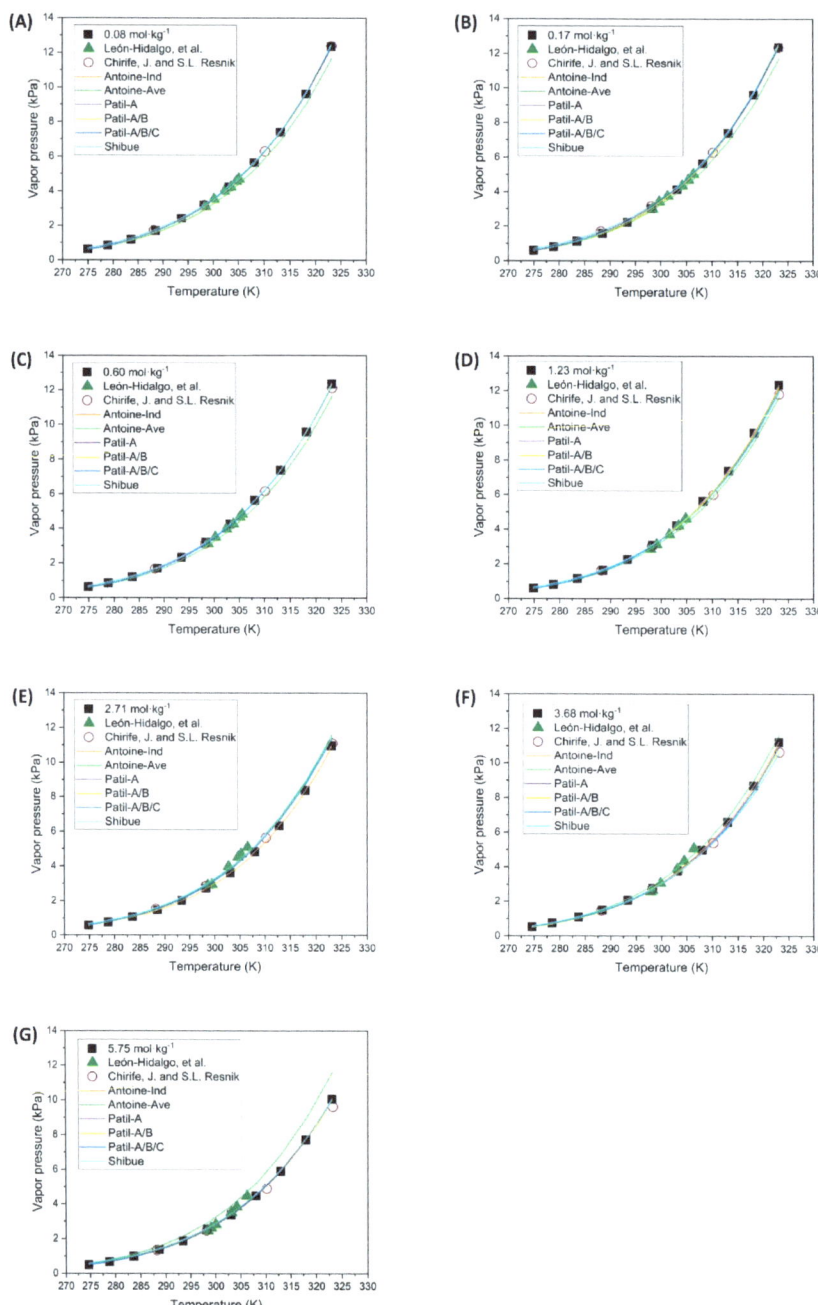

Figure 5. Vapor pressure of unsaturated NaCl solutions and different expressions for parameter fitting (■ Experimental data; ▲ León-Hidalgo, et al. [41]; ○ Chirife, J. and S.L. Resnik [46]; — Individual fitting; — Average fitting; — Patil-A; — Patil-A/B; — Patil-A/B/C; — Shibue).

Patil's expression, with the fitting parameters represented as cubic polynomials in terms of molality, has been fitted sequentially. This approach avoids complicating the fitting method and ensures the fitting values have acceptable confidence limits for the parameters. The first approach to fitting the parameters involved using Equation (4) but only considering parameter A as a function of molality. The results of the fitting parameters are shown in column $A(m)$ of Table 5. The theoretical curves obtained from these parameters are illustrated in Figure 5, closely matching the experimental points for all the concentrations tested. Observing the values of the fitting parameters in Table 5, they are considered accurate as their confidence intervals are close to the estimated values, with the exception of parameters B_0 and C_0. Parameters B_0 and C_0 likely do not have acceptable confidence intervals because they do not significantly influence the vapor pressures in Equation (4). Comparing these results with those provided by Patil for the tested solutions, they are of the same order of magnitude and are thus deemed correct.

Table 5. Parameter fitting to all experimental NaCl vapor pressure data for Equation (4).

Parameters	$A(m)$	$A(m) - B(m)$	$A(m) - B(m) - C(m)$
A_0	5.63 ± 1.96	5.35 ± 2.08	4.48 ± 4.21
A_1	$-2.30 \times 10^{-3} \pm 1.10 \times 10^{-2}$	$-1.41 \times 10^{-1} \pm 4.34 \times 10^{-1}$	$4.00 \times 10^{-1} \pm 9.19$
A_2	$-5.80 \times 10^{-3} \pm 5.00 \times 10^{-3}$	$7.40 \times 10^{-2} \pm 1.98 \times 10^{-1}$	$3.50 \times 10^{-1} \pm 3.84$
A_3	$6.00 \times 10^{-4} \pm 1.00 \times 10^{-3}$	$-1 \times 10^{-2} \pm 2.30 \times 10^{-2}$	$-6.00 \times 10^{-2} \pm 5.20 \times 10^{-1}$
B_0	$-6.72 \times 10^2 \pm 1.21 \times 10^3$	$-4.83 \times 10^2 \pm 1.28 \times 10^3$	$24.8 \pm 2.55 \times 10^3$
B_1		$4.66 \times 10^1 \pm 1.37 \times 10^2$	$-2.54 \times 10^2 \pm 6.17 \times 10^3$
B_2		-26.5 ± 62.6	$-2.11 \times 10^2 \pm 2.81 \times 10^3$
B_3		3.48 ± 7.19	$37.3 \pm 2.48 \times 10^2$
C_0	$-2.57 \times 10^5 \pm 1.87 \times 10^5$	$-2.88 \times 10^5 \pm 1.97 \times 10^5$	$-3.62 \times 10^5 \pm 4.02 \times 10^5$
C_1			$3.99 \times 10^4 \pm 8.73 \times 10^5$
C_2			$3.10 \times 10^4 \pm 3.72 \times 10^5$
C_3			$-5.53 \times 10^3 \pm 4.97 \times 10^4$
Residuals	1.08	1.14	1.04

The second approach involves considering both parameters A and B of Equation (4) as functions of molality, with the estimated values recorded in column $A(m) - B(m)$ in Table 5. In the third step, all three parameters (A, B and C) have been assumed to vary as functions of molality, and the results are shown in column $A(m) - B(m) - C(m)$ of Table 5. The corresponding theoretical curves are depicted in Figure 5.

Based on the results obtained from the various parameterizations of Equation (4), it can be concluded that increasing the number of fitting parameters leads to poorer confidence intervals, despite minimal changes in residuals. This trend suggests that using A, B and C simultaneously as functions of molality is not justified, especially given the significant errors observed for parameters B and C. Upon reviewing Figure 5, it can be concluded that the simplest approach is the most appropriate and the parameters presented in column $A(m)$ of Table 5 are the most suitable for determining vapor pressures as functions of temperature and concentration.

Finally, Equation (8) proposed by Shibue [30] was used as the last fitting expression, employing either 3 (a_1, a_2, a_3) or 5 (a_1, a_2, a_3, b_1, b_2) fitting parameters depending on whether the diluted or concentrated region was considered. The estimated parameters obtained with the equations corresponding to the diluted region are shown in Table 6 and the theoretical curves in Figure 5. These parameters present values with highly acceptable confidence limits. Moreover, this fitting expression yields a considerably lower error than other tested expressions, and notably, it serves effectively as a fitting model for diluted and concentrated regions. However, attempts to fit the parameters for the equations associated with the concentrated region were unsuccessful. Additionally, using Shibue's expression with a quadratic difference in logarithmic vapor pressures also did not yield a successful fit. The fit of parameters in the dilute region is considered accurate, as they are similar to those obtained by Shibue, except for a_3, which in the current investigation is an order of magnitude lower than that reported by Shibue.

Table 6. Fitting parameters for all experimental values of NaCl vapor pressure for Equation (8).

Parameter	Value	Residual
a_1	$2.66 \times 10^{-1} \pm 3.60 \times 10^{-2}$	
a_2	$-2.25 \pm 4.62 \times 10^{-1}$	2.63×10^{-6}
a_3	-43.5 ± 14.2	

In conclusion, the best fits are those with Equations (8)–(11) shown by Shibue. With just three fitting parameters, this approach perfectly explains the behavior of the vapor pressures of the solutions across the entire range of concentrations and temperatures tested. Next, Patil's expression modified using only the combination of Equations (4) and (5), where only the fitting parameter A is assumed as a function of molality, is also considered effective. With only 6 fitting parameters, this expression results in a better fit with more restrictive confidence intervals. Finally, while Antoine's expression (3) is deemed correct, its limitation is that the fitted values are only valid for the specific concentrations of the solution used, as it does not account for the effect of concentration on the fitting parameters.

3.2.2. $CuSO_4 \cdot 5H_2O$ Solutions

The results obtained for the unsaturated $CuSO_4 \cdot 5H_2O$ solutions were carried out up to a molal concentration of 1.23 mol·kg^{-1}, because its solubility is highly dependent on temperature. At a molal concentration of 2.71 mol·kg^{-1} the solution is in a saturated state, so the tests were conducted for the first four molal concentrations studied. The results of these tests for the range of working temperatures are depicted in Figure 6.

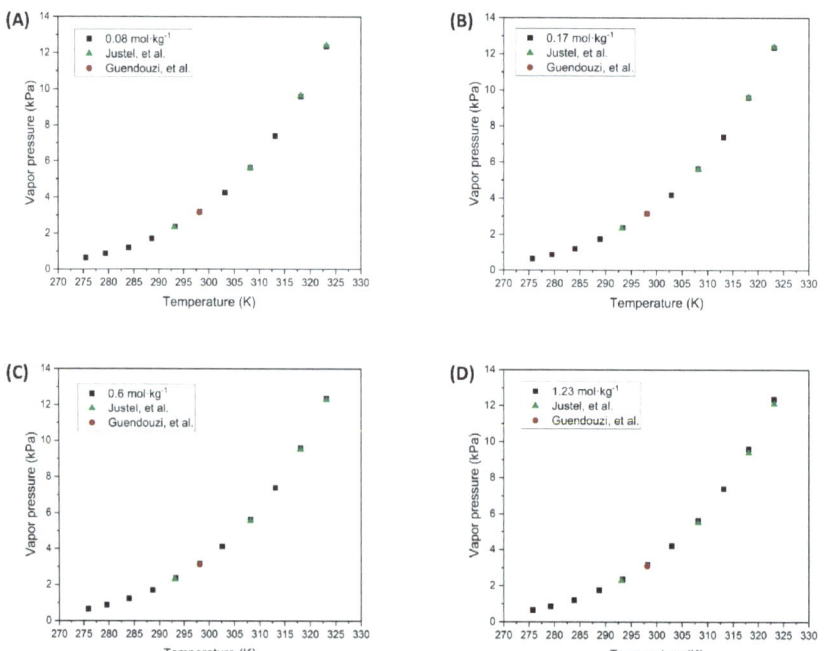

Figure 6. Vapor pressure of unsaturated $CuSO_4 \cdot 5H_2O$ solutions at different molalities. (■ Experimental data; ▲ Justel, et al. [47]; ● Guendouzi, et al. [48]).

Based on the results obtained, it can be seen that the experimental data from this investigation closely align with literature values within the working temperature range. However, a slight deviation is observed at temperatures around 50 °C, corresponding to the experimental values reported by Justel et al. [47]. As previously mentioned, this drift in the experimental measurement may be due to the vapor pressure expression of pure water used. Conversely, the data at 25 °C and lower temperatures agree with the experimental points reported by Guendouzi et al. [48].

In the tests carried out in the present investigation, it was observed that the relative humidity of the air, measured with the hygrometer, showed 100% saturation when the equilibrium temperature of 20 °C was reached, regardless of the solution concentration. This observation can be directly related to the stability of the salt with hydrated pairs, as reported by Glasser [49]. This may be because as the salt has a larger hydration sphere and its solubility increases with temperature, it is easier for these water molecules to dissociate from the main molecule and evaporate. Consequently, the $CuSO_4$ salt forms smaller hydrated pairs resulting in the environment reaching 100% relative humidity.

The parameter fitting for the vapor pressure expressions at different concentrations has been performed similarly to the NaCl solutions, starting with the Antoine Equation (3) for each concentration separately and for all concentrations simultaneously. The results of this parameter fitting are presented in Table 7, and the corresponding theoretical curves are shown in Figure 7.

Table 7. The fitted parameters of Antoine's Equation (3) for each molality separately and also for all molalities together with the $CuSO_4 \cdot 5H_2O$ solution.

Molality (mol·kg^{-1})	Parameters			Residual
	A	B	C	
0.08	13.9 ± 1.24	$2.66 \times 10^3 \pm 5.43 \times 10^2$	-89.7 ± 22.3	5.00×10^{-3}
0.17	13.5 ± 1.45	$2.47 \times 10^3 \pm 6.12 \times 10^2$	-97.7 ± 26.0	7.00×10^{-3}
0.60	13.7 ± 1.30	$2.57 \times 10^3 \pm 5.58 \times 10^2$	-93.7 ± 23.3	5.00×10^{-3}
1.23	14.0 ± 1.42	$2.70 \times 10^3 \pm 6.25 \times 10^2$	-88.0 ± 25.5	6.00×10^{-3}
All Exp. data	$13.8 \pm 5.30 \cdot 10^{-1}$	$2.60 \times 10^3 \pm 2.30 \times 10^2$	-92.3 ± 9.55	2.40×10^{-2}

As seen in the results obtained, the fitting parameters for the different concentrations are not drastically affected, and even the estimated parameters for all concentrations fall within the range of values of the other settings. In contrast, observing the fitting error and the theoretical curves, it can be concluded that the fit is accurate. This simple expression effectively describes all concentrations of this solution, even though the concentration is not explicitly included in the fitting expression.

The Patil model has been used to validate a more complex expression that incorporates the molality of the solutions tested in this research. The fitting procedure was conducted similarly to that for the NaCl solution, beginning with a simpler expression where only parameter A depends on the concentration and subsequently applying equations where all parameters depend on the concentration. The results of the fitted parameters are presented in the corresponding column of Table 8. The theoretical curves obtained are illustrated in Figure 7.

According to the parameter values obtained, the same trend observed with the NaCl solution can be seen: as the number of fitting parameters increases their corresponding confidence intervals also expand, but the residuals are not improved. Additionally, it is notable that even with the simplest expression it is possible to accurately describe the vapor pressures that would be obtained in any case.

Observing the results obtained from the fitting expressions and the theoretical curves in Figure 7, it was decided not to use the fitting of Equation (8) proposed by Shibue, since the previous estimations have already produced theoretical curves that faithfully represent the experimental data of the present investigation, as seen in Figure 7. The theoretical curves from each fitting are practically superimposed, with minimal differences. Therefore, it can be concluded that for the different concentrations of the $CuSO_4 \cdot 5H_2O$ solution, the Antoine-type Equation (3) for the average values shown in Table 7 is sufficient to predict the vapor pressures of this solution. Conversely, the fittings made using Equation (4) are not considered adequate due to the large errors in the confidence intervals, despite the curves overlapping the experimental points.

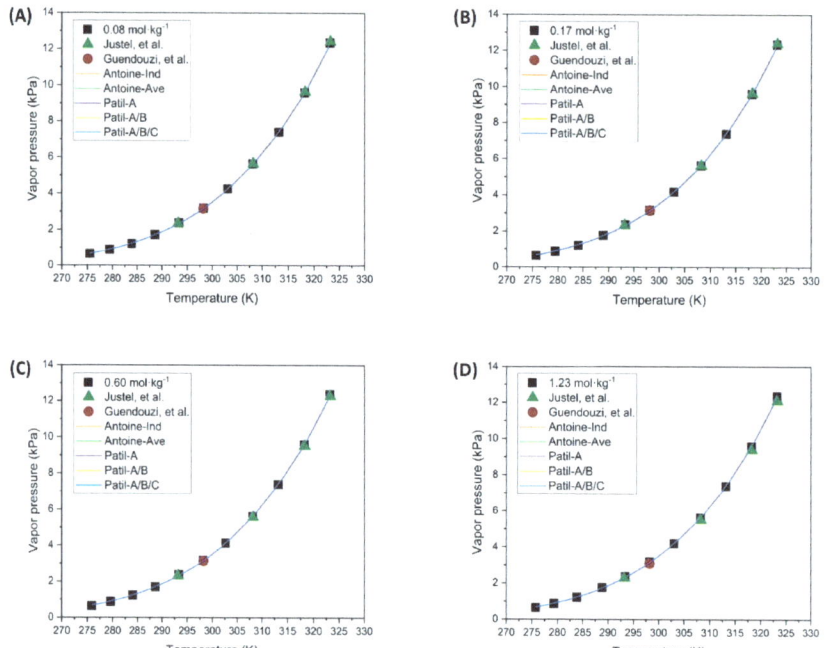

Figure 7. Vapor pressure of unsaturated $CuSO_4 \cdot 5H_2O$ solutions and different expressions for parameter fitting (■ Experimental data; ▲ Justel, et al. [47]; ● Guendouzi, et al. [48]; ─── Individual fitting; ─── Average fitting; ─── Patil-A; ─── Patil-A/B; ─── Patil-A/B/C).

Table 8. Parameter fitting to all experimental $CuSO_4 \cdot 5H_2O$ vapor pressure data for Equation (4).

Parameters	$A(m)$	$A(m) - B(m)$	$A(m) - B(m) - C(m)$
A_0	$5.02 \pm 5.25 \times 10^{-1}$	$5.00 \pm 5.60 \times 10^{-1}$	5.68 ± 3.12
A_1	$-1.6 \times 10^{-3} \pm 3.50 \times 10^{-2}$	$2.36 \times 10^{-1} \pm 1.34$	-7.94 ± 33.2
A_2	$3.20 \times 10^{-3} \pm 7.30 \times 10^{-2}$	$-4.70 \times 10^{-1} \pm 2.78$	16.1 ± 36.8
A_3	$-1.60 \times 10^{-3} \pm 3.80 \times 10^{-2}$	$2.32 \times 10^{-1} \pm 1.45$	-8.04 ± 35.9
B_0		$-3.25 \times 10^2 \pm 3.39 \times 10^2$	$-7.44 \times 10^2 \pm 1.92 \times 10^3$
B_1		$-75.1 \pm 4.23 \times 10^2$	$4.98 \times 10^3 \pm 1.06 \times 10^4$
B_2		$1.50 \times 10^2 \pm 8.79 \times 10^2$	$-1.01 \times 10^4 \pm 4.26 \times 10^4$
B_3		$-74.0 \pm 4.58 \times 10^2$	$5.04 \times 10^3 \pm 1.21 \times 10^4$
C_0	$-3.04 \times 10^5 \pm 5.00 \times 10^4$	$-3.04 \times 10^5 \pm 5.20 \times 10^4$	$-.39 \times 10^5 \pm 2.97 \times 10^5$
C_1			$-7.80 \times 10^5 \pm 3.17 \times 10^6$
C_2			$1.58 \times 10^6 \pm 3.42 \times 10^6$
C_3			$-7.89 \times 10^5 \pm 3.42 \times 10^6$
Residuals	2.62×10^{-2}	2.59×10^{-2}	2.56×10^{-2}

3.3. Evaporation of Saturated Salts on Different Free Evaporating Surfaces

Figure 8 presents the results for the natural evaporation of a saturated NaCl solution across different evaporation surfaces: S_1 (141.2 mm^2), S_2 (1683.7 mm^2), and S_3 (3452.4 mm^2). The data show that surface S_1 exhibits distinct behavior in evaporation per unit area and requires a longer time to fully evaporate the water compared to S_2 and S_3, with S_1 requiring approximately 180 days to complete the process. For surfaces S_2 and S_3, the difference in evaporation per unit area becomes noticeable only after the complete evaporation of the water, as indicated by the plateau in the curves. Initially, the slopes for S_2 and S_3 are identical, indicating that the evaporation surface area does not significantly impact the evaporation rate. Instead, increasing the surface area results in a higher total mass of water evaporated, but does not alter the evaporation per unit area.

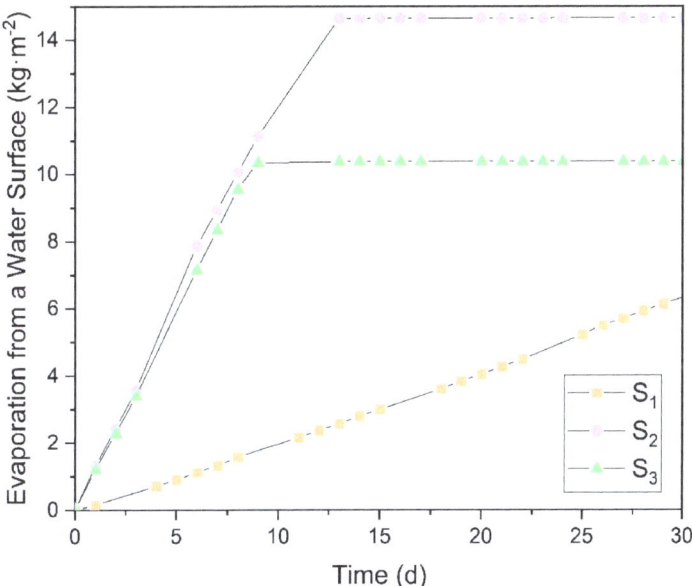

Figure 8. Evaporation of water in saturated NaCl solution for three evaporating surfaces. (S_1 (141.2 mm^2); S_2 (1683.7 mm^2); S_3 (3452.4 mm^2)).

The other salts studied for natural evaporation exhibited similar behavior to that shown in Figure 8, so this figure is presented primarily to illustrate the experimental test conducted. To compare the results of the different saturated salt solutions, the slopes of the evaporation curves during the initial linear evaporation phase, corresponding to the first days of evaporation, were analyzed. These results are displayed in Figure 9, which compares the evaporation rates of all solutions tested in this research, including distilled water.

The results indicate no significant difference in the evaporation rates between surfaces S_2 and S_3, as their values are nearly identical across all study cases, with any minor differences likely attributable to measurement errors. However, a notable difference is observed between these larger surfaces (S_2 and S_3) and surface S_1, with a reduction in the evaporation rate ranging from 82.7% to 92.2%. In contrast, despite having twice the evaporation area, S_2 and S_3 exhibit similar evaporation rates, suggesting that there may be a critical surface area at which the evaporation rate stabilizes regardless of further increases in surface area.

Assouline et al. [50] experimented to determine water evaporation in two tanks. A black polypropylene plate covered the surface of one of the tanks with evenly distributed square perforations. This design allowed the use of different evaporation areas in the experiment. Assouline et al. demonstrated that with a very small evaporation surface, the evaporation rate deviates from the ideal trend. As the surface area increases, it approaches the maximum evaporation rate. Therefore, it can be concluded that there is a limiting surface area, beyond which further increases do not affect the evaporation rate. This effect may explain the behavior observed for surface S_1, which deviates from the ideal evaporation trend. In contrast, surfaces S_2 and S_3 exhibit the maximum possible evaporation rate.

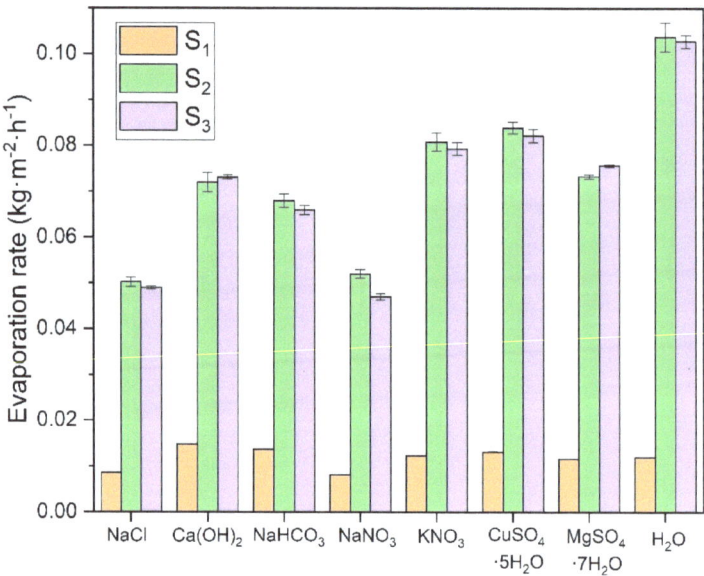

Figure 9. Evaporation rates of saturated solutions for three evaporating surfaces. (■ S_1 (141.2 mm^2); ■ S_2 (1683.7 mm^2); ■ S_3 (3452.4 mm^2)).

The different salts' evaporation rates depend mainly on their vapor pressure in equilibrium with the air temperature. In Section 3.1, the vapor pressures of these saturated solutions are determined. Comparing the vapor pressure results with the evaporation rate experiments confirms a good correlation between both variables. The $CuSO_4 \cdot 5H_2O$ and $MgSO_4 \cdot 7H_2O$ solutions exhibit the highest evaporation rates due to their hydration equilibrium state, which allows them to eliminate excess water molecules from the solution at a higher rate regardless of their vapor pressure. It should be noted that when calculating the evaporation rates in the linear evaporation interval, these solutions have the highest values. However, as they reach equilibrium with the ambient humidity, they do not completely evaporate the water from their structure and can even gain mass if the environmental humidity increases. This effect is evident when these solutions have almost no apparent liquid left in the vessel and salt deposits become noticeable. In contrast, other solutions with high vapor pressures dried out completely during the experiment, maintaining a constant evaporation rate throughout the trial. The evaporation rate of these hydrated salts drops drastically once they reach equilibrium with the environment. Genceli, F. E. et al. [51] investigated $MgSO_4 \cdot nH_2O$ salt in different hydration states, determining that at room temperature and depending on the relative humidity it can overhydrate up to 11 times. This demonstrates that these hydrated salts can retain water in their crystalline structure.

KNO_3 shows a slightly higher evaporation rate than $NaHCO_3$ and $Ca(OH)_2$, while lower evaporation rates characterize NaCl and $NaNO_3$. Comparing all these results with those of distilled water, it can be seen that this shows the highest evaporation rate since it evaporates on a free film, and its rate is influenced primarily by environmental conditions such as temperature and relative humidity.

This experimental test is used to determine the evaporation rates of salts in the saturated state, representing the worst possible concentration conditions for the evaporation of the water present in the solutions. Gilron et al. [25] determined an evaporation rate for the volume reduction of waste from brackish water, where they obtained values of evaporation rates from the free surface of water of 0.0796 kg·m^{-2}·h^{-1}. This value approximates the results obtained for the solutions in the present investigation. More recent studies by Gilron et al. [52], investigating evaporation rates for brines concentrated to a state of ZLD or almost no liquid, found an approximate evaporation rate of 0.0459 kg·m^{-2}·h^{-1}, for the water-free film. This is similar to the values obtained in the present study for the saturated NaCl solution, with an average evaporation rate of 0.0496 kg·m^{-2}·h^{-1}.

3.4. Evaporation of NaCl and $CuSO_4 \cdot 5H_2O$ Saturated Solutions in Fabrics

As previously mentioned, evaporation studies on natural and synthetic fabrics have been carried out using saturated NaCl and $CuSO_4 \cdot 5H_2O$ solutions on samples with a 50 mm × 75 mm surface area and an effective evaporation area of 50 mm × 60 mm. These two salts were chosen for determining evaporation rates due to their distinct properties: NaCl, with the lowest vapor pressure, and $CuSO_4 \cdot 5H_2O$, which is a structurally larger salt in its hydrated state. Figure 10 shows the evaporation process for the Ara sample in NaCl solution across different evaporation cycles.

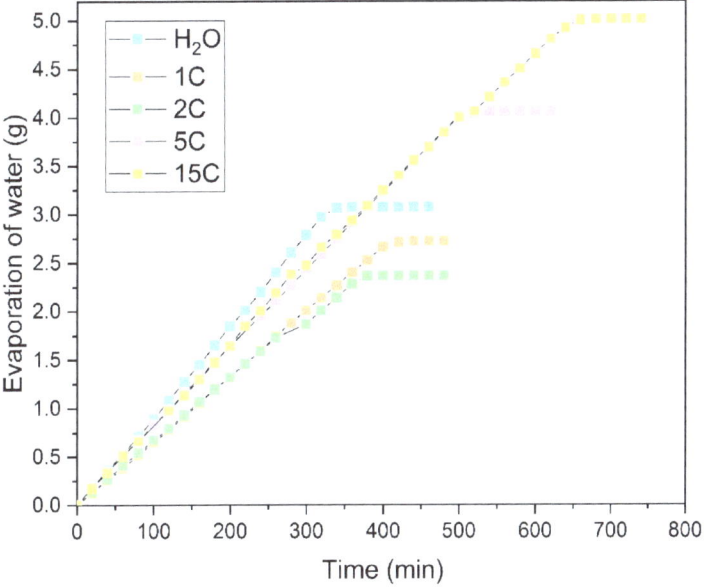

Figure 10. Evaporation of water from the saturated NaCl solution in the Ara sample. (● Distilled water; Evaporation cycles ■ 1C; ■ 2C; ■ 5C; ■ 15C).

In these tests, only the evaporation of water was accounted for, with the salts precipitated on the fabric as the water evaporated being discounted. The graphs compare the evaporation per cycle of each saturated solution with that measured in distilled water. In the NaCl solution, shown in Figure 10, it can be observed that during the first few operation cycles (1C and 2C), the water absorbed by the fabric and measured in evaporation is similar to that of distilled water. This suggests that salt precipitation does not initially affect evaporation. As the number of evaporation cycles increases, NaCl precipitates within the fabric structure. This effect is illustrated in Figure 11, showing NaCl crystals growing perpendicularly to the fabric surface, particularly at the edges of the sample. This salt precipitation causes an increase in water absorption due to the combined effects of the fabric's inherent absorption and the hydration of the precipitated crystals. This results in increased liquid absorption and, consequently, higher total water evaporation, as seen in Figure 10.

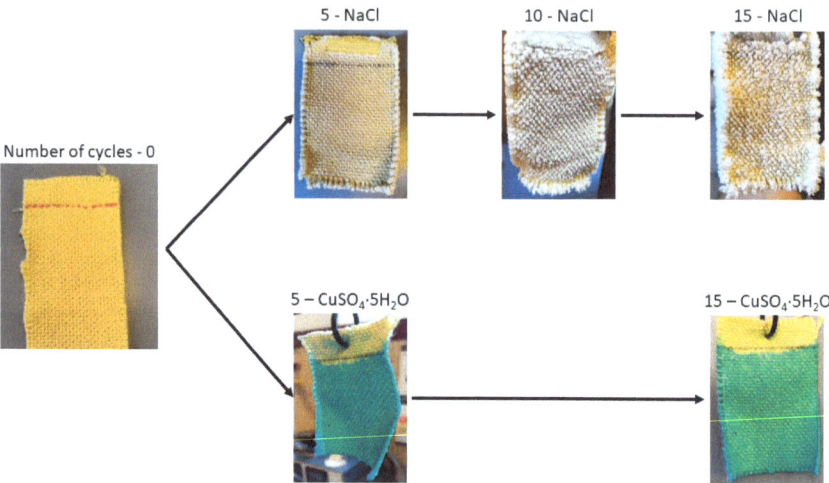

Figure 11. Precipitation of NaCl and $CuSO_4·5H_2O$ salts on Ara fabric along 15 cycles of evaporation.

Figure 12 shows the tests conducted with $CuSO_4·5H_2O$ salt and highlights a completely different behavior compared to the NaCl salt observed previously. As can be seen, compared to evaporation with distilled water, the absorption capacity reduces from the first cycle onwards and continues to decline with subsequent evaporation cycles. This reduction is likely due to the pentahydrated nature of the salt and to the formation of $CuSO_4$ crystals. As shown in Figure 11, these crystals do not grow in the same manner as NaCl crystals, with little difference between the 5C and 15C cycles. The precipitation of $CuSO_4$ forms a dense, solid layer that prevents the solution from penetrating into the fabric structure during each operation cycle and inhibits the fabric's ability to retain water. From cycle 5C onwards, the mass of salt precipitated on the fabric remains consistent at approximately 0.1 g, indicating that the saturated $CuSO_4·5H_2O$ solution is retained only on the $CuSO_4·5H_2O$ crystals already precipitated in the previous cycle.

Only the graphs showing the water evaporation over time for the Ara sample are presented, as its trend mirrors that of the other fabrics tested. Next, the evaporation rates for all samples are analyzed during each work cycle, focusing on the range where the evaporation rate is at its maximum. In this zone, a linear evaporation trend with a constant slope has been observed, as indicated in the previous figures, until complete evaporation is achieved.

The results for the saturated NaCl solution are shown in Figure 13. The general trend observed is consistent with that of the Ara sample. Both natural and synthetic fabrics exhibit an increase in evaporation rate as the number of cycles increases, with some fabrics even surpassing the evaporation rate of distilled water (PES and Jut). Two natural fabrics, Bam and WL-T, display different behaviors. The Bam sample, with a very low fabric weight of 0.2193 g, shows minimal salt retention, making the effect of additional hydration as evaporation cycles increase less noticeable. In contrast, the WL-T fabric exhibits a decreasing trend in evaporation rate as the cycles progress. This reduction may be attributed to the interaction between the saturated NaCl solution and the palm fibers, which could occlude voids between the fibers and increasingly restrict water evaporation from the fabric to the environment.

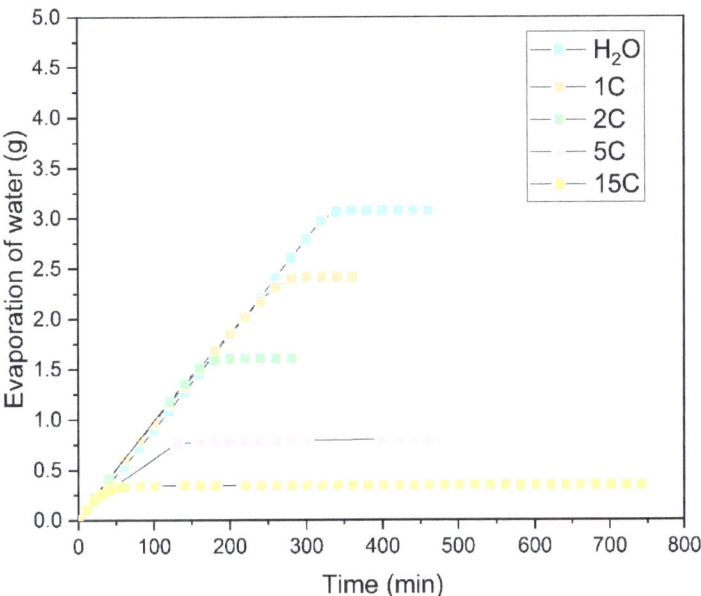

Figure 12. Evaporation of water from the saturated CuSO$_4$·5H$_2$O solution in the sample Ara. (■ Distilled water; Evaporation cycles ■ 1C; ■ 2C; ■ 5C; ■ 15C).

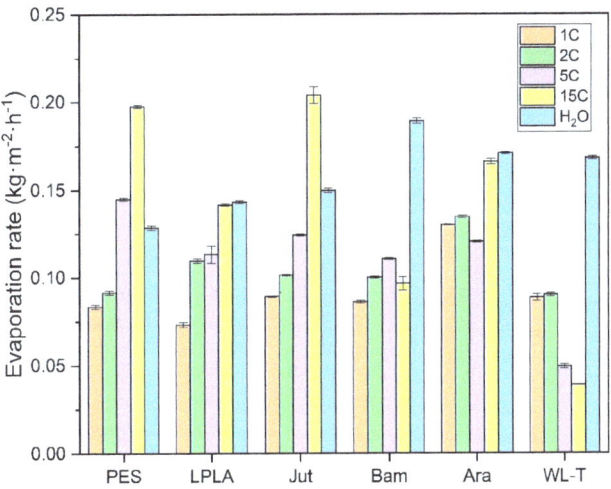

Figure 13. Evaporation rate in saturated NaCl solution for the different fabrics in the zone of maximum evaporation. (Evaporation cycles ■ 1C; ■ 2C; ■ 5C; ■ 15C; ■ Distilled water).

Observing the results obtained in these initial experiments with the NaCl saturated solution, the most promising fabrics for long-term evaporation cycles are PES, LPLA, Jut, and Ara. At around 15 evaporation cycles, these fabrics exhibit an increased evaporation rate, often surpassing that of distilled water due to the hydration of NaCl crystals deposited on their structure. Conversely, the Bam fabric stands out for applications requiring low salt retention. Despite its stable evaporation rate throughout the cycles, its salt retention is significantly lower than that of the other tested fabrics.

In the present investigation, it has been observed that evaporation rates for distilled water range between 0.1285 to 0.1892 kg·m^{-2}·h^{-1} under ambient laboratory conditions, with a temperature range of 20 ± 2 °C, relative humidity of 40 ± 5 % and zero air velocity. Authors such as Shokri Kuehni et al. [53] state that environmental conditions of temperature and relative humidity significantly affect the evaporation rate from porous media. Shokri Kuehni et al. conducted tests at a constant temperature with increasing relative humidity and observed a decrease in the evaporation rate. Conversely, evaporation is favored at constant relative humidity with an increase in temperature.

Liu et al. [54] determined evaporation rates for polyester fabrics under conditions simulating human skin, at a temperature of 36 °C, relative humidity of 40%, and a wind speed of 0.3 m·s^{-1}. Because these conditions are more favorable than those tested in the present study, evaporation rates of 0.2544 to 0.3078 kg·m^{-2}·h^{-1} were obtained, approximately twice the values reported in this paper. This difference is explained by the temperature and wind speed variations at which the experimental tests were conducted.

Some authors who have worked under similar conditions include Fourt et al. [55], who studied hand-made fabrics at a temperature of 21 °C and a relative humidity of 65%, obtaining evaporation rates of 0.045 to 0.072 kg·m^{-2}·h^{-1}. On the other hand, Gurudatt et al. [56] carried out tests in ambient conditions of 25 °C and relative humidity of 64%, obtaining evaporation rates of 0.1062 to 0.1158 kg·m^{-2}·h^{-1}. In the present work, slightly higher evaporation rates than those shown by the other authors have been measured at a lower relative humidity of 40%.

Alternatively, the studies of Wu et al. [57], who investigated evaporation in saline media with surface-treated cotton fabrics, are worth mentioning. They worked at temperatures of 25 °C and relative humidities of 35 to 55%, obtaining evaporation rates of 0.166 kg·m^{-2}·h^{-1}. These values are more similar to those obtained in the present research, with a difference of 12%, although they are different solutions.

In the previous cases analyzed, only the environmental factors of temperature and relative humidity, which critically affect the evaporation rate, have been compared. Other factors influencing the evaporation rate may include the weight of the fabrics, as indicated by Chau et al. [58]. These authors suggest that as the grammage of a fabric increases, the evaporation rate is reduced. Crow and Osczevski [59] showed that fabrics can absorb more liquids if their thickness is increased and that the drying time of the samples depends on the grammage and thickness, as these factors directly influence the fabric's ability to retain more liquids in their structure.

Figure 13 also shows that during the evaporation cycles for the saturated solution of NaCl, some of the fabrics exhibit an evaporation rate higher than that of water. This effect sounds counterintuitive, as the vapor pressure of this salt is quite low and, therefore, should not exceed that of distilled water. Shokri-Kuehni et al. [60] demonstrated in their experiments that the mass loss over days in a porous medium for a concentrated NaCl solution was lower than when it was more diluted. In their research, Norouzi Rad and Shokri [61] observed that adding NaCl in a porous medium up to a concentration of approximately 1.5 mol·kg^{-1} decreased the evaporation rate. However, above this point, increasing the concentration caused an increase in the evaporation rate. This effect of increasing the evaporation rate in a porous medium was also demonstrated by Sghaier and Prat [62], who observed that the evaporation rate for a saturated NaCl solution was higher than that of pure water in the first hours of evaporation, until the crystallization phase of the salt crystals was reached.

Subsequently, the same type of fabrics were used for evaporation tests in a saturated $CuSO_4·5H_2O$ solution. The values of the evaporation rates determined in the zone of maximum evaporation can be seen in Figure 14. Although the time for complete evaporation after five cycles of operation increases dramatically compared to the saturated NaCl solution tests, the evaporation rate shows higher average values than the NaCl solution. This effect is mainly due to the vapor pressure of each solution at the temperature at which each test was performed. In this case, the vapor pressure associated with $CuSO_4·5H_2O$

is higher than that of NaCl at the same temperature. This effect causes the evaporation rate of the water to be faster until it reaches the crystallization state of the salts, where the time for complete evaporation of the $CuSO_4 \cdot 5H_2O$ salt is higher than that of NaCl at the same temperature.

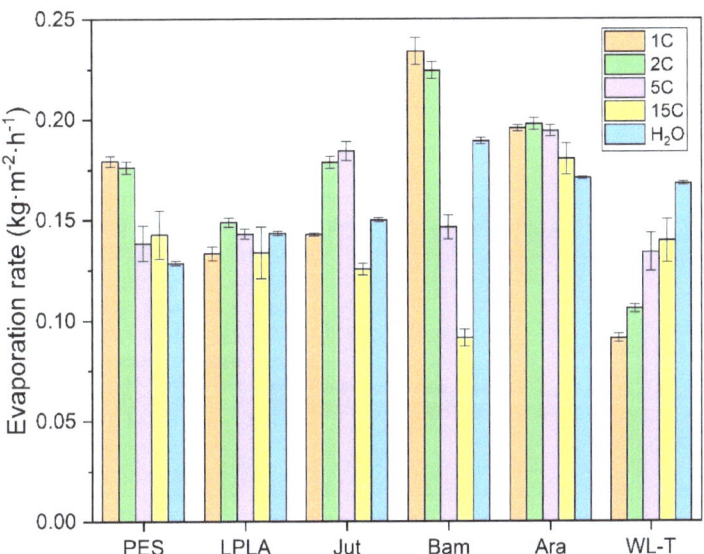

Figure 14. Evaporation rate in saturated $CuSO_4 \cdot 5H_2O$ solution for the different fabrics in the zone of maximum evaporation. (Evaporation cycles ■ 1C; ■ 2C; ■ 5C; ■ 15C; ■ Distilled water).

As seen in the results obtained, the maximum evaporation rate obtained in the first cycles of the tests carried out for the $CuSO_4 \cdot 5H_2O$ salt is not drastically affected, like the NaCl salt, showing a reduction as the evaporation cycles increase. The Bam sample stands out due to its small weight, causing the $CuSO_4 \cdot 5H_2O$ salt to crystallize within the fabric structure and completely block the gaps between the fibers, which reduces its evaporation rate. This is because it absorbs little solution and quickly moves to the crystallization phase. On the other hand, the WL-T fabric slightly increases its evaporation rate because the fibers on its surface are not completely blocked and the surface roughness is able to retain some more solution in its structure without reaching the crystallization phase.

Although the drying times for the fabrics in the presence of the saturated $CuSO_4 \cdot 5H_2O$ solution are longer, we focused on the initial evaporation phase, where the salt has not yet precipitated, and the evaporation rates remain stable initially, decreasing in the later cycles. This is due to the high vapor pressure of $CuSO_4 \cdot 5H_2O$, which is similar to that of pure water at temperatures of 20–25 °C until the nucleation phase. De Castelnuovo et al. [63] and Saig et al. [64] explain the dehydration mechanisms of $CuSO_4 \cdot 5H_2O$ in equilibrium with $CuSO_4 \cdot 3H_2O + 2H_2O$, highlighting that at high temperatures, excess water converts to gas and reabsorbs to maintain equilibrium. This could explain the slow drying observed in experiments with high evaporation cycles. Additionally, Donkers et al. [65] demonstrated through thermal decomposition by thermogravimetric analysis (TGA) that $CuSO_4 \cdot 5H_2O$ is highly stable with its five water molecules, indicating that excess water can be eliminated while decomposition does not occurs, achieving high evaporation rates.

It was decided to compare the evaporation rates between cycles and saturated salts used in the different fabrics in the first cycle (1C), the last cycle (15C), and in comparison with distilled water (Table 9).

Table 9. Comparison of evaporation cycles between NaCl and $CuSO_4 \cdot 5H_2O$ solutions using different fabrics.

Fabrics	NaCl ($kg \cdot m^{-2} \cdot h^{-1}$)		$CuSO_4 \cdot 5H_2O$ ($kg \cdot m^{-2} \cdot h^{-1}$)		H_2O ($kg \cdot m^{-2} \cdot h^{-1}$)
	1C	15C	1C	15C	
PES	8.40×10^{-2}	1.97×10^{-1}	1.79×10^{-1}	1.43×10^{-1}	1.29×10^{-1}
LPLA	7.30×10^{-2}	1.42×10^{-1}	1.33×10^{-1}	1.34×10^{-1}	1.43×10^{-1}
Jut	8.90×10^{-2}	2.04×10^{-1}	1.43×10^{-1}	1.26×10^{-1}	1.50×10^{-1}
Bam	8.60×10^{-2}	9.70×10^{-2}	2.34×10^{-1}	9.10×10^{-2}	1.89×10^{-1}
Ara	1.30×10^{-1}	1.66×10^{-1}	1.95×10^{-1}	1.80×10^{-1}	1.71×10^{-1}
WL-T	8.90×10^{-2}	3.90×10^{-2}	9.10×10^{-2}	1.40×10^{-1}	1.68×10^{-1}

As can be observed, the tendency in the NaCl cycles is to increase the evaporation rate due to the increase in water absorption, even surpassing the evaporation rate of the $CuSO_4 \cdot 5H_2O$ solution despite having a lower vapor pressure than this solution. In contrast, the $CuSO_4 \cdot 5H_2O$ solution shows a slight reduction in its evaporation rate over the fifteen cycles calculated in the zone of the highest evaporation rate.

In view of the results obtained from the evaporation tests using natural and synthetic fabrics, a summary of the conditions that most affect the evaporation rate has been compiled. These factors are listed in Table 10 in order of importance, considering that the evaporation tests were conducted with negligible or no air velocity.

Table 10. Evaporation determinant factors in fabrics.

Determinant Factor	Effect on Evaporation
Temperature	Higher temperature: increases evaporation rate
Relative humidity	Increases relative humidity: decreases evaporation
Salt to treat	The crystal formation can enhance the evaporation
Solution	Vapor pressure affects the evaporation
Grammage	High grammage increases liquid absorption but reduces evaporation
Thickness	Less liquid retention favors lower salt precipitation
Porosity	A more porous fabric will make it more difficult for liquid to escape from inside the fabric and evaporate.

When treating wastewater or saline water with a significant presence of salts, the factors mentioned in Table 10 should be considered to select the fabric that best performs the process under study.

4. Conclusions

The determination of vapor pressures of saturated solutions has shown similar trends and results to those obtained by other authors mentioned in this research, with slight deviations at elevated temperatures in the range of 35 to 50 °C. These deviations are probably due to the fitting expression for the vapor pressure of pure water. However, there is no available information from other researchers on the $Ca(OH)_2$ and $NaHCO_3$ solutions. The trend shown by these two solutions is similar, as both have an equilibrium relative humidity higher than 95% from a temperature of 15 °C, indicating a vapor pressure closely that of pure water.

On the other hand, the determination of vapor pressures for unsaturated NaCl solutions has shown the expected trend: as the concentration of the solution increases, the vapor pressure decreases, making evaporation more difficult in open environments. For $CuSO_4 \cdot 5H_2O$ solutions at concentrations below saturation, all tested molal concentrations have shown, starting at a temperature of 20 °C, a relative humidity of 100%.

Parameter fitting conducted for unsaturated NaCl and $CuSO_4 \cdot 5H_2O$ solutions at various concentrations has shown that the Antoine-type expression needs to be fitted for each solution separately. However, for $CuSO_4 \cdot 5H_2O$, an average fit can accurately represent the theoretical results. On the other hand, Patil's expression is noteworthy because it can accurately describe the results with a simpler form across all concentrations. Finally, Shibue's expression is particularly useful for NaCl concentrations. In the dilute region, using three fitting parameters, the behavior was effectively captured across the entire range of concentrations studied.

The evaporation of saturated solutions of the salts used in this research from free water surfaces has shown the expected effect: they all reduce vapor pressure, resulting in slower evaporation rates compared to distilled water. NaCl exhibited the slowest evaporation rate among these salts due to its lowest vapor pressure. Conversely, $CuSO_4 \cdot 5H_2O$ and $MgSO_4 \cdot 7H_2O$ showed higher evaporation rates when considering the maximum evaporation rate. However, both salts do not completely evaporate the water occluded in their structure once they reach equilibrium with the environment. This drastically reduces their evaporation rate and can even increase their mass if the environmental humidity is high. Additionally, the effect of the evaporation surface indicates that there is a critical surface area beyond which it does not influence the evaporation rate, as observed in the differences between surface S_1 and surfaces S_2 and S_3 in all cases.

Finally, studies conducted to determine the evaporation of water on natural and synthetic fabrics using saturated NaCl and $CuSO_4 \cdot 5H_2O$ solutions have shown different behavior than on free water surfaces. On the one hand, for the saturated NaCl solution on the fabrics, it was observed that as the evaporation cycles increase and more salt deposits on the fabrics, the more water evaporates from the fabrics, the greater the evaporation rate and the higher the salt deposited on the fabrics. In some cases, such as the PES and Jut samples, the evaporation rate in cycle 15C exceeds that of distilled water, despite NaCl having a lower vapor pressure than pure water at the working temperature. Conversely, the $CuSO_4 \cdot 5H_2O$ salt exhibited an opposite effect. As the evaporation cycles increase, the absorption capacity of the fabric decreases because the salt precipitates on the fabric structure, blocking its surface and internal structure. This obstruction makes it difficult for the liquid solution to interact with the fabric, causing the solution to remain in contact with the salt deposited from previous cycles. The maximum evaporation rate determined for $CuSO_4 \cdot 5H_2O$ is sometimes higher than that of water due to its overhydration effect, having more than five water molecules in its structure. When evaporation begins, this excess water starts to precipitate on the fabric, reaching equilibrium and drastically reducing the evaporation rate, resulting in very long periods for complete evaporation.

The findings presented in this study establish a foundation for the development of mathematical models of evaporation, particularly those involving natural and synthetic fabrics, as well as the evaporation of water masses in free surface films. This research has systematically examined the vapor pressures of both saturated solutions and solutions at varying concentrations, enabling the precise calibration of vapor pressure models across a broad temperature spectrum. These parameters are crucial for advancing our understanding and will serve as a basis for the development of both simple and complex evaporation models that utilize vapor pressure as the driving gradient.

In addition, the investigation into evaporation using natural and synthetic fabrics has provided valuable insights into the behavior of different materials under extreme conditions in the context of saturated solution treatment. These findings will facilitate the selection of fabrics with optimal performance characteristics for wastewater treatment, with potential implications for industrial-scale applications.

Author Contributions: Conceptualization, A.L.-B., J.L.-G., V.F., S.C.C. and M.-F.L.-P.; Investigation, V.F.; Methodology, A.L.-B., V.F., S.C.C. and M.-F.L.-P.; Supervision, J.L.-G., V.F., S.C.C. and M.-F.L.-P.; Writing—original draft, A.L.-B.; Writing—review & editing, A.L.-B., V.F., S.C.C. and M.-F.L.-P. All authors have read and agreed to the published version of the manuscript.

Funding: This research work was funded by the Ministry of Science and Innovation, "Retos de la Sociedad". Project reference: PID2020-119142RA-I00.

Institutional Review Board Statement: Not applicable.

Data Availability Statement: The data presented in this study are available upon request from the corresponding author.

Acknowledgments: The authors are grateful to the Ministry of Science and Innovation for supporting the "Retos de la Sociedad" project. Project reference: PID2020-119142RA-I00.

Conflicts of Interest: The authors declare no conflict of interest.

References

1. Afroz, R.; Masud, M.M.; Akhtar, R.; Duasa, J.B. Water pollution: Challenges and future direction for water resource management policies in Malaysia. *Environ. Urban. ASIA* **2014**, *5*, 63–81. [CrossRef]
2. Dwivedi, A.K. Researches in water pollution: A review. *Int. Res. J. Nat. Appl. Sci.* **2017**, *4*, 118–142.
3. Kumar, S.; Islam, A.R.M.T.; Islam, H.T.; Hasanuzzaman, M.; Ongoma, V.; Khan, R.; Mallick, J. Water resources pollution associated with risks of heavy metals from Vatukoula Goldmine region, Fiji. *J. Environ. Manag.* **2021**, *293*, 112868. [CrossRef] [PubMed]
4. Parris, K. Impact of agriculture on water pollution in OECD countries: Recent trends and future prospects. *Int. J. Water Res. Dev.* **2011**, *27*, 33–52. [CrossRef]
5. Chaudhry, F.N.; Malik, M. Factors affecting water pollution: A review. *J. Ecosyst. Ecography* **2017**, *7*, 225–231.
6. Halder, J.N.; Islam, M.N. Water pollution and its impact on the human health. *J. Environ. Hum.* **2015**, *2*, 36–46. [CrossRef]
7. Haseena, M.; Malik, F.M.; Javed, A.; Arshad, S.; Asif, N.; Zulfiqar, S.; Hanif, J. Water pollution and human health. *Environ. Risk Assess. Remediat.* **2017**, *1*, 16–19. [CrossRef]
8. Naidoo, S.; Olaniran, A.O. Treated wastewater effluent as a source of microbial pollution of surface water resources. *Int. J. Environ. Res. Public Health* **2014**, *11*, 249–270. [CrossRef]
9. Marrot, B.; Barrios-Martinez, A.; Moulin, P.; Roche, N. Industrial wastewater treatment in a membrane bioreactor: A review. *Environ. Prog.* **2004**, *23*, 59–68. [CrossRef]
10. Obotey Ezugbe, E.; Rathilal, S. Membrane technologies in wastewater treatment: A review. *Membranes* **2020**, *10*, 89. [CrossRef]
11. de Nicolás, A.P.; Molina-García, Á.; García-Bermejo, J.T.; Vera-García, F. Desalination, minimal and zero liquid discharge powered by renewable energy sources: Current status and future perspectives. *Renew. Sustain. Energy Rev.* **2023**, *187*, 113733. [CrossRef]
12. Panagopoulos, A.; Haralambous, K.-J. Minimal Liquid Discharge (MLD) and Zero Liquid Discharge (ZLD) strategies for wastewater management and resource recovery–Analysis, challenges and prospects. *J. Environ. Chem. Eng.* **2020**, *8*, 104418. [CrossRef]
13. Tong, T.; Elimelech, M. The global rise of zero liquid discharge for wastewater management: Drivers, technologies, and future directions. *Environ. Sci. Technol.* **2016**, *50*, 6846–6855. [CrossRef] [PubMed]
14. Yaqub, M.; Lee, W. Zero-liquid discharge (ZLD) technology for resource recovery from wastewater: A review. *Sci. Total Environ.* **2019**, *681*, 551–563. [CrossRef] [PubMed]
15. Panagopoulos, A. Techno-economic assessment of zero liquid discharge (ZLD) systems for sustainable treatment, minimization and valorization of seawater brine. *J. Environ. Manag.* **2022**, *306*, 114488. [CrossRef] [PubMed]
16. Katzir, L.; Volkmann, Y.; Daltrophe, N.; Korngold, E.; Mesalem, R.; Oren, Y.; Gilron, J. WAIV-Wind aided intensified evaporation for brine volume reduction and generating mineral byproducts. *Desalination Water Treat.* **2010**, *13*, 63–73. [CrossRef]
17. Ahmad, J.; Zhou, Z. Mechanical properties of natural as well as synthetic fiber reinforced concrete: A review. *Constr. Build. Mater.* **2022**, *333*, 127353. [CrossRef]
18. Jagadeesh, P.; Puttegowda, M.; Boonyasopon, P.; Rangappa, S.M.; Khan, A.; Siengchin, S. Recent developments and challenges in natural fiber composites: A review. *Polym. Compos.* **2022**, *43*, 2545–2561. [CrossRef]
19. Seki, Y.; Selli, F.; Erdoğan, Ü.H.; Atagür, M.; Seydibeyoğlu, M.Ö. A review on alternative raw materials for sustainable production: Novel plant fibers. *Cellulose* **2022**, *29*, 4877–4918. [CrossRef]
20. Jariwala, H.; Jain, P. A review on mechanical behavior of natural fiber reinforced polymer composites and its applications. *J. Reinf. Plast. Compos.* **2019**, *38*, 441–453. [CrossRef]
21. Kerni, L.; Singh, S.; Patnaik, A.; Kumar, N. A review on natural fiber reinforced composites. *Mater. Today Proc.* **2020**, *28*, 1616–1621. [CrossRef]
22. Elfaleh, I.; Abbassi, F.; Habibi, M.; Ahmad, F.; Guedri, M.; Nasri, M.; Garnier, C. A comprehensive review of natural fibers and their composites: An eco-friendly alternative to conventional materials. *Results Eng.* **2023**, *19*, 101271. [CrossRef]
23. López-Borrell, A.; Lora-García, J.; Fombuena, V.; Cardona, S.C.; López-Pérez, M.-F. Characterization of Natural and Synthetic Fabrics for the Treatment of Complex Wastes. *Polymers* **2023**, *16*, 84. [CrossRef]
24. Patil, V.; Subeshan, B.; Asmatulu, E.; Asmatulu, R. Enhancing saline water evaporation rates via floatable, conductive nanoparticles embedded in superhydrophobic cotton gauze at air-water interface. *Diam. Relat. Mater.* **2023**, *136*, 110047. [CrossRef]
25. Gilron, J.; Folkman, Y.; Savliev, R.; Waisman, M.; Kedem, O. WAIV—Wind aided intensified evaporation for reduction of desalination brine volume. *Desalination* **2003**, *158*, 205–214. [CrossRef]
26. Panagopoulos, A.; Haralambous, K.-J.; Loizidou, M. Desalination brine disposal methods and treatment technologies—A review. *Sci. Total Environ.* **2019**, *693*, 133545. [CrossRef] [PubMed]
27. Huang, J. A simple accurate formula for calculating saturation vapor pressure of water and ice. *J. Appl. Meteorol. Clim.* **2018**, *57*, 1265–1272. [CrossRef]

28. Thomson, G.W. The Antoine equation for vapor-pressure data. *Chem. Rev.* **1946**, *38*, 1–39. [CrossRef]
29. Patil, K.R.; Tripathi, A.D.; Pathak, G.; Katti, S.S. Thermodynamic properties of aqueous electrolyte solutions. 2. Vapor pressure of aqueous solutions of sodium bromide, sodium iodide, potassium chloride, potassium bromide, potassium iodide, rubidium chloride, cesium chloride, cesium bromide, cesium iodide, magnesium chloride, calcium chloride, calcium bromide, calcium iodide, strontium chloride, strontium bromide, strontium iodide, barium chloride, and barium bromide. *J. Chem. Eng. Data* **1991**, *36*, 225–230.
30. Shibue, Y. Vapor pressures of aqueous NaCl and CaCl2 solutions at elevated temperatures. *Fluid Phase Equilibria* **2003**, *213*, 39–51. [CrossRef]
31. Apelblat, A. The vapour pressures of saturated aqueous lithium chloride, sodium bromide, sodium nitrate, ammonium nitrate, and ammonium chloride at temperatures from 283 K to 313 K. *J. Chem. Thermodyn.* **1993**, *25*, 63–71. [CrossRef]
32. Apelblat, A. The vapour pressures of saturated aqueous solutions of potassium bromide, ammonium sulfate, copper (II) sulfate, iron (II) sulfate, and manganese (II) dichloride, at temperatures from 283 K to 308 K. *J. Chem. Thermodyn.* **1993**, *25*, 1513–1520. [CrossRef]
33. Apelblat, A.; Korin, E. The vapour pressures of saturated aqueous solutions of sodium chloride, sodium bromide, sodium nitrate, sodium nitrite, potassium iodate, and rubidium chloride at temperatures from 227 K to 323 K. *J. Chem. Thermodyn.* **1998**, *30*, 59–71. [CrossRef]
34. Apelblat, A.; Korin, E. Vapour pressures of saturated aqueous solutions of ammonium iodide, potassium iodide, potassium nitrate, strontium chloride, lithium sulphate, sodium thiosulphate, magnesium nitrate, and uranyl nitrate fromT=(278 to 323) K. *J. Chem. Thermodyn.* **1998**, *30*, 459–471. [CrossRef]
35. Apelblat, A.; Manzurola, E. Solubilities and vapour pressures of saturated aqueous solutions of sodium tetraborate, sodium carbonate, and magnesium sulfate and freezing-temperature lowerings of sodium tetraborate and sodium carbonate solutions. *J. Chem. Thermodyn.* **2003**, *35*, 221–238. [CrossRef]
36. O'brien, F. The control of humidity by saturated salt solutions. *J. Sci. Instrum.* **1948**, *25*, 73–76. [CrossRef]
37. Diesnis, M. The determination of the critical hygrometric state. *Ann. Chim.* **1937**, *7*, 5–69.
38. Adams, J.R.; Merz, A.R. Hygroscopicity of fertilizer materials and mixtures. *Ind. Eng. Chem.* **1929**, *21*, 305–307. [CrossRef]
39. Wexler, A.; Hasegawa, S. Relative humidity-temperature relationships of some saturated salt solutions in the temperature range 0 to 50 C. *J. Res. Natl. Bur. Stand.* **1954**, *53*, 19. [CrossRef]
40. Ozcan, O.; Miller, J. Flotation of sodium carbonate and sodium bicarbonate salts from their saturated brines. *Miner. Eng.* **2002**, *15*, 577–584. [CrossRef]
41. León-Hidalgo, M.; Gozálvez-Zafrilla, J.; Lora-García, J.; Arnal-Arnal, J. Study of the vapour pressure of saturated salt solutions and their influence on evaporation rate at room temperature. *Desalination Water Treat.* **2009**, *7*, 111–118. [CrossRef]
42. Knuutila, H.; Hessen, E.T.; Kim, I.; Haug-Warberg, T.; Svendsen, H.F. Vapor–liquid equilibrium in the sodium carbonate–sodium bicarbonate–water–CO2-system. *Chem. Eng. Sci.* **2010**, *65*, 2218–2226. [CrossRef]
43. Duchesne, J.; Reardon, E. Measurement and prediction of portlandite solubility in alkali solutions. *Cem. Concr. Res.* **1995**, *25*, 1043–1053. [CrossRef]
44. Pallagi, A.; Tasi, Á.; Gácsi, A.; Csáti, M.; Pálinkó, I.; Peintler, G.; Sipos, P. The solubility of Ca (OH) 2 in extremely concentrated NaOH solutions at 25 C. *Cent. Eur. J. Chem.* **2012**, *10*, 332–337. [CrossRef]
45. Konno, H.; Nanri, Y.; Kitamura, M. Crystallization of aragonite in the causticizing reaction. *Powder Technol.* **2002**, *123*, 33–39. [CrossRef]
46. Chirife, J.; Resnik, S.L. Unsaturated solutions of sodium chloride as reference sources of water activity at various temperatures. *J. Food Sci.* **1984**, *49*, 1486–1488. [CrossRef]
47. Justel, F.J.; Taboada, M.E.; Jimenez, Y.P. Thermodynamic study of the Cu-Na-H-SO4-Cl-HSO4-H2O system for the solubility of copper sulfate in acid seawater at different temperatures. *J. Mol. Liq.* **2018**, *249*, 702–709. [CrossRef]
48. Guendouzi, M.E.; Mounir, A.; Dinane, A. Water activity, osmotic and activity coefficients of aqueous solutions of Li2SO4, Na2SO4, K2SO4,(NH4) 2SO4, MgSO4, MnSO4, NiSO4, CuSO4, and ZnSO4 at T = 298.15 K. *J. Chem. Thermodyn.* **2003**, *35*, 209–220. [CrossRef]
49. Glasser, L. Thermodynamics of inorganic hydration and of humidity control, with an extensive database of salt hydrate pairs. *J. Chem. Eng. Data* **2014**, *59*, 526–530. [CrossRef]
50. Assouline, S.; Narkis, K.; Or, D. Evaporation from partially covered water surfaces. *Water Resour. Res.* **2010**, *46*. [CrossRef]
51. Genceli, F.E.; Lutz, M.; Spek, A.L.; Witkamp, G.J. Crystallization and characterization of a new magnesium sulfate hydrate MgSO4· 11H2O. *Cryst. Growth Des.* **2007**, *7*, 2460–2466. [CrossRef]
52. Gilron, J.; Ramon, E.; Assaf, N.; Kedem, O. Wind-aided intensified evaporation (WAIV): An environmentally sustainable method for brine management. In *Current Trends and Future Developments on (Bio-) Membranes*; Elsevier: Amsterdam, The Netherlands, 2019; pp. 215–241.
53. Shokri-Kuehni, S.M.; Rad, M.N.; Webb, C.; Shokri, N. Impact of type of salt and ambient conditions on saline water evaporation from porous media. *Adv. Water Resour.* **2017**, *105*, 154–161. [CrossRef]
54. Liu, S.-R.; Dai, X.-Q.; Hong, Y. Prediction of the water evaporation rate of wet textile materials in a pre-defined environment. *Int. J. Cloth. Sci. Technol.* **2020**, *32*, 356–365. [CrossRef]
55. Fourt, L.; Sookne, A.M.; Frishman, D.; Harris, M. The rate of drying of fabrics. *Text. Res. J.* **1951**, *21*, 26–33. [CrossRef]

56. Gurudatt, K.; Nadkarni, V.M.; Khilar, K.C. A study on drying of textile substrates and a new concept for the enhancement of drying rate. *J. Text. Inst.* **2010**, *101*, 635–644. [CrossRef]
57. Wu, Y.-G.; Xue, C.-H.; Guo, X.-J.; Huang, M.-C.; Wang, H.-D.; Ma, C.-Q.; Wang, X.; Shao, Z.-Y. Highly efficient solar-driven water evaporation through a cotton fabric evaporator with wettability gradient. *Chem. Eng. J.* **2023**, *471*, 144313. [CrossRef]
58. Chau, K.-H.; Tang, K.-P.M.; Kan, C.-W. Constant temperature drying rate tester: Real-time water evaporation measurement of fabrics. *IEEE Trans. Instrum. Meas.* **2018**, *67*, 2635–2648. [CrossRef]
59. Crow, R.M.; Osczevski, R.J. The interaction of water with fabrics. *Text. Res. J.* **1998**, *68*, 280–288. [CrossRef]
60. Shokri-Kuehni, S.M.; Vetter, T.; Webb, C.; Shokri, N. New insights into saline water evaporation from porous media: Complex interaction between evaporation rates, precipitation, and surface temperature. *Geophys. Res. Lett.* **2017**, *44*, 5504–5510. [CrossRef]
61. Norouzi Rad, M.; Shokri, N. Nonlinear effects of salt concentrations on evaporation from porous media. *Geophys. Res. Lett.* **2012**, *39*. [CrossRef]
62. Sghaier, N.; Prat, M. Effect of efflorescence formation on drying kinetics of porous media. *Transp. Porous Media* **2009**, *80*, 441–454. [CrossRef]
63. De Castelnuovo, S.; Harness, J.; McColm, I. The role of liquid water in crystalline hydrate dehydration: Copper sulphate pentahydrate. *J. Therm. Anal. Calorim.* **2001**, *63*, 233–247. [CrossRef]
64. Saig, A.; Danon, A.; Finkelstein, Y.; Kimmel, G.; Koresh, J. A continuous polymorphic transition of coordinating water molecules in CuSO4· 5H2O. *J. Phys. Chem. Solids* **2003**, *64*, 701–706. [CrossRef]
65. Donkers, P.; Pel, L.; Adan, O. Hydration/dehydration cycles of salt hydrates–studied with NMR–. In Proceedings of the Conference ISES-Eurosun, Aix-les-Bains, France, 16–19 September 2014.

Disclaimer/Publisher's Note: The statements, opinions and data contained in all publications are solely those of the individual author(s) and contributor(s) and not of MDPI and/or the editor(s). MDPI and/or the editor(s) disclaim responsibility for any injury to people or property resulting from any ideas, methods, instructions or products referred to in the content.

Article

Effective Removal of Glyphosate from Aqueous Systems Using Synthesized PEG-Coated Calcium Peroxide Nanoparticles: Kinetics Study, H₂O₂ Release Performance and Degradation Pathways

Fan Li [1], Thomas Shean Yaw Choong [1,2,*], Luqman Chuah Abdullah [1,2], Siti Nurul Ain Md. Jamil [3,4] and Nurul Nazihah Amerhaider Nuar [3]

1. Department of Chemical and Environmental Engineering, Faculty of Engineering, Universiti Putra Malaysia, Serdang 43400, Selangor, Malaysia
2. Institute of Tropical Forestry and Forest Product (INTROP), Universiti Putra Malaysia, Serdang 43400, Selangor, Malaysia
3. Department of Chemistry, Faculty of Science, Universiti Putra Malaysia, Serdang 43400, Selangor, Malaysia
4. Centre of Foundation Studies for Agricultural Science, Universiti Putra Malaysia, Serdang 43400, Selangor, Malaysia
* Correspondence: csthomas@upm.edu.my

Citation: Li, F.; Choong, T.S.Y.; Abdullah, L.C.; Md. Jamil, S.N.A.; Amerhaider Nuar, N.N. Effective Removal of Glyphosate from Aqueous Systems Using Synthesized PEG-Coated Calcium Peroxide Nanoparticles: Kinetics Study, H₂O₂ Release Performance and Degradation Pathways. *Polymers* **2023**, *15*, 775. https://doi.org/10.3390/polym15030775

Academic Editors: Marta Otero and Ricardo N. Coimbra

Received: 20 December 2022
Revised: 25 January 2023
Accepted: 26 January 2023
Published: 3 February 2023

Copyright: © 2023 by the authors. Licensee MDPI, Basel, Switzerland. This article is an open access article distributed under the terms and conditions of the Creative Commons Attribution (CC BY) license (https://creativecommons.org/licenses/by/4.0/).

Abstract: Glyphosate (N-phosphonomethyl glycine) is a non-selective, broad-spectrum organophosphate herbicide. Its omnipresent application with large quantity has made glyphosate as a problematic contaminant in water. Therefore, an effective technology is urgently required to remove glyphosate and its metabolites from water. In this study, calcium peroxide nanoparticles (nCPs) were functioned as an oxidant to produce sufficient hydroxyl free radicals (·OH) with the presence of Fe^{2+} as a catalyst using a Fenton-based system. The nCPs with small particle size (40.88 nm) and high surface area (28.09 m²/g) were successfully synthesized via a co-precipitation method. The synthesized nCPs were characterized using transform infrared spectroscopy (FTIR), X-ray diffractometry (XRD), Brunauer–Emmett–Teller analysis (BET), dynamic light scattering (DLS), and field emission scanning electron microscopy (FESEM) techniques. Under the given conditions (pH = 3.0, initial nCPs dosage = 0.2 g, Ca^{2+}/Fe^{2+} molar ratio = 6, the initial glyphosate concentration = 50 mg/L, RT), 99.60% total phosphorus (TP) removal and 75.10% chemical oxygen demand (COD) removal were achieved within 75 min. The degradation process fitted with the Behnajady–Modirshahla–Ghanbery (BMG) kinetics model. The H_2O_2 release performance and proposed degradation pathways were also reported. The results demonstrated that calcium peroxide nanoparticles are an efficient oxidant for glyphosate removal from aqueous systems.

Keywords: calcium peroxide; nanoparticles; AOPs; kinetics study; H_2O_2 release; agricultural wastewater

1. Introduction

Water pollution caused by herbicides is one of major global concerns due to the potential negative effects of these compounds on ecosystems and human health (e.g., cardiac and respiratory impacts) [1]. Particularly, glyphosate containing wastewater has drawn considerable attention to public. Glyphosate, known as PMG or Roundup (commercial formulation), is globally used as an organophosphorus post-emergence herbicide to manage the development of weeds, with an annual production volume of ~700,000 tons [2]. Its widespread application in agriculture in massive quantities could bring about environmental issues, considering the biosafety concerns [3]. Hence, it can be widely detected in surface water (originated from agricultural fields) and municipal wastewater [4]. For instance, the concentration of glyphosate in surface or groundwater was found to be in the range of 2–430 μg/L in the USA [5], which is higher than that in Europe: 0.59–165 μg/L [6,7]. Owing to the three polar functional groups (-PO₃H₂, -COOH and -NH-) in molecular glyphosate (shown in Figure 1), glyphosate has proven to be a more challenging water contaminant to effectively remove [8]. Thus, there is an urgent need to develop an effective method to remove glyphosate and its metabolites, such as aminomethylphosphonic acid (AMPA) and sarcosine, from the aqueous environment.

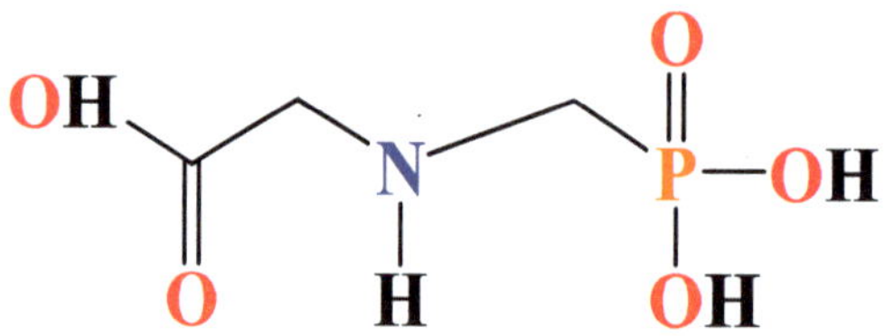

Figure 1. Chemical structure of glyphosate.

Several methods have been employed to treat wastewater contaminated with glyphosate, mainly including adsorption, biological oxidation, reverse osmosis, and advanced oxidation processes (AOPs) [9]. Among these, physicochemical processes (adsorption, coagulation, and reverse osmosis) are non-destructive, and post-treatments of adsorbent materials or solid wastes are required and costly [10,11]. Biological processes (bacteria and fungi) are eco-friendly but limited to lab scale. Besides, another drawback of the above methods is that the metabolites (e.g., AMPA, sarcosine) with higher toxicity cannot be entirely eliminated [12]. In the past years, AOPs (photolysis-based, Fenton-based, electrochemical and ozonation oxidation) have acquired widespread acceptance as potential effective methods for treating glyphosate containing wastewater [13]. AOPs typically operate at or near ambient temperature and pressure conditions involving the production of powerful oxidizing agents in sufficient quantity for water purification. The main advantages of AOPs are attributed to the shorter treatment time, higher removal efficiency (>90%), less limitation, and a better capacity to remove intractable compounds or metabolites. Fenton-based oxidation is preferentially utilized among AOPs for wastewater treatment [14]. Due to the high standard potentials (2.8 V vs. NHE) and oxidation capacity in acidic media, the hydroxyl radical (·OH) is significantly crucial in AOPs [15]. The large amount of free hydroxyl radicals (·OH) are generated from the reaction between Fenton reagents, which are H_2O_2 and Fe^{2+}, and the reaction equation is shown in the Equation (1).

$$H_2O_2 + Fe^{2+} \rightarrow \cdot OH + OH^- + Fe^{3+} \tag{1}$$

Calcium peroxide (CaO_2, CP) is considered as a liquid form of H_2O_2 to generate ·OH with higher stability and environmental friendliness [16]. In fact, CaO_2 can dissolve in water to form H_2O_2 in acidic condition through Equation (3), liberating a maximum of 0.47 g H_2O_2/g CaO_2 [17]. However, in alkaline condition, CaO_2 is directly converted to molecular oxygen instead of H_2O_2 as described in the Equation (2).

$$CaO_2 + H_2O \rightarrow 0.5O_2 + Ca(OH)_2 \tag{2}$$

$$CaO_2 + 2H_2O \rightarrow H_2O_2 + Ca(OH)_2 \tag{3}$$

Assorted studies have assessed the efficacy of calcium peroxide on pollutants removal. For instance, the removal of fluticasone propionate and clobetasol propionate [18], trichloroethylene (TCE) [19,20], methylene blue [21], and 2,4-dichlorophenol [22] by CP has been reported. In the recent years, nanoparticles CaO_2 (nCPs) have acquired the researcher's attentions due to the noteworthy and preferable performance than the commercial CP [16,17]. Results from previous research suggested that the reaction rate of synthesized nCPs was more rapid than commercial CP with silver nanoparticles [23]. Besides, the surface stabilizers used during the synthesis process, such as polyethylene glycol (PEG), polyvinyl pyrrolidone (PVP), polyvinyl alcohol (PVA), dextran, and diethylene glycol monomethyl ether (DEGMME), also dominated the removal efficiency [24]. These surface stabilizers adsorb on nanoparticles and prevent irreversible agglomeration, resulting in stable dispersion with higher steric repulsion [19]. The previous studies have confirmed that nCPs preformed effectively in the removal of organic micropollutants and inorganic heavy metals [25–27], acid blue 161 [28], acrylic acid [29,30], benzene [31], and doxycycline [14]. Therefore, the excellent properties of CaO_2 to generate ·OH at a controlled rate drew an increasing number of researchers to apply CaO_2 in wastewater industry.

To this end, the objectives of the present study were to synthesize nCPs and to evaluate its performance on glyphosate removal from aqueous systems. The synthesis process was based on a co-precipitation technique, using calcium chloride ($CaCl_2$) as a precursor and polyethylene glycol 200 (PEG 200) as a surface stabilizer. The synthesize nCPs were characterized by transform infrared spectroscopy (FTIR), X-ray diffractometry (XRD), Brunauer–Emmett–Teller analysis (BET), dynamic light scattering (DLS), potassium permanganate titration method, and field emission scanning electron microscopy (FESEM) techniques. The removal process of glyphosate from aqueous systems was investigated, utilizing nCPs as an effective oxidant in Fenton-based system. The fitting of different kinetics models was conducted, and the H_2O_2 release as well as proposed degradation pathways were also assessed.

2. Materials and Methods

2.1. Chemical Reagents

Roundup (commercial grade, 41%, containing 360 g/L glyphosate) was diluted to prepare a 1000 mg/L glyphosate stock solution with a natural pH of 4.9–5.9. Calcium peroxide (CaO_2, 65%), ammonium molybdate ($H_8MoN_2O_4$, 99%), and ascorbic acid ($C_6H_8O_6$, 99%) was purchased from Alfa Aesar, Ward Hill, MA, USA. Calcium chloride ($CaCl_2$, $\geq$99.5%), hydrogen peroxide (H_2O_2, 30%), ammonia solution (NH_4OH, 25 wt%), ferrous sulfate ($FeSO_4 \cdot 7H_2O$, $\geq$98%), and ethyl alcohol (C_2H_5OH, 95%) were supplied from R&M Chemicals Sdn. Bhd. (Semenyih, Malaysia). Polyethylene glycol 200 [$HO(C_2H_4O)nH$, PEG 200], antimony potassium tartrate [$K_2Sb_2(C_4H_2O_6)_2$, 99%], sodium sulfite (Na_2SO_3, $\geq$98%, 2.0 M), sodium hydroxide (NaOH), and sulfuric acid (H_2SO_4, 95%) were procured from Sigma Aldrich (St Louis, MO, USA). Cerium (IV) sulfate tetrahydrate ($CeO_8S_2 \cdot 4H_2O$, $\geq$98%) was purchased from Acros Organics, Geel, Belgium. All the chemical reagents used in this study were of analytical reagent grade. Distilled water was used throughout the experiments to prepare the solutions and the stock solutions were freshly prepared every week. The pH of the solutions was adjusted by the addition of 0.1 M H_2SO_4 solution and 0.1 M NaOH solution.

2.2. Synthesis of Calcium Peroxide Nanoparticles (nCPs)

In this study, calcium peroxide nanoparticles were synthesized utilizing co-precipitation technique with minor modifications [23]. Initially, 3.0 g of $CaCl_2$ was dissolved in the 30 mL distilled water and heated to 50 °C to accomplish complete dissolution. 15 mL NH_4OH (1 M) and 120 mL of PEG 200 solution were added into the stirring mixture. The PEG 200 solution functioned as a steric stabilizer and inhibitor of irreversible agglomeration of nanoparticles throughout the precipitation. After 20 min of agitation, 15 mL of 30% H_2O_2 was added to the stirring mixture dropwise with the burette at the rate of 20 drops per minute. The above mixture was constantly stirred for 2 h until a beige to yellowish solution was obtained. The agitation speed throughout the synthesis process was 350 rpm at the ambient temperature. Subsequently, to precipitate the product, 0.1 M NaOH solution was added dropwise until the pH of solution was at 11.50. The appearance of beige to yellowish precipitates indicated the formation of nCPs. The beige to yellowish colored precipitates were separated by centrifugation and after the centrifugation process the powder was washed three times with ethanol. Eventually, two additional washes by distilled water were carried out. The resultant precipitate was dried at 80 °C for 24 h in an evacuated oven to obtain the nCPs. The evacuated oven was used to avoid contamination which may cause impurity in samples. It also permitted the drying process of nCPs in the lower temperature. Equation (4) describes the overall reaction involved in the synthesis process of nCPs.

$$CaCl_2 \cdot 6H_2O + 2H_2O + H_2O_2 + 2NH_3 \rightarrow CaO_2 \cdot 8H_2O + 2NH_4Cl \quad (4)$$

A schematic diagram for synthesis process of nCPs is displayed in Figure 2.

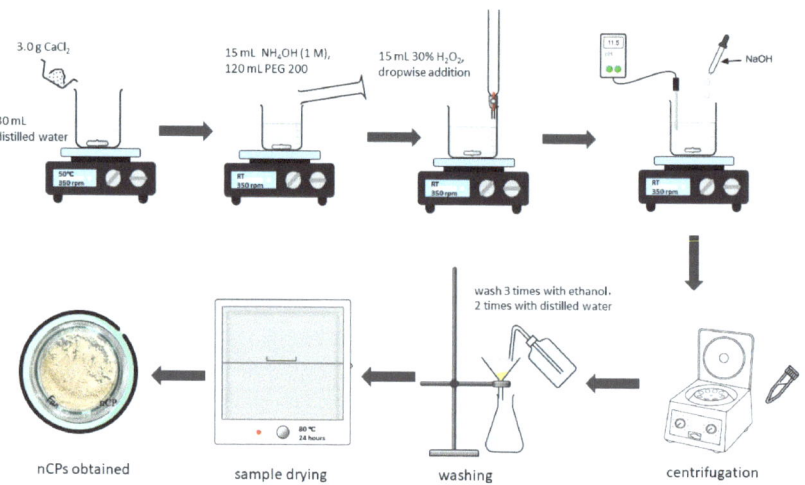

Figure 2. A schematic diagram of synthesis of nCPs.

2.3. Fenton-Based Reaction Experiment Procedure

In a typical experiment, the stock glyphosate-containing solution (1000 mg/L) was diluted to 50 mg/L and placed in the 250 mL conical flasks. The equal initial dosage (0.2 g) of nCPs and CP were weighed out and added into the solutions, respectively. The required amount of FeSO$_4$ was calculated and added, based on our previous research [13] where Ca^{2+}/Fe^{2+} molar ratio was kept at 6. The pH of the solutions was adjusted to 3. The experiments were equipped in a water bath at the agitation speed of 150 rpm, at ambient temperature. The excess of 2.0 M Na$_2$SO$_3$, as an inhibitor to terminate the Fenton reaction, was added to the samples, assuring the elimination of the remaining H$_2$O$_2$ in the solutions [32]. The contact time was 2 h, and the absorbance values were monitored at an interval of 15 min. At the first 15 min, sampling was carried out every 5 min. The precipitated supernatant was taken and filtered with 0.22 μm membrane filters before analyses. Two parameters, the total phosphorus (TP) value and chemical oxygen demand (COD) value, were utilized to evaluate the glyphosate removal process. All experiments were conducted in triplicate, and the mean values were reported.

2.4. Analytical Method

The pH of the solutions was measured using a digital pH meter (Model: Sartorius PB-10). The ammonium molybdate spectrophotometry method was employed to determine TP values with an ultraviolet-visible spectrophotometer (Dynamica, HALO DB-20). The absorbance of each sample was determined at λ = 880 nm and the calibration curve of standard phosphorus (R^2 = 0.9980) was obtained. The total removal efficiency was calculated via the Equation (5).

$$\text{Total removal efficiency (\%)} = \left(1 - \frac{C_t}{C_0}\right) \times 100 \quad (5)$$

where C_0 and C_t are the concentration of total phosphate at the initial and a given contact time t, respectively.

The dichromate digestion-colorimetric method (Lovibond thermoreactor RD125, DR/890 Portable colorimeter) was adopted to analyze the COD values before and after treatment. The total COD removal percentage was calculated using Equation (6), and the flowchart of COD determination is shown in Figure 3.

$$\text{COD removal (\%)} = \left(\frac{\text{COD}_{\text{before treatment}} - \text{COD}_{\text{after treatment}}}{\text{COD}_{\text{before treatment}}}\right) \times 100 \tag{6}$$

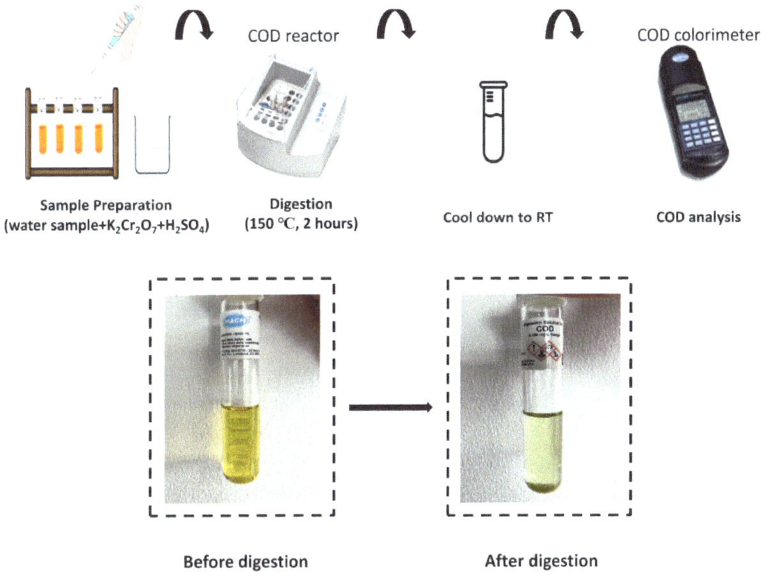

Figure 3. The flowchart of COD determination.

The potassium permanganate titration method was used to determine the purity of the nCPs sample. The purity was determined using the Equation (7) below.

$$\text{Purity of calcium peroxide (\%)} = \frac{5 \times C(\text{KMnO}_4) \times V(\text{KMnO}_4) \times 72.08}{2 \times m(\text{CaO}_2) \times 100} \tag{7}$$

where $C(\text{KMnO}_4)$ represents the concentration of standard KMnO$_4$ solution, which was 0.02 M in the present study. $V(\text{KMnO}_4)$ indicates the volume (mL) of consumed KMnO$_4$ solutions. $m(\text{CaO}_2)$ indicates the mass of sample to be analyzed, which, for each sample in the present study, was 0.05 g. Titration was run in triplicate and the mean value was reported.

The FTIR spectra of CP and nCPs were analyzed via a compact FTIR Spectrometer (Bruker ALPHA II, MA, USA) in the wavelength range of 4000–500 cm^{-1}. X-ray diffraction (XRD) analysis was conducted to identify the structure and phase analysis of CP and the synthesized nCPs by X-ray diffractometer (PHILIPS PW 3040/60 MPD X'Pert high Pro-PAN analytical). The software applied to analyze the XRD data was X'Pert HighScore Plus. The surface morphology of the CP and nCPs was characterized by Nova NanoSEM 230 FE-SEM. The particle size distribution from FESEM results were obtained using the software, SPSS. The textural properties of CP and nCPs were acquired by Brunauer–Emmet–Teller (BET) analysis (degassed at 200 °C for 2 h in a nitrogen environment at 77 K) using the Autosorb-1 (Quantachrome Co., Hampshire, UK). The average particle size and polydispersity index

(PDI) of the CP and nCPs (dispersant: ethanol) were measured using a nanosizer (Nano S, Malvern Instruments Ltd., Malvern, UK). The release of H_2O_2 from CP and nCPs was analyzed via the cerium sulfate spectrophotometry method using a UV spectrophotometer (Dynamica, HALO DB-20). The wavelength used in this study was 480 nm and the calibration curve of $Ce(SO_4)_2$ was prepared and saved for further study (R^2 = 0.9993). Equation (8) was used to calculate the instantaneous H_2O_2 concentration.

$$H_2O_2 \text{ concentration} = \frac{C_0 V_0 - C_t \left(V_0 + V_{sample}\right) \times 0.017}{V_{sample}} \times 1000 \quad (8)$$

where C_0 and C_t represent the initial $CeSO_4$ concentration and that of at the given time, t, respectively. V_0 indicates the volume (mL) of the cerium sulfate standard solution and Vsample represents the volume (mL) of the unknown water sample.

2.5. Kinetic Study

The reaction kinetics of glyphosate removal by CP and nCPs were examined by pseudo-zero order, pseudo-first order, pseudo-second order, and BMG models. The BMG model was initially proposed by Behnajady, Modirshahla, and Ghanbery in 2007 [33] and was fitted with degradation data in representative research [34–36]. The equations are listed in Table 1.

Table 1. The equations of kinetic equations.

Order	Equation Applied	Linear Form by Integration
Zero-order	$\frac{dC_t}{dt} = -k_0$	$C_t = C_0 - k_0 \cdot t$
First-order	$\frac{dC_t}{dt} = -k_1 \cdot C_t$	$\ln C_t = \ln C_0 - k_1 \cdot t$
Second-order	$\frac{dC_t}{dt} = -k_2 \cdot (C_t)^2$	$\frac{1}{C_t} = \frac{1}{C_0} + k_2 \cdot t$
BMG model	$\frac{C_t}{C_0} = 1 - \left[\frac{t}{(m + b \cdot t)}\right]$	$\frac{t}{1 - \left(\frac{C_t}{C_0}\right)} = m + b \cdot t$

Note: k_0, k_1, and k_2 are apparent kinetic rate constants of zero-, first-, and second-order models, respectively. t is reaction time, and C_t is the concentration at a given time t. where m and b are two constants concerning initial degradation rate and maximum oxidation capacity, respectively.

3. Results
3.1. Characterization
3.1.1. FTIR Spectra of CP and nCPs

The FTIR spectra of calcium peroxide and synthesized nCPs were acquired within the range 4000 to 400 cm^{-1} and presented in Figure 4. The spectra were examined for the functional groups for further confirmation of nCPs. The common and characteristic absorption peaks from two samples were denoted by dashed lines or boxes with numbers I–IV. The common peaks present at 855–880 cm^{-1} (dashed line I) together with the range of 710–750 cm^{-1} corresponded to the O–O bridge of CaO_2 [37], and that of 520–565 cm^{-1} referred to the O–Ca–O vibrations [19]. An intense peak near 1450–1490 cm^{-1} (dashed line III) was detected for both CP and synthesized nCPs samples, manifesting the O–Ca–O stretching in CaO_2. A sharp absorption peak near 1100 cm^{-1} (dashed line II) in nCPs sample was identified as the presence of the carbonate ion (CO_3^{2-}) [38], while this signal was weak in CP sample. This finding implied that the synthesized nCPs contained calcite ($CaCO_3$) as an impurity. Another weak peak detected near 1660 cm^{-1} (dashed box IV) was attributed to C=C stretch in nCPs sample [38]. Additionally, a broad absorption peak (dashed box V) at the range of 3000–3650 cm^{-1} was identified as the O-H vibration from the surface stabilizer and water molecules [39].

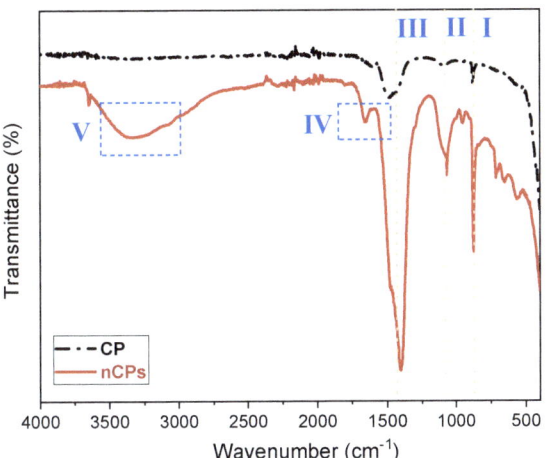

Figure 4. FTIR spectra of CP and nCPs.

3.1.2. XRD Patterns of CP and nCPs

X-ray diffraction (XRD) analysis of nCPs samples was carried out in the 2θ range from 20° to 80° using a CuKα radiation (λ = 0.1540 nm, 30 mA, 40 kV) to ascertain the chemical composition and calculate the crystallite size. As indicated in Figure 5, the five dominant peaks were detected in synthesized calcium peroxide nanoparticles at 2θ values of 30.18, 35.90, 47.20, 51.90, and 60.94, consistent with standard XRD card data of calcium peroxide (JCPDS-00-003-0865) [40]. In addition, calcite is the sole calcium product precipitated throughout the synthesis other than CP, and no other distinctive peak was detected in the XRD pattern [19]. The primary cause of calcite formation was the carbonation of calcium hydroxide, which is generated because of hydrolysis of the precipitated calcium peroxide during the synthesis [41]. This conclusion accorded with our earlier observations from FTIR results and the purity result from the latter part (Section 3.1.3). Table 2 compares the 2θ and d-spacing values of standard CP with the synthesized nCPs, and the analogy validates its chemical composition. The XRD results firmly confirmed that nCPs was successfully synthesized by this method.

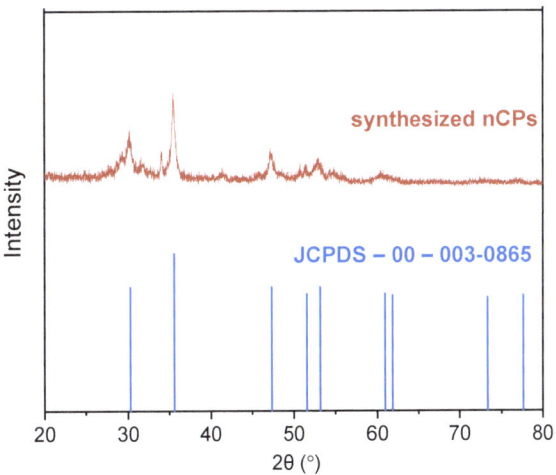

Figure 5. XRD spectra of CP and nCPs.

Table 2. Comparison of 2θ and d-spacing values of the standard CaO_2 with the synthesized nCPs.

Peak Number	CaO_2 (JCPDS-00-003-0865)			Synthesized nCPs		
	2θ (Degree)	d-Spacing (Å)	Miller Indices (h k L)	2θ (Degree)	d-Spacing (Å)	Miller Indices (h k L)
1	30.27	2.95	0 0 2	30.18	2.95	0 0 2
2	35.60	2.52	1 1 0	35.90	2.52	1 1 0
3	47.31	1.92	1 1 2	47.20	1.92	1 1 2
4	51.60	1.77	2 0 0	51.90	1.77	2 0 0
5	60.89	1.52	2 0 2	60.94	1.52	2 0 2

The Debye–Scherrer equation is displayed as Equation (9) and was utilized to calculate the average particle size of nCPs.

$$D = \frac{K\lambda}{\beta cos\theta} \quad (9)$$

where k is the Debye–Scherrer constant (k = 0.9), λ is the wavelength of the incident X-ray radiation (λ = 0.1540 nm), θ is the Bragg's angle in radians, and β is the full width at half maximum (FWHM). The most intense peak at 2θ = 35.90 was used to calculate the average particle size. The calculated nanocrystalline size of nCPs was 38.69 nm.

3.1.3. Physicochemical Properties of CP and nCPs

Table 3 presents the physicochemical properties of CP and nCPs, including the results of potassium permanganate titration, BET and DLS analysis. The BET isotherm and the result of DLS analysis are shown in Figure 6. The purity of Ca^{2+} in calcium peroxide samples increased from 65.0% to 75.1%, which reduced the production of byproducts and contaminants. Thus, the main impurity was found to be $CaCO_3$ at about 24.90%. Of note, the surface area drastically improved for synthesized CaO_2 nanoparticles. The surface area of the typically 65% CP was only 3.1079 m^2/g and of synthesized nCPs was 28.0860 m^2/g. This remarkable improvement would greatly increase the reaction rate and strengthen the performance of removal. The higher surface area of nCPs corresponded to the larger pore size (23.1324 nm) and larger pore volume (1.3248 cm^3/g) and the smaller average particle size (40.88 nm) obtained from DLS analysis. Notably, the result of average particle size was relatively in agreement with the XRD result. The PDI value (0.228) indicated that the synthesized PEG-coated nCPs were uniformly dispersed without much agglomeration while the PDI value of CP (0.592) was greatly larger. As shown in Figure 6b, the width of peaks also certified the PDI result. The pore size (23.1324 nm) demonstrated that the synthesized PEG-coated nCPs were well-defined mesoporous (between 2–50 nm). The higher ratio of surface-volume of nCPs samples also confirmed that it would perform better in terms of removing the pollutants. It could be observed from Figure 6a that the limiting gas uptake over high P/P_0 (0.4–1.0) may be caused by capillary condensation that occurred in pores with monolayer–multilayer adsorption [19]. Likewise, this feature revealed that the adsorption process of nCPs belonged to IUPAC type IV isotherm [42]. Collectively, the enhanced physicochemical properties would be conducive to the greater removal performance.

Table 3. Physicochemical properties of CaO_2 and nCPs.

Sample	Purity (%)	Surface Area (m^2/g)	Pore Size (nm)	Pore Volume (cm^3/g)	Surface-Volume Ratio	Average Size (nm)	PDI
CP	65.0%	3.1079	2.365	0.3644	13.70	220.16	0.592
nCPs	75.1%	28.0860	23.1324	1.3248	21.25	40.88	0.228

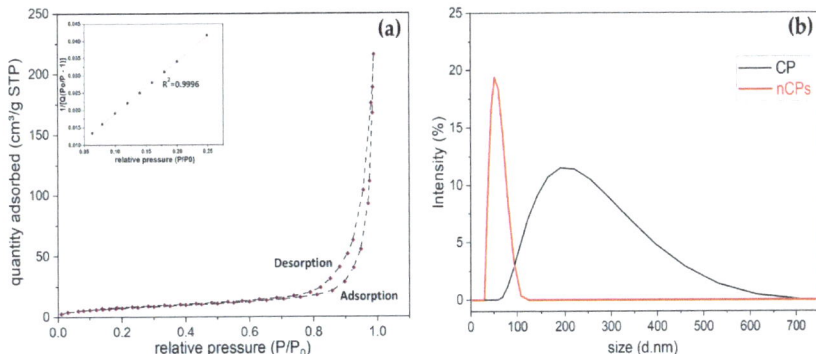

Figure 6. (a) BET nitrogen adsorption isotherm plot. (b) DLS results of CP and nCPs.

3.1.4. FESEM Results of CP and nCPs

Figure 7 presents the morphology of samples determined via FESEM with the same magnification (100,000 X) and the particle size distribution of CP and nCPs. Figure 7a shows that conventional CP appeared as irregular flakes of particles in pieces with excessive agglomeration. This would significantly reduce the surface area and removal efficiency. The minimum size was approximately 98.99 nm whereas the maximum size was relatively 1.19 µm due to the exceeding agglomeration. The particle size distribution of CP was mostly between 90 and 400 nm, shown in Figure 7c, as correlated with the DLS results (average particle size: 220.16 nm; PDI: 0.592). In contrast, observed from Figure 7b, the heterogeneous surface morphology with regular mostly spherical shape and smaller particle size could be detected on nCPs, contributing to the higher removal performance. Apparently, the CP nanoparticles were uniformly distributed, and the agglomeration was mitigated. The minimum and maximum particle sizes of nCPs were approximately 10 nm and 94.28 nm, respectively. The particle size distribution of nCPs was mainly at the range of 15–60 nm, illustrated by Figure 7d. Remarkably, PEG 200, which was an effective stabilizer and dispersant to mitigate agglomeration, was attributed to the smaller particle size and better distribution of nCPs [19]. This finding was in great agreement with the previous studies [24,29], which reported the heterogeneous surface morphology of nCPs.

3.2. Glyphosate Removal Process Using Fenton-Based System

Figure 8 indicates the glyphosate removal process based on CP and nCPs using a Fenton-based system. The optimal experimental conditions obtained from our prior investigation were applied to the present study, with minor modifications [13]. The investigation was carried out at the initial pH = 3.0, the initial calcium peroxide (CP and nCPs, respectively) dosage = 0.2 g, Ca^{2+}/Fe^{2+} molar ratio = 6, the initial glyphosate concentration = 50 mg/L under room temperature at the stirring speed of 150 rpm. The smaller particle size with larger surface area and higher purity of calcium peroxide nanoparticles significantly enhanced the removal process. Thus, the higher removal efficiency was achieved compared to the CP. This clearly indicated that the degradation of glyphosate was largely dependent on particle size. It is notable that within 75 min (marked as a vertical dashed line), up to 99.60% removal was reached using nCPs while at the same contact time the removal efficiency using CP was only 69.48%. Within contact time = 75 min, the total phosphorus in the solution was removed effectively by the powerful free hydroxyl radicals (·OH) generated from the Fenton reagents. However, between contact time = 75 min and 120 min, the removal efficiency of CP was still climbing from 69.48% to 73.53% and the equilibrium was detected around 90 min. The reason was that the surface area of conventional CP was quite smaller than nCPs, resulting in the slower removal rate and longer removal time. Additionally, the Fenton reaction involving Fe^{2+} as a catalyst often proceeds in two stages, a fast one and a slower one [35]. The faster stage is caused by an interaction between Fe^{2+} and H_2O_2, whereas the slower stage is attributed to the accumulation of Fe^{3+} and the limited recovery of Fe^{2+} via H_2O_2 [43]. This behavior was observed in the previous research [39,44] and was also supported by the finding of Figure 8a. The total removal efficiency is displayed in Figure 8b. The TP and COD removal of glyphosate-containing solution using nCPs were considerably more effective compared to the conventional CP. Under the same condition, total 75.10% COD removal was detected while only 48.50% COD removal using conventional CP. Hence, the nCPs is preferred in terms of COD removal because it is an important parameter in the treatment of wastewater with a high phosphorus content. Comparing our total removal efficiencies with the previous studies under the same mechanism and similar experimental conditions [13,45,46], the TP removal efficiency improved by 3.9–5.4% and 6.5–13.0% improvement was detected for COD removal.

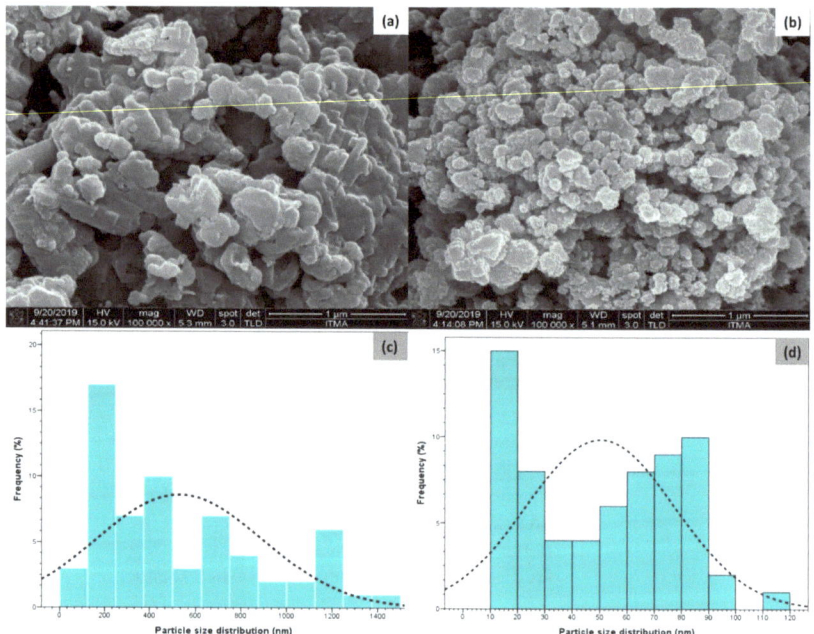

Figure 7. FESEM images of (**a**) CP (**b**) nCPs; Particle size distribution of (**c**) CP (**d**) nCPs.

3.3. Effect of Initial Glyphosate Concentration

To assess the significance and reliability of our method, the batch degradation study was carried out with the initial glyphosate concentration ranging from 50 ppm to 200 mg/L. The efficiency of TP removal and COD removal was investigated at the initial pH = 3.0, initial nCPs dosage = 0.2 g, Ca^{2+}/Fe^{2+} molar ratio = 6, contact time = 75 min and agitation speed = 150 rpm under ambient temperature. The results of total removal efficiency before and after treatment are displayed in Figure 9. It can be observed that the TP and COD removal efficiency typically reduced with the increased initial glyphosate concentration from 50 mg/L to 200 mg/L. The TP and COD removal decreased from 99.60% to 71.40% and from 75.10% to 40.30%, respectively. The lower removal efficiency at higher initial glyphosate concentration can be explained due to the greater amount of generated metabolites such as AMPA or sarcosine, which compete with glyphosate for Fenton reaction with the stationary concentration of ·OH [47,48]. Therefore, the removal efficiency is strictly restricted by the amount of free hydroxyl radicals (·OH) generated. To sum up the main points, this method was quite effective, and the initial glyphosate concentration was one of the crucial factors that had a significant impact on the glyphosate removal. As has been noted, considering implementing this process at full-scale, the strict control of operational conditions (e.g., acidic medium) and the efficiency of mineralization are worth further studying.

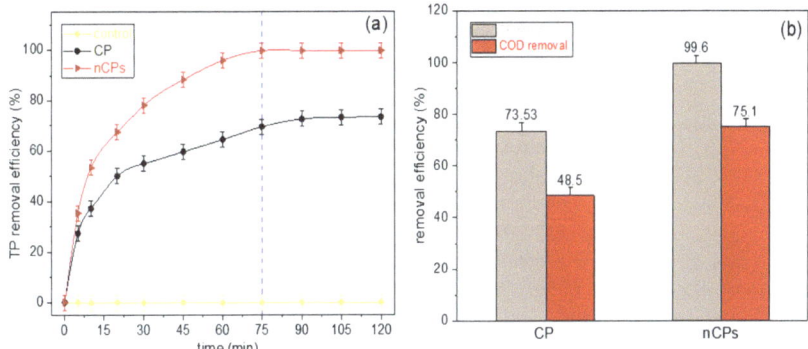

Figure 8. (a) The effect of contact time on TP removal based on nCPs using Fenton-based system (Condition: pH, 3.0; the initial dosage of CP and nCPs, 0.2 g; Ca^{2+}/Fe^{2+} molar ratio, 6; initial glyphosate concentration, 50 mg/L; RT; 150 rpm). (b) Comparison of total removal efficiency of CP and nCPs.

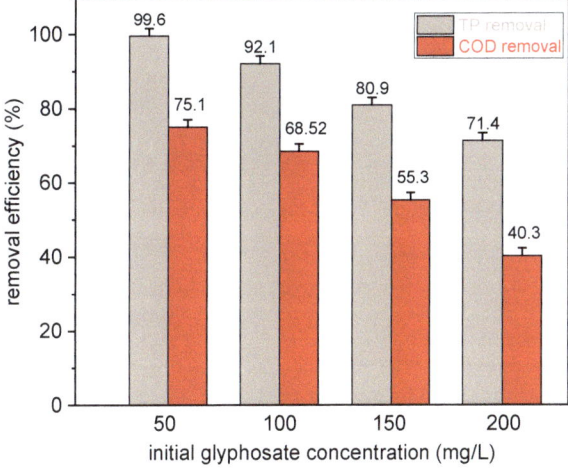

Figure 9. Effect of initial glyphosate concentration on TP and COD removal (Condition: pH, 3.0; the initial nCPs dosage, 0.2 g; Ca^{2+}/Fe^{2+} molar ratio, 6; contact time, 75 min; RT; 150 rpm).

3.4. H_2O_2 Release

The investigation on H_2O_2 release by CP and nCPs was conducted in a 250 mL conical flask equipped with the agitation speed of 150 rpm at room temperature. 0.5 g of CP and nCPs was weighed out and dissolved in 200 mL distilled water before adding in the flasks, respectively. To monitor the instantaneous concentration of H_2O_2 generated from calcium peroxide and the instantaneous pH of solutions, 2 mL of sample was collected from each flask within 5 h. For the first 2 h, absorbance was monitored at intervals of 15 min. The results of the controlled release and the change of pH with time are displayed in Figure 10. The theoretical H_2O_2 release was calculated according to molecular formula. It could be observed that the rate of H_2O_2 released for both CP and nCPs was remarkably fast in the initial 60 min, which proved the first stage of Fenton reaction as confirmed by the previous study [44]. It is worth noting that up to 70.90% of CP and 72.20% of nCPs were converted the liquid H_2O_2 within the initial hour. The CP and nCPs samples continued to actively release H_2O_2 until the equilibrium was reached, which was expected to be around 3 h (CP) and 2 h (nCPs). Though the nCPs posed smaller particle size and larger surface area, nCPs released H_2O_2 relatively slowly compared to CP at the first of 30 min, due to the PEG 200 coated during the synthesis. This was because the involvement of PEG 200 functioning as a membrane on the surface of nCPs could result in a slower rate of H_2O_2 release [24]. The conversion percentage of nCPs (94.50%) was much higher than that of CP (89.30%) within 5 h, suggesting that nanoscale calcium peroxide performed better than conventional CP in water regarding releasing H_2O_2. The remainder of CP (10.70%) and nCPs (5.50%) was related to decomposition in water and oxygen generation [19]. When it came to pH changes, an increase of 1.97 for CP and 1.26 for nCPs was observed due to increased H_2O_2 release in the nCPs system. To conclude, nanoparticles calcium peroxide is of great efficiency as an oxidant owing to the high dispersion and smaller particle size [44].

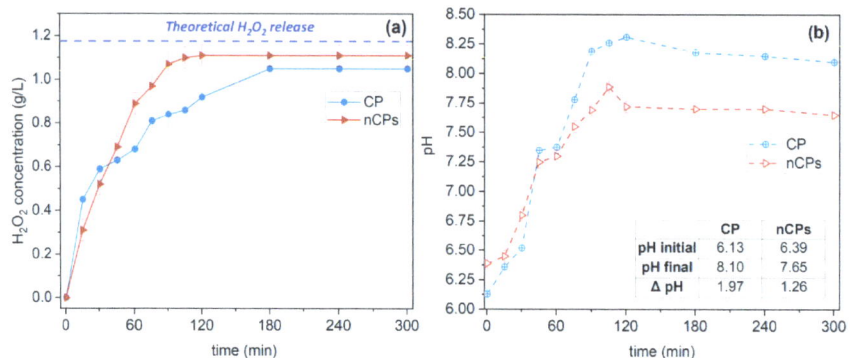

Figure 10. (a) Release of H_2O_2 by CP and nCPs under stirring condition (150 rpm, RT), [CaO_2] = [nCPs] = 2.5 g/L. (b) The change of pH with time of CP and nCPs.

3.5. Kinetic Study

In the kinetic study, four models were utilized to study the degradation kinetic of the glyphosate based on calcium peroxide using Fenton-based system. The obtained kinetic parameters are illustrated in Figure 11 and Table 4. For second-order kinetic study, the contact time of 60 min was selected to fit with degradation data due to the feasibility of data fitting. Overall, the most appropriate model to describe glyphosate removal was the BMG model reaction kinetics (average R^2 = 0.9965), followed by first-order kinetics (average R^2 = 0.9336). Due to the low average correlation coefficient (average R^2 < 0.9), the zero-order (0.7598) as well as the second order (0.8920) could not fit with the degradation data. Regarding parameter 1/m (initial degradation rate) acquired from BMG model, the initial degradation rate increased to four times in nCPs system (1/m = 0.4916), leading to the faster removal rate. That is to say, the first stage of the degradation process made the greatest contributions to the overall removal efficiency because, as was discussed before, it is a rapid stage of the Fenton reaction. By considering the parameter 1/b (maximum oxidation capacity), the same finding can be observed from the Table 4. The maximum oxidation capacity in nCPs system was 1.1799 whereas that of the CP system was only 0.9981, suggesting that the higher removal efficiency could be observed in nCPs system. When it came to the first-order kinetics model, it can be observed that the reaction rate coefficient (k_1) in nCPs system was even six times greater than the CP system. Therefore, the reaction rate of nCPs system is considerably rapid than the CP system. The result of total removal efficiency proved this finding as well. These basic findings are consistent with previous research [13,21,35] showing that BMG model is the most appropriate model to describe the degradation data by AOPs.

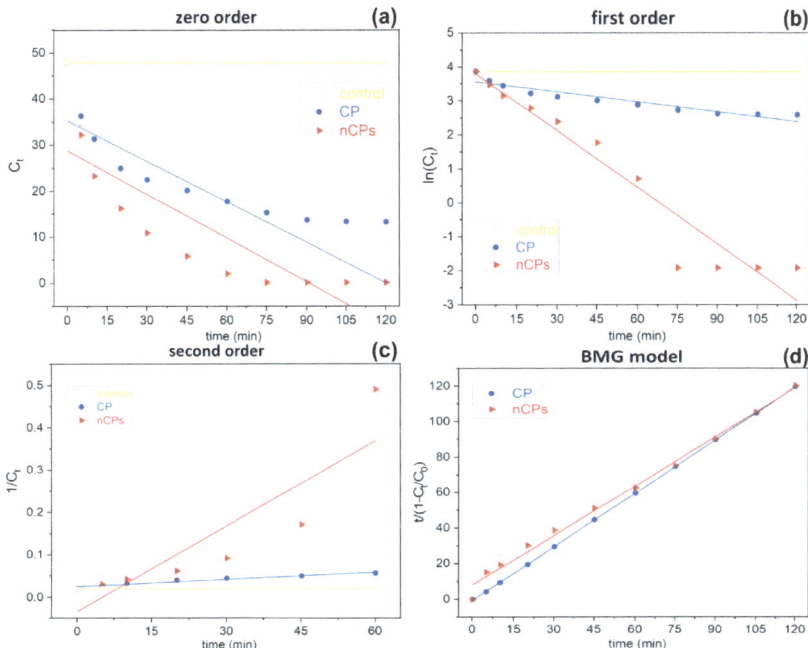

Figure 11. Degradation data of glyphosate using different kinetic models.

Table 4. Comparison of kinetic models for different samples (CP and nCPs).

Sample	Removal (%)	Zero-Order		First-Order		Second-Order		BMG Model		
		k_0	R^2	k_1	R^2	k_2	R^2	$1/m$	$1/b$	R^2
CP	73.53	0.2993	0.8248	0.0097	0.9385	0.0067	0.7449	0.1218	0.9981	0.9999
nCPs	99.60	0.3163	0.6949	0.0555	0.9286	5.5700	0.9411	0.4916	1.1799	0.9930
Average R^2			0.7598		0.9336		0.8430			0.9965

3.6. Proposed Degradation Pathways

According to the literature [8], glyphosate degradation and mineralization by AOPs are linked to the cleavage of C-P and C-N bonds by hydroxyl radicals (·OH). Two oxidation pathways via AOPs were reported, which are AMPA pathway and sarcosine pathway [9]. Furthermore, it is common to detect these two pathways can exist separately or coexist during the degradation process [49]. The possible degradation mechanisms in the present study are displayed in Figure 12. Initially, glyphosate is attacked by hydroxyl radicals (·OH) to yield AMPA and glycolic acid (AMPA pathway), or sarcosine and PO_4^{3-} (sarcosine pathway). Two metabolites, which are AMPA and glycolic acid, can be produced via AMPA pathway. Through the cleavage of C-P bond by means of ·OH, AMPA can be further converted to NH_4^+, NO_3^-, and PO_4^{3-} and glycolic acid can be finally oxidized to CO_2. On the other hand, glyphosate can be degraded into sarcosine and PO_4^{3-} via sarcosine pathway. Subsequently, sarcosine can be further oxidized to N-contained metabolites, which can be eventually converted to NH_4^+, NO_3^-, and CO_2. The possible mechanisms also implied that ·OH can oxidize and decompose organic matter until its total mineralization to CO_2, H_2O, and its corresponding inorganic salt [15,50,51]. As a result, AOPs are viewed as playing a critical role in the treatment of glyphosate containing wastewater.

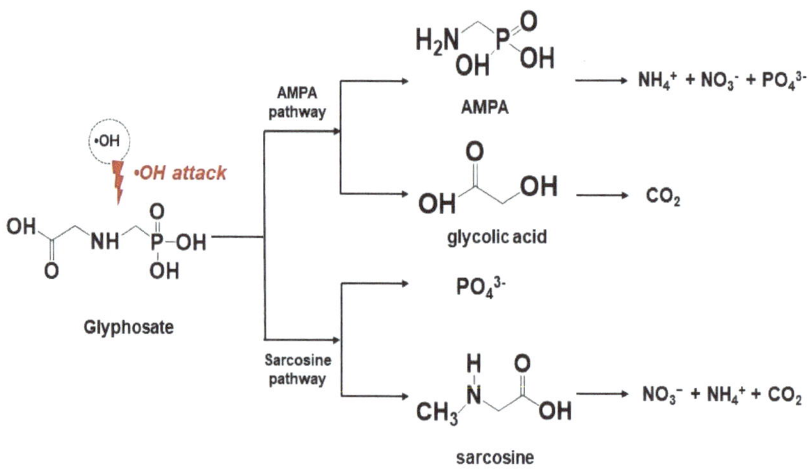

Figure 12. Proposed glyphosate degradation pathways via AOPs.

4. Conclusions

In the present study, PEG-coated calcium peroxide nanoparticles were successfully synthesized, and the effective removal efficiency was successfully achieved. The average particle size of synthesized nCPs was approximately 40.88 nm with larger surface area of 28.09 m^2/g and a uniform distribution. The maximum TP removal efficiency (99.60%) and total COD removal efficiency (75.10%) could be acquired at the given conditions (pH = 3.0, the initial nCPs dosage = 0.2 g, Ca^{2+}/Fe^{2+} molar ratio = 6, the initial glyphosate concentration = 50 mg/L, RT). The degradation rate of glyphosate in aqueous system was rapid and equilibrium was achieved within 75 min. The degradation process of glyphosate could be most effectively described by the BMG model. The particle size and initial glyphosate concentration revealed a significant impact on glyphosate removal. Moreover, the effectiveness of the H$_2$O$_2$ release and the outcome of the initial glyphosate concentration study demonstrated the reliability and potential of this material. To summarize the findings of this work, calcium peroxide nanoparticles represent an efficient oxidant for glyphosate removal from aqueous systems.

Author Contributions: Conceptualization, F.L., T.S.Y.C., L.C.A., S.N.A.M.J. and N.N.A.N.; methodology, F.L., T.S.Y.C., L.C.A., S.N.A.M.J. and N.N.A.N.; software, F.L., T.S.Y.C., L.C.A., S.N.A.M.J., and N.N.A.N.; validation, F.L., T.S.Y.C., L.C.A., S.N.A.M.J. and N.N.A.N.; formal analysis, F.L., T.S.Y.C., L.C.A., S.N.A.M.J. and N.N.A.N.; investigation, F.L., T.S.Y.C., L.C.A., S.N.A.M.J. and N.N.A.N.; resources, F.L., T.S.Y.C., L.C.A., S.N.A.M.J. and N.N.A.N.; data curation, F.L., T.S.Y.C., L.C.A., S.N.A.M.J. and N.N.A.N.; writing—original draft preparation, F.L., T.S.Y.C., L.C.A., S.N.A.M.J. and N.N.A.N.; writing—review and editing, F.L., T.S.Y.C., L.C.A., S.N.A.M.J. and N.N.A.N.; visualization, F.L., T.S.Y.C., L.C.A., S.N.A.M.J. and N.N.A.N.; supervision, F.L., T.S.Y.C., L.C.A., S.N.A.M.J. and N.N.A.N.; project administration, F.L., T.S.Y.C., L.C.A., S.N.A.M.J. and N.N.A.N.; funding acquisition, T.S.Y.C., L.C.A. and S.N.A.M.J. All authors have read and agreed to the published version of the manuscript.

Funding: This research was funded by the Ministry of Higher Education (KPT), Malaysia under the Fundamental Research Grant Scheme FRGS/1/2020/TK0/UPM/01/2 (03 01 20 2250FR).

Institutional Review Board Statement: Not applicable.

Data Availability Statement: The data presented in this study are available on demand from the corresponding author or the first author.

Acknowledgments: The authors would like to acknowledge the Department of Chemistry, Faculty of Science, Universiti Putra Malaysia (UPM) and Center of Sustainable Research Department of Chemical and Environmental Engineering, Faculty of Engineering, Universiti Putra Malaysia (UPM), for the research facilities to conduct this study. The authors thank the grant FRGS/1/2020/TK0/UPM/01/2 (03 01 20 2250FR) received from Ministry of Higher Education (KPT), Malaysia, to carry out the project.

Conflicts of Interest: No conflict of interest exists in the submission of this manuscript and all authors approve this manuscript for publication. The authors confirm here that this manuscript has not been previously published in whole and is not under consideration by any other journals.

References

1. Nicolopoulou-Stamati, P.; Maipas, S.; Kotampasi, C.; Stamatis, P.; Hens, L. Chemical Pesticides and Human Health: The Urgent Need for a New Concept in Agriculture. *Front. Public Health* **2016**, *4*, 148. [CrossRef]
2. Sustainable Water Solutions. *Nat. Sustain.* **2020**, *3*, 73. [CrossRef]
3. Jia, D.; Liu, M.; Xia, J.; Li, C. Effective Removal of Aqueous Glyphosate Using $CuFe_2O_4$@biochar Derived from Phragmites. *J. Chem. Technol. Biotechnol.* **2020**, *95*, 196–204. [CrossRef]
4. Poiger, T.; Keller, M.; Buerge, I.J.; Balmer, M.E. Behavior of Glyphosate in Wastewater Treatment Plants. *Chimia* **2020**, *74*, 156–160. [CrossRef]
5. Mahler, B.J.; Van Metre, P.C.; Burley, T.E.; Loftin, K.A.; Meyer, M.T.; Nowell, L.H. Similarities and Differences in Occurrence and Temporal Fluctuations in Glyphosate and Atrazine in Small Midwestern Streams (USA) during the 2013 Growing Season. *Sci. Total Environ.* **2017**, *579*, 149–158. [CrossRef]
6. Villeneuve, A.; Larroudé, S.; Humbert, J.F. Herbicide Contamination of Freshwater Ecosystems: Impact on Microbial Communities. *Pestic. - Formul. Eff. Fate* **2011**. [CrossRef]
7. Tóth, G.; Háhn, J.; Szoboszlay, S.; Harkai, P.; Farkas, M.; Radó, J.; Göbölös, B.; Kaszab, E.; Szabó, I.; Urbányi, B.; et al. Spatiotemporal Analysis of Multi-Pesticide Residues in the Largest Central European Shallow Lake, Lake Balaton, and Its Sub-Catchment Area. *Environ. Sci. Eur.* **2022**, *34*. [CrossRef]
8. Espinoza-Montero, P.J.; Vega-Verduga, C.; Alulema-Pullupaxi, P.; Fernández, L.; Paz, J.L. Technologies Employed in the Treatment of Water Contaminated with Glyphosate: A Review. *Molecules* **2020**, *25*, 5550. [CrossRef]
9. Feng, D.; Soric, A.; Boutin, O. Treatment Technologies and Degradation Pathways of Glyphosate: A Critical Review. *Sci. Total Environ.* **2020**, *742*, 140559. [CrossRef]
10. Li, Y.; Zhao, C.; Wen, Y.; Wang, Y.; Yang, Y. Adsorption Performance and Mechanism of Magnetic Reduced Graphene Oxide in Glyphosate Contaminated Water. *Environ. Sci. Pollut. Res.* **2018**, *25*, 21036–21048. [CrossRef]
11. Li, F.; Wang, Y.; Yang, Q.; Evans, D.G.; Forano, C.; Duan, X. Study on Adsorption of Glyphosate (N-Phosphonomethyl Glycine) Pesticide on MgAl-Layered Double Hydroxides in Aqueous Solution. *J. Hazard. Mater.* **2005**, *125*, 89–95. [CrossRef] [PubMed]
12. Zhan, H.; Feng, Y.; Fan, X.; Chen, S. Recent Advances in Glyphosate Biodegradation. *Appl. Microbiol. Biotechnol.* **2018**, *102*, 5033–5043. [CrossRef]
13. Li, F.; Choong, T.S.Y.; Soltani, S.; Chuah Abdullah, L.; Siti Nurul, S.N.A.; Amerhaider Nuar, N.N. Investigation of Glyphosate Removal from Aqueous Solutions Using Fenton-like System Based on Calcium Peroxide. *Processes* **2022**, *10*, 2045. [CrossRef]
14. Amerhaider Nuar, N.N.; Siti Nurul, S.N.A.; Li, F.; Mat Azmi, I.D.; Chiang, P.C.; Choong, T.S.Y. Synthesis of Controlled-Release Calcium Peroxide Nanoparticles Coated with Dextran for Removal of Doxycycline from Aqueous System. *Polymers* **2022**, *14*, 3866. [CrossRef] [PubMed]
15. Wang, J.L.; Xu, L.J. Advanced Oxidation Processes for Wastewater Treatment: Formation of Hydroxyl Radical and Application. *Crit. Rev. Environ. Sci. Technol.* **2012**, *42*, 251–325. [CrossRef]
16. Lu, S.; Zhang, X.; Xue, Y. Application of Calcium Peroxide in Water and Soil Treatment: A Review. *J. Hazard. Mater.* **2017**, *337*, 163–177. [CrossRef] [PubMed]
17. Xue, Y.; Sui, Q.; Brusseau, M.L.; Zhou, W.; Qiu, Z.; Lyu, S. Insight into CaO_2-Based Fenton and Fenton-like Systems: Strategy for CaO_2-Based Oxidation of Organic Contaminants. *Chem. Eng. J.* **2019**, *361*, 919–928. [CrossRef]
18. Zhang, A.; Wang, J.; Li, Y. Performance of Calcium Peroxide for Removal of Endocrine-Disrupting Compounds in Waste Activated Sludge and Promotion of Sludge Solubilization. *Water Res.* **2015**, *71*, 125–139. [CrossRef]
19. Ali, M.; Farooq, U.; Lyu, S.; Sun, Y.; Li, M.; Ahmad, A.; Shan, A.; Abbas, Z. Synthesis of Controlled Release Calcium Peroxide Nanoparticles (CR-NCPs): Characterizations, H_2O_2 Liberate Performances and Pollutant Degradation Efficiency. *Sep. Purif. Technol.* **2020**, *241*, 116729. [CrossRef]
20. Sun, Y.; Lyu, S.; Brusseau, M.L.; Tang, P.; Jiang, W.; Gu, M.; Li, M.; Lyu, Y.; Qiu, Z.; Sui, Q. Degradation of Trichloroethylene in Aqueous Solution by Nanoscale Calcium Peroxide in the Fe(II)-Based Catalytic Environments. *Sep. Purif. Technol.* **2019**, *226*, 13–21. [CrossRef]
21. Li, F.; Shean, T.; Choong, Y.; Soltani, S.; Chuah, L.; Nurul, S.; Jamil, A. Kinetic Study of Fenton-Like Degradation of Methylene Blue in Aqueous Solution Using Calcium Peroxide. *Pertanika J. Sci. Technol.* **2022**, *30*. [CrossRef]

22. Wang, H.; Zhao, Y.; Su, Y.; Li, T.; Yao, M.; Qin, C. Fenton-like Degradation of 2,4-Dichlorophenol Using Calcium Peroxide Particles: Performance and Mechanisms. *RSC Adv.* **2017**, *7*, 4563–4571. [CrossRef]
23. Khodaveisi, J.; Banejad, H.; Afkhami, A.; Olyaie, E.; Lashgari, S.; Dashti, R. Synthesis of Calcium Peroxide Nanoparticles as an Innovative Reagent for in Situ Chemical Oxidation. *J. Hazard. Mater.* **2011**, *192*, 1437–1440. [CrossRef] [PubMed]
24. Rastinfard, A.; Nazarpak, M.H.; Moztarzadeh, F. Controlled Chemical Synthesis of CaO_2 Particles Coated with Polyethylene Glycol: Characterization of Crystallite Size and Oxygen Release Kinetics. *RSC Adv.* **2018**, *8*, 91–101. [CrossRef]
25. Massalimov, I.A.; Il'yasova, R.R.; Samsonov, M.R.; Mustafin, A.G.; Medvedev, Y.A. Recovery of Heavy Metal Ions with Calcium Peroxide Microparticles. *Russ. J. Appl. Chem.* **2016**, *89*, 360–366. [CrossRef]
26. Garnjanagoonchorn, W.; Changpuak, A. Preparation and Partial Characterization of Eggshell Calcium Chloride. *Int. J. Food Prop.* **2007**, *10*, 497–503. [CrossRef]
27. Huang, M.R.; Li, X.G. Highly Sensing and Transducing Materials for Potentiometric Ion Sensors with Versatile Applicability. *Prog. Mater. Sci.* **2022**, *125*, 100885. [CrossRef]
28. Trovó, A.G.; Senivs, P.; Palmiste, Ü.; Sillanpää, M.; Tang, W.Z. Decolorization Kinetics of Acid Blue 161 by Solid Peroxides Catalyzed by Iron in Aqueous Solution. *Desalin. Water Treat.* **2016**, *57*, 19344–19356. [CrossRef]
29. Madan, S.S.; Wasewar, K.L. Synthesis of CaO_2 Nanoparticles for Environmental Remediation. *EnergyEnviron. Sel. Proc. ICWEES-2016* **2018**, 69–76. [CrossRef]
30. Madan, S.S.; Upwanshi W, A.; Wasewar, K.L. Adsorption of α-Toluic Acid by Calcium Peroxide Nanoparticles. *Desalin. Water Treat.* **2016**, *57*, 16507–16513. [CrossRef]
31. Mosmeri, H.; Alaie, E.; Shavandi, M.; Dastgheib, S.M.M.; Tasharrofi, S. Benzene-Contaminated Groundwater Remediation Using Calcium Peroxide Nanoparticles: Synthesis and Process Optimization. *Environ. Monit. Assess.* **2017**, *189*. [CrossRef] [PubMed]
32. Souza, D.R.D.; Trovó, A.G.; Filho, N.R.A.; Silva, M.A.A.; Machado, A.E.H. Degradation of the Commercial Herbicide Glyphosate by Photo-Fenton Process: Evaluation of Kinetic Parameters and Toxicity. *J. Braz. Chem. Soc.* **2013**, *24*, 1451–1460. [CrossRef]
33. Behnajady, M.A.; Modirshahla, N.; Ghanbary, F. A Kinetic Model for the Decolorization of C.I. Acid Yellow 23 by Fenton Process. *J. Hazard. Mater.* **2007**, *148*, 98–102. [CrossRef] [PubMed]
34. Ertugay, N.; Acar, F.N. Removal of COD and Color from Direct Blue 71 Azo Dye Wastewater by Fenton's Oxidation: Kinetic Study. *Arab. J. Chem.* **2017**, *10*, S1158–S1163. [CrossRef]
35. Santana, C.S.; Ramos, M.D.N.; Velloso, C.C.V.; Aguiar, A. Kinetic Evaluation of Dye Decolorization by Fenton Processes in the Presence of 3-Hydroxyanthranilic Acid. *Int. J. Environ. Res. Public Health* **2019**, *16*, 1602. [CrossRef]
36. Bougdour, N.; Sennaoui, A.; Bakas, I.; Assabbane, A. Experimental Evaluation of Reactive Yellow 17 Degradation Using UV Light and Iron Ions Activated Peroxydisulfate: Efficiency and Kinetic Model. *Sci. Technol. Mater.* **2018**, *30*, 157–165. [CrossRef]
37. Zegliński, J.; Piotrowski, G.P.; Piekoś, R. A Study of Interaction between Hydrogen Peroxide and Silica Gel by FTIR Spectroscopy and Quantum Chemistry. *J. Mol. Struct.* **2006**, *794*, 83–91. [CrossRef]
38. Matuana, L.M.; Balatinecz, J.J.; Sodhi, R.N.S.; Park, C.B. Surface Characterization of Esterified Cellulosic Fibers by XPS and FTIR Spectroscopy. *Wood Sci. Technol.* **2001**, *35*, 191–201. [CrossRef]
39. Li, X.; Xie, Y.; Jiang, F.; Wang, B.; Hu, Q.; Tang, Y.; Luo, T.; Wu, T. Enhanced Phosphate Removal from Aqueous Solution Using Resourceable Nano-CaO_2/BC Composite: Behaviors and Mechanisms. *Sci. Total Environ.* **2020**, *709*, 136123. [CrossRef]
40. Madan, S.S.; Wasewar, K.L.; Ravi Kumar, C. Adsorption Kinetics, Thermodynamics, and Equilibrium of α-Toluic Acid onto Calcium Peroxide Nanoparticles. *Adv. Powder Technol.* **2016**, *27*, 2112–2120. [CrossRef]
41. Yuan, D.; Zhang, C.; Tang, S.; Li, X.; Tang, J.; Rao, Y.; Wang, Z.; Zhang, Q. Enhancing CaO_2 Fenton-like Process by Fe(II)-Oxalic Acid Complexation for Organic Wastewater Treatment. *Water Res.* **2019**, *163*, 114861. [CrossRef]
42. Rahman, M.M.; Shafiullah, A.Z.; Pal, A.; Islam, M.A.; Jahan, I.; Saha, B.B. Study on Optimum Iupac Adsorption Isotherm Models Employing Sensitivity of Parameters for Rigorous Adsorption System Performance Evaluation. *Energies* **2021**, *14*, 7478. [CrossRef]
43. Barreto, F.; Santana, C.S.; Aguiar, A. Behavior of Dihydroxybenzenes and Gallic Acid on the Fenton-Based Decolorization of Dyes. *Desalin. Water Treat.* **2016**, *57*, 431–439. [CrossRef]
44. Xie, Y.; Yang, X.; Li, J.; Wu, T.; Huang, J.; Xu, F. Enhanced Removal of Glyphosate from Aqueous Solution by Nano-CaO_2/AS Composite: Oxidation and Precipitation. *Sep. Purif. Technol.* **2022**, *288*, 120349. [CrossRef]
45. Liao, H.; Tan, B.; Ke, M.; Li, Z.; Lu, J. Pretreatment of Glyphosate Wastewater with Fenton Reagent. *Technol. Chem. Ind.* **2009**, *38*, 48–50.
46. Zhai, J.; Jiang, C.H. Synthesis of Calcium Peroxide Microparticles and Its Application in Glyphosate Wastewater Pretreatment. *Adv. Mater. Res.* **2014**, *881–883*, 1139–1143. [CrossRef]
47. Balci, B.; Oturan, M.A.; Oturan, N.; Sires, I. Decontamination of Aqueous Glyphosate, (Aminomethyl) Phosphonic Acid, and Glufosinate Solutions by Electro-Fenton-Iike Process with Mn^{2+} as the Catalyst. *J. Agric. Food Chem.* **2009**, *57*, 4888–4894. [CrossRef] [PubMed]
48. Wang, H.; Zhao, Y.; Li, T.; Chen, Z.; Wang, Y.; Qin, C. Properties of Calcium Peroxide for Release of Hydrogen Peroxide and Oxygen: A Kinetics Study. *Chem. Eng. J.* **2016**, *303*, 450–457. [CrossRef]
49. Jönsson, J.; Camm, R.; Hall, T. Removal and Degradation of Glyphosate in Water Treatment: A Review. *J. Water Supply Res. Technol. AQUA* **2013**, *62*, 395–408. [CrossRef]

50. De, W.C.; Cui, Q.I.; Zhao, L.I. Evaluation of Calcium Peroxide as a New Water Purifying and Oxygen Releasing Agent in Aquaculture. *Life Sci. Res.* **2012**. [CrossRef]
51. Andreozzi, R.; Caprio, V.; Insola, A.; Marotta, R. Advanced Oxidation Processes (AOP) for Water Purification and Recovery. *Catal. Today* **1999**, *53*, 51–59. [CrossRef]

Disclaimer/Publisher's Note: The statements, opinions and data contained in all publications are solely those of the individual author(s) and contributor(s) and not of MDPI and/or the editor(s). MDPI and/or the editor(s) disclaim responsibility for any injury to people or property resulting from any ideas, methods, instructions or products referred to in the content.

Article

Removal of Sulfide Ions from Kraft Washing Effluents by Photocatalysis with N and Fe Codoped Carbon Dots

Hao Luo [1], Hao Liu [1,2,*] and Chengwu Sun [2]

[1] State Key Laboratory of Pulp and Paper Engineering, South China University of Technology (SCUT), Guangzhou 510640, China
[2] Bengbu-SCUT Research Center for Advanced Manufacturing of Biomaterials, Bengbu 233010, China
* Correspondence: feliuh@scut.edu.cn; Tel.: +86-20-2223-6028

Abstract: N and Fe codoped carbon dots (N,Fe-CDs) were fabricated from citric acid, L-glutamic acid and ferric chloride via a hydrothermal method for the photocatalytic removal of S^{2-} from kraft washing effluents (KWE). The N,Fe-CDs were fluorescent nanoparticles (average size of 3.18 nm) and catalyzed the oxidation of S^{2-} following a first-order kinetic model with an activation energy of 33.77 kJ/mol. The N,Fe-CDs tolerated elevated temperatures as high as 80 °C without catalyst deactivation. The N,Fe-CDs catalysts were reusable for at least four cycles, preserving over 90% of the activity. In the treatment of KWE from the kraft pulping of eucalyptus, the concentration of S^{2-} was decreased by the N,Fe-CDs from 1.19 to 0.41 mmol/L in 6 h. Consequently, near complete remediation was obtained in 24 h. In addition, half of the chemical oxygen demand was removed after treatment with 500 mg/L of the N,Fe-CDs. In addition, the present photocatalyst was safe within a concentration of 200 mg/L, as indicated by the acetylcholinesterase inhibition test. Our findings may help develop a cleaner production process for kraft brownstock washing.

Keywords: N and Fe codoped carbon dots (N,Fe-CDs); sulfide ions (S^{2-}); photocatalysis; kraft washing effluents (KWE)

Citation: Luo, H.; Liu, H.; Sun, C. Removal of Sulfide Ions from Kraft Washing Effluents by Photocatalysis with N and Fe Codoped Carbon Dots. *Polymers* **2023**, *15*, 679. https://doi.org/10.3390/polym15030679

Academic Editors: Marta Otero and Ricardo N. Coimbra

Received: 22 December 2022
Revised: 21 January 2023
Accepted: 25 January 2023
Published: 29 January 2023

Copyright: © 2023 by the authors. Licensee MDPI, Basel, Switzerland. This article is an open access article distributed under the terms and conditions of the Creative Commons Attribution (CC BY) license (https:// creativecommons.org/licenses/by/ 4.0/).

1. Introduction

Kraft cooking is currently the dominant chemical pulping method contributing to about 90% of the global production of brownstock pulp [1]. In both batch and continuous kraft processes, wood chips are delignified at a given H factor with white liquor, composed of NaOH and Na_2S, to reach a desired kappa number [2]. After cooking, brownstock slurry is delivered into a blow tank and subsequently sent to pulp washers [3]. About 90% of spent cooking liquor, or so-called black liquor, separated from the washing process is subjected to evaporators and boilers for the recovery of heat and chemicals [4]. The rest is primarily diluted with washing water and pumped to an effluent treatment system (ETS) [5]. In the kraft pulping of non-woody biomass, the extraction percentage of black liquor is usually lower than 80% [6], which means 20% of the spent chemicals enter the subsequent units. Sulfide ions (S^{2-}, or HS^- under alkaline pH) are one of the major hazardous chemicals in kraft washing effluents (KWE) [7]. They impact the ETS by declining the performance of activated sludge due to their high oxygen demand and strong biological toxicity [8]. There is urgent need to remediate S^{2-} prior to the active sludge stages.

A routine solution is to add microbial promoters, e.g., sulfur-oxidizing bacteria, together with a bio-energizer to restore the efficiency of activated sludge to a normal level [9]. However, this approach usually takes a few days, not to mention the significant increase in operation costs. Conventional chemical sulfide oxidation employs ClO_2 or Fenton's reagents [10,11]. The former technology requires additional equipment for the generation of ClO_2 [10]. The latter requires acidic conditions and generates a considerable amount of ferric sludge [11]. It is desired to develop biosourced recyclable catalysts for the green remediation of S^{2-} in KWE.

Carbon dots (CDs) are a new type of photocatalysts with a small particle size, high chemical stability and excellent catalytic performance [12]. CDs have been widely studied in pollution control and waste valorization since they were first synthesized in 2004 [13]. In general, CDs particles consist of carbon cores with outer layers rich in functional groups (-COOH, -OH, -NH$_2$) as well as codoped metal and/or non-metal atoms if designed as such [14]. Light irradiation could effectively excite the catalytic activity of CDs by promoting the migration of electrons, which finally accelerates the degradation of pollutants [15]. However, to date, CDs have not been studied for sulfide removal from industrial effluents. Instead, there have been many reports on the quantification of S^{2-} and sulfide compounds using CDs as detection probes [16–18]. For instance, CD/AgI hybrid probes were fabricated using citric acid as a precursor, enabling the measurement of the S^{2-} concentration in a range of 1~10 μM. A minimum detection limit of 0.48 μM was obtained suggesting a very high sensitivity [19]. In another report, dual-emission fluorescent probes were prepared by doping copper nanoclusters in CDs for the determination of trace sulfides with an accuracy of 1 ppm [20]. Novel stable CD/MnO$_2$ nanosheets were developed by Liu et al., which could serve as a turn-on fluorescent sulfide probe [21].

The objective of this work is to fabricate N,Fe-CDs specifically for the removal of S^{2-} from KWE. The doping of Fe aims to enhance the local electron density of the material matrix and promote intermolecular electron transfer [22]. Doping of N facilitates the entry of photogenerated electrons into the conduction band [23]. To the best of our knowledge, this article provides the first detailed assessment of CDs for sulfide remediation in pulping effluents. The catalytic performance of the N,Fe-CDs was evaluated. A preliminary study of safety was performed as well.

2. Materials and Methods

2.1. Chemicals

Elemental iodine, sodium thiosulfate, ferric chloride, soluble starch, L-glutamic acid and citric acid were purchased from Macklin Chemical Reagent Co., Ltd. (Shanghai, China). Other chemicals were of analytical grade and were supplied by Guangzhou Chemical Reagent Factory (Guangzhou, China).

2.2. Synthesis of N,Fe-CDs

The fabrication of N,Fe-CDs is illustrated in Figure 1. A total of 2.0 g of citric acid, 1.5 g of L-glutamic acid and 1.0 g of ferric chloride were dissolved sequentially in 70 mL of distilled water at 25 °C. The solution was transferred to a JGF-100 high pressure reactor (Jiean, Shanghai, China) in which a hydrothermal reaction was carried out at 200 °C for 4 h. Then, the reaction mixtures were cooled and centrifuged (5500 rpm, 25 °C) for 20 min to remove precipitates. The supernatant was vacuum-filtered through a 0.22 μm membrane. It was finally dialyzed against distilled water in a cellulose acetate dialysis tubing (MWCO 500 Da) for 48 h.

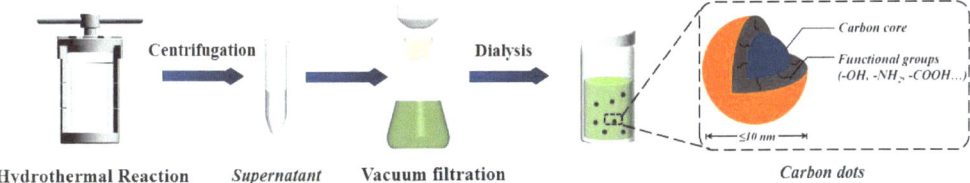

Figure 1. Fabrication diagram of N,Fe-CDs.

2.3. Kraft Washing Effluent (KWE)

Eucalyptus wood chips were cooked in a horizontal rotary digester (KRK 2611, Japan) under the following conditions: effective alkali charge of 30% (on basis of NaOH), sulfidity of 15%, solid-to-liquid ratio of 1:4, cooking temperature of 170 °C and total H factor of

1555. After kraft cooking, the spent black liquor was separated with a 200 mesh nylon sieve bag. The solids were subsequently washed with 2 L of deionized water filtered through the nylon bag to produce the kraft washing effluent (KWE). The pH of the KWE was 12.7 ± 0.1, and the concentration of S^{2-} was 1.19 ± 0.10 mmol/L.

2.4. Characterization of N,Fe-CDs

UV-Vis spectra of the N,Fe-CDs were recorded using a UV-Vis 1900 spectrophotometer (Shimadzu, Kyoto, Japan). Fluorescence emission and excitation spectra were taken on a FluoroMax-4 fluorescence spectrometer (Horiba, Kyoto, Japan). The excitation wavelengths were 280, 300, 320, 340, 360, 380, 400 and 420 nm. Both excitation and emission slits were 5 nm. The N,Fe-CDs were also characterized using FT-IR spectroscopy on a Tensor spectrometer (Bruker Optics, Karlsruhe, Germany). All dried samples were diluted with KBr and pressed into tablets. An X-ray photoelectron spectroscopic (XPS) study was performed on a Kratos AXIS Ultra DLD spectrometer (Shimadzu, Tokyo, Japan).

The particle size distribution of the N,Fe-CDs was measured using a Nano Zetasizer (Malvern, Malvern, UK). For the transmission electron microscopy (TEM) examination, the N,Fe-CDs were deposited on a piece of carbon-coated copper grid and air-dried. The morphology was scanned on a JEM-2100F microscope (JEOL, Tokyo, Japan). Tapping mode atomic force microscopy of the N,Fe-CDs was carried out on a Multimode-8 scanning probe microscope (Bruker, Karlsruhe, Germany). Samples were prepared by depositing dilute N,Fe-CDs solution on fresh mica, followed with drying using N_2. All data were processed with the Nanoscope analysis software.

2.5. Catalytic Oxidation of S^{2-} by N,Fe-CDs

Na_2S solutions (30~90 mg/L) were prepared using dissolved oxygen-free water to minimize the auto-oxidation of S^{2-}. A total of 20~80 mg of the N,Fe-CDs was added into 200 mL of Na_2S solution in a photoreaction vessel at a certain temperature (20~80 °C) and under light irradiation. To record the time-dependent catalytic oxidation of S^{2-} by the N,Fe-CDs, the reaction mixtures were pumped from the photoreaction vessel into a 10 mm quartz flow cuvette (Figure S1). Real-time variation curves of absorbance values at λ_{230nm} (S^{2-}) and λ_{215nm} ($S_2O_3^{2-}$) were recorded and denoted as A_{230} and A_{215}. Molar concentrations of S^{2-} and $S_2O_3^{2-}$ were derived from A_{230} and A_{215} according to Lambert–Beer law. Their respective molar absorption coefficients were 2.28×10^3 and 2.22×10^3 L/mol/cm. Reaction rate constants (k) were obtained by simulating the kinetics data measured under varied temperature and light conditions, following Equation (1).

$$\ln C_0 - \ln C_A = kt \quad (1)$$

where C_0 and C_A were concentrations of S^{2-} at the starting time and time t, respectively.

For evaluation of the recyclability of catalysts, 60 mg of N,Fe-CDs were added to the reaction vessel with 200 mL of the Na_2S solution (50 mg/L). The reaction was maintained at 25 °C for 2 h, monitored with the flowing spectroscopic method as described above. The reactants were dialyzed against water in a cellulose acetate dialysis tubing at the end of first run reaction and subsequently subjected to 200 mL of fresh Na_2S solution (50 mg/L) for the next run test. A total of six run cycles were tested. The concentration of S^{2-} was determined every 6 min in each cycle.

2.6. Catalytic Oxidation of S^{2-} in KWE by N,Fe-CDs

The applicability of photocatalytic oxidation was studied by mixing 10 mg of N,Fe-CDs in 200 mL of the KWE. The treatment was carried out at 25 °C, at 2000 Lux (white LED), for 12 h. Water was used as reference. The concentrations of S^{2-} and $S_2O_3^{2-}$ were measured using the titration method because the deep-colored KWE could not be precisely quantified through the direct measurement of UV absorbance [24]. In flask A, 100 mL of the KWE after treatment with N,Fe-CDs was mixed with 50 mL of I_2 standard solution

(0.1 mol/L) and one drop of soluble starch solution (1 wt%). The mixture was titrated with a standard Na$_2$S$_2$O$_3$ solution (0.05 mol/L) until the blue color disappeared. In flask B, 100 mL of the KWE which had undergone treatment with N,Fe-CDs was precipitated by ZnCl$_2$. The reactants were centrifuged at 5500 rpm for 10 min to remove solids. The supernatant obtained was mixed with one drop of starch indicator (0.5 wt%). Standard I$_2$ solution (0.05 mol/L) was used to titrate the mixture solution until it turned blue.. The concentration of S$_2$O$_3^{2-}$ and S^{2-} were deduced from Equations (2) and (3), respectively.

$$S^{2-} + I_2 \rightarrow 2I^- + S$$

$$S_2O_3^{2-} + 4I_2 + 5H_2O \rightarrow 2SO_4^{2-} + 8I^- + 10H^+$$

$$\text{In flask B, } C\left(S_2O_3^{2-}\right) = \frac{C_B(I_2) \times V_B(I_2)}{4V_w} \tag{2}$$

where, C_B (I$_2$) was 0.05 mol/L; V_B (I$_2$) was the titration volume, L; and V_w, the volume of ZnCl$_2$-treated KWE, was 0.1 L.

$$\text{In flask A, } C\left(S^{2-}\right) = \frac{C(I_2) \times V(I_2) - 4C_A\left(S_2O_3^{2-}\right) \times V_A\left(S_2O_3^{2-}\right) - C_B(I_2) \times V_B(I_2)}{V_w} \tag{3}$$

where C_A(S$_2$O$_3^{2-}$) was 0.05 mol/L; V_A(S$_2$O$_3^{2-}$) was the titration volume, L; and V_w was the KWE volume, L.

The determination of the SO$_4^{2-}$ ions was performed by precipitating the KWE with BaCl$_2$ solution followed by the collection, drying and weighing of the generated BaSO$_4$. The concentration of SO$_4^{2-}$ was calculated according to Equation (4).

$$C\left(SO_4^{2-}\right) = \frac{m(BaSO_4)}{233.39 \times V_w} \tag{4}$$

where m represents mass, g; and 223.39 is the relative molecular mass of BaSO$_4$, g/mol.

The chemical oxygen demand (COD) of the KWE was measured using a COD determination kit (Huaikai, Guangzhou, China) on the basis of the potassium dichromate method [25]. Potassium hydrogen phthalate was used as the standard for COD determination. The absorbance at 420 nm (A$_{420}$) was measured and the COD was calculated according to Equation (5).

$$A_{420} = -0.0033 \times COD + 0.5853, \ R^2 = 0.9983, \ n = 7 \tag{5}$$

2.7. Safety Assessment of N,Fe-CDs

Acetylcholinesterase (AchE) inhibition by the N,Fe-CDs was quantitatively determined using a commercial assay kit (HPER Scientific Instrument Co., Ltd., Guangzhou, China) [26]. In a 4-mL cuvette, 0.14 mL of phosphate buffer (20 mM, pH 7.6), 0.05 mL of acetylthiocholine iodide (5 mM) and 0.01 mL of the N,Fe-CDs solution (water as reference) were pipetted sequentially. The mixture was incubated at 38 °C for 10 min. At the end of reaction, 1.8 mL of 5,5'-dithiobis-2-nitrobenzoic acid/phosphate ethanol reagent was added. The change in absorbance at 412 nm in 3 min (ΔA_{412}) was recorded from which the relative activity of AchE (RA, %) was reported according to Equation (6).

$$RA = \frac{\Delta A_{412,Sample}}{\Delta A_{412,Water}} \times 100\% \tag{6}$$

3. Results

3.1. Characterization of N,Fe-CDs

The prepared N,Fe-CDs solution was apparently brownish-yellow colored having an absorption spectrum covering the UV region (Figure 2a). No specific peak at around 260 nm was found, suggesting Fe^{3+} was not abundantly bound or perhaps in a chelated state. Fluorescence spectra show the N,Fe-CDs were more readily excited by long wavelength UV lights (340~420 nm) than short ones (280~340 nm) to produce fluorescence. A maximum emission wavelength (E_m) of 410 nm was obtained when the excitation wavelength (E_x) was 340 nm (Figure 2b). The previously reported N,Fe-CDs by Wu et al. [27] were derived from ethanediamine and citric acid and had an E_m at 450 nm after being excited by 360 nm UV light. Their longer wavelength fluorescence was probably due to the structure of the precursors. On the other hand, fluorescence spectral examinations show the present N,Fe-CDs did not precipitate S^{2-} through the formation of Fe_2S_3 because the excited fluorescence produced by the N,Fe-CDs was not influenced by the presence of S^{2-} (0.05~10 mmol/L, Figure S2).

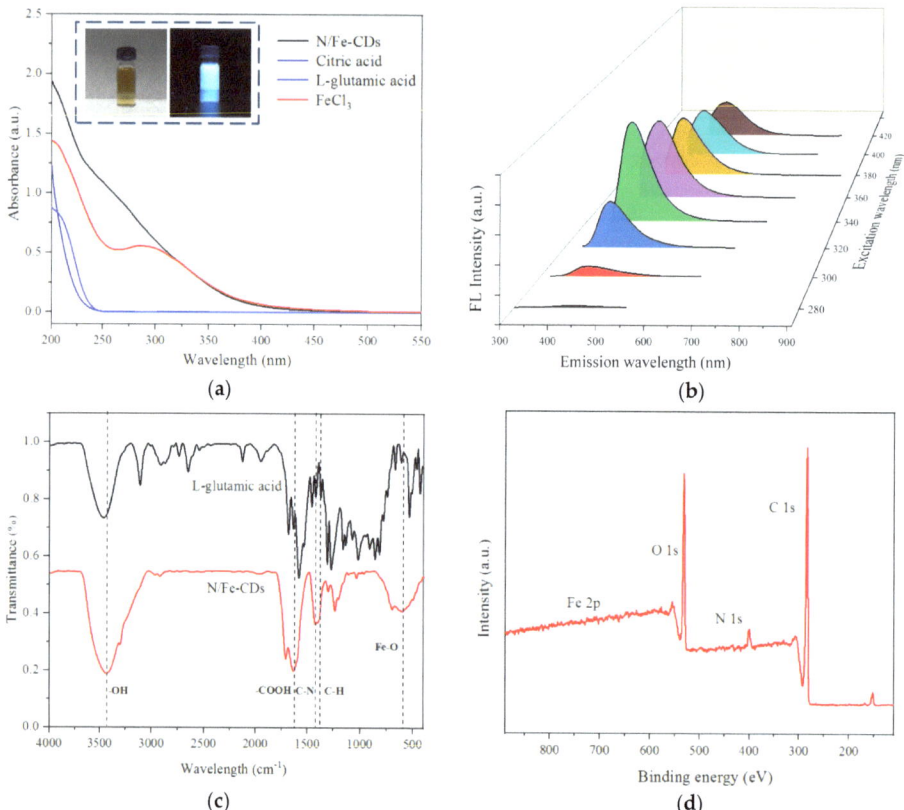

Figure 2. Spectral characteristics of N,Fe-CDs, (**a**) UV-Vis spectra, (**b**) Fluorescence spectra, (**c**) FI-IR spectra, and (**d**) XPS spectra.

FT-IR spectra showed the N,Fe-CDs had strong transmittance peaks at 1750, 3460 and 1346 cm^{-1} (Figure 2c), corresponding to $-COOH$, $-OH$ and $C-H$ stretching vibrations, respectively. Bands at around 640 cm^{-1} corresponded to the stretching vibration of Fe$-$O [28]. The peaks at 1452 cm^{-1} and 1161 cm^{-1} were attributed to the N$-$H bending vibration and C$-$N stretching vibration, respectively. Both were present in L-glutamic acid but disappeared in the N,Fe-CDs, showing the amino groups were destroyed in the hydrothermal reaction.

The XPS spectra verified the incorporation of N and Fe in the N,Fe-CDs (Figure 2d). The contents of C, O, N and Fe were 80.5%, 15.7%, 3.8% and 0.1%, respectively. Peak fitting of C spectra resulted in three peaks attributed to C−C (284.7 eV), C−OR (285.8 eV) and C=O (287.7 eV, Figure S3). N elements mostly exist in (398.2 eV) a pyridine-like structure, consistent with the FT-IR results. The binding amount of Fe was quite low. The XPS peaks of Fe included Fe 2p 3/2 (706.75 eV), Fe 2p1/2 (707.91 eV), Fe 2p3/2 (710.96 eV) and Fe 2p1/2 (711.55 eV), respectively (Figure S3). Doping of Fe ions could promote the local electron density of nanoparticles and thereby enhance the intermolecular electron transfer for catalytic activity [29].

TEM shows the N,Fe-CDs granular nanoparticles dispersed uniformly in aqueous solution without obvious agglomeration (Figure 3a). The particle size of the N,Fe-CDs was 1.5~2.5 nm according to the TEM images which was smaller than the average data, 3.18 nm, measured by zetasizer (Figure 3b). Additional information provided by zetasizer was that all the N,Fe-CDs had size dimensions within a range of 2~4.5 nm. Those particles with size of 2.7~3.65 nm accounted for 65% of the total. AFM offered 3D images of the N,Fe-CDs deposited on mica and the average vertical size of the N,Fe-CDs was 2.3 nm. Data from microscopic measurements lower than the results by zetasizer could be explained by the drying treatment in sample preparation. Our N,Fe-CDs have an average particle size close to the amino-acid-based CDs (3.19 nm) prepared by Pandit et al. [30].

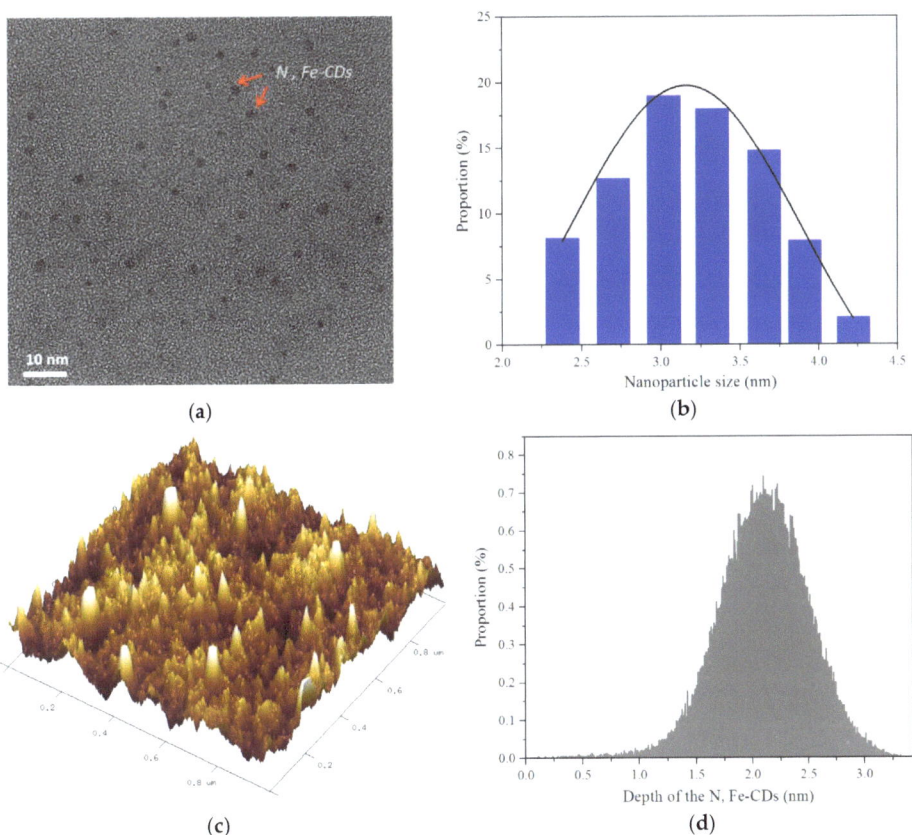

Figure 3. Morphology and size distribution of N,Fe-CDs, (**a**) TEM morphology, (**b**) Size distribution by zetasizer, (**c**) 3D AFM morphology, and (**d**) Size distribution by AFM.

3.2. Catalytic Properties of N,Fe-CDs for Oxidation of S^{2-}

The catalytic oxidation of S^{2-} by the N,Fe-CDs was kinetically measured under varied conditions (Figure 4). S^{2-} was stable in the absence of dissolved oxygen (DO). In the presence of 6 mg/L of DO, S^{2-} was oxidized slowly. Its content was decreased by 19.4% in 2 h. The N,Fe-CDs accelerated the oxidation even in darkness, eliminating 55.4% of S^{2-} in 2 h. Under white LED irradiation, the N,Fe-CDs were activated and remediated 67.8% of S^{2-} (Figure 4a). Obviously efficient catalysis by the N,Fe-CDs was dependent on O_2 and light. In a previous report, N,Fe-CDs derived from diethylenetriamine, pentaacetic acid and Fe(NO)$_3$ showed peroxidase-like properties, depending on the participation of H_2O_2 for the formation of reactive oxygen species [31]. The present catalyst did not require H_2O_2 but utilized O_2 for S^{2-} oxidation.

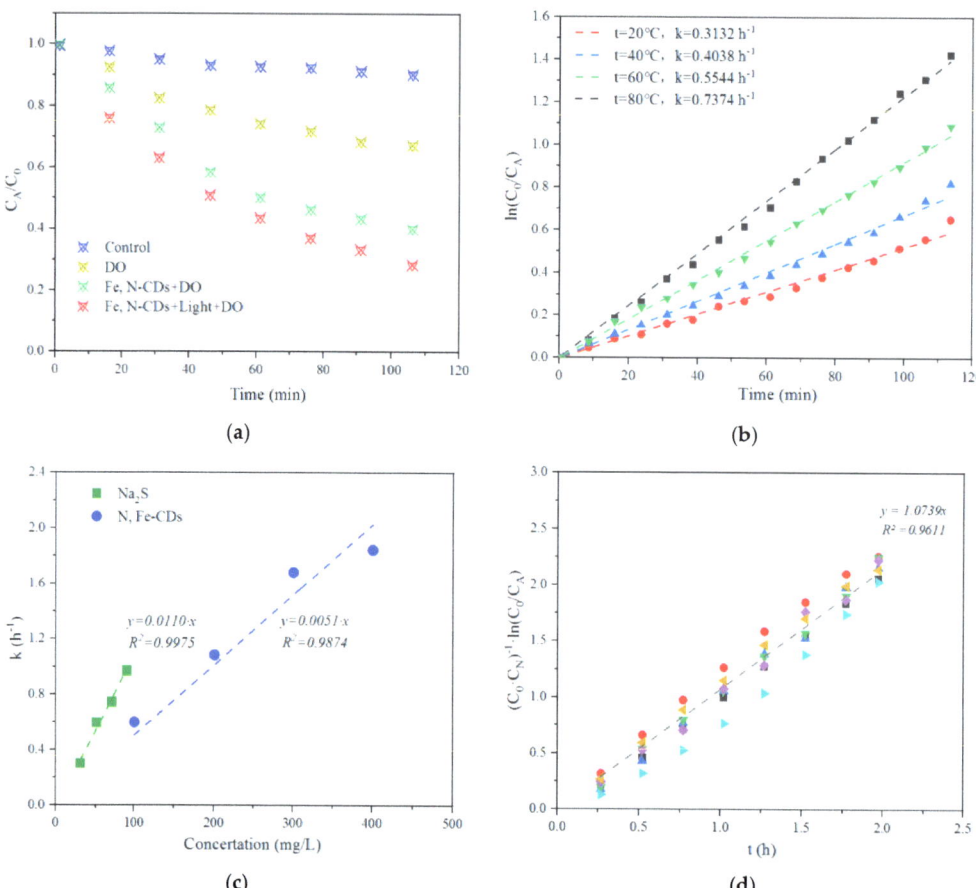

Figure 4. Kinetics of S^{2-} oxidation catalyzed by Fe,N-CDs. (**a**) Time-dependent oxidation of S^{2-} by 100 mg/L Fe,N-CDs with light and DO at 25 °C. (**b**) Kinetic curves at varied temperatures. (**c**) Correlation of k with concentrations of S^{2-} and N,Fe-CDs. (**d**) Fitting of data in Figure 4c with Equation (9).

Kinetic curves of S^{2-} oxidation catalyzed by 100 mg/L of the N,Fe-CDs at varied temperatures are shown in Figure 4b. All curves could be fitted well with the first-order kinetic model (all $R^2 > 0.97$). The values of constant k were increased with elevation of temperatures, i.e., from 20 to 80 °C. A total of 75.6% of S^{2-} was oxidized at 80 °C after 120 min. Cui et al. have developed CDs doped with N and S for the probing of Fe^{3+}. They found 25 °C was a preferred working temperature, and fluorescence intensity decreased linearly with temperatures [32]. The present N,Fe-CDs are more active at elevated temperatures, which matches well the brownstock washing conditions, optimally 62~68 °C in a pulp mill [33]. On the other hand, according to the Arrhenius equation (Equation (7)), the activation energy (E_a) of the process was calculated to be 33.77 kJ/mol.

$$k = Ae^{\frac{-E_a}{RT}} \rightarrow \ln\frac{k_2}{k_1} = \frac{-E_a}{R}\left(\frac{1}{T_2} - \frac{1}{T_1}\right) \tag{7}$$

where E_a is the activation energy, kJ/mol; k is the apparent reaction constant, 1/h; A is the Arrhenius constant, 1/h; T is temperature, K; and R is the gas constant, 8.314 kJ/mol/K.

The constant k was linearly correlated with the concentrations of S^{2-} (C_0) and N,Fe-CDs (C_N) according to the data measured under varied C_0 and C_N conditions (Figure 4c). Therefore, Equation (1) could be rewritten as Equations (8) and (9). Data in Figure 4c were re-arranged by graphing $\ln(C_0/C_A)/(C_0C_N)$ vs. t according to Equation (9). The linear relationship ($R^2 = 0.9611$) was valid as shown in Figure 4d.

$$\ln\frac{C_0}{C_A} = C_0 \cdot C_N \cdot k't \tag{8}$$

$$\frac{1}{C_0 \cdot C_N} \ln\frac{C_0}{C_A} = k't \tag{9}$$

$$\text{where, } k = C_0 C_N k' \tag{10}$$

Recyclability is a crucial parameter of a catalyst for practical application in industrial processes. The recycling of the N,Fe-CDs was performed by encapsulating the suspension of the catalysts in a dialysis tubing. In the first run, the tubing containing 300 mg/L of the N,Fe-CDs was immersed with fresh Na_2S solution. Catalytic oxidation was carried out under conditions of DO of 6 mg/L, an LED light density of 2000 lux, a temperature of 60 °C and duration time of 120 min. The spent N,Fe-CDs in dialysis tubing were re-immersed in fresh Na_2S solution in the next run. Results in Figure 5 show that a high photocatalytic activity towards S^{2-} (removal ratio > 90%) was maintained after being recycled four times. The catalytic activity was not changed in the first three cycles. In the fourth run, the activity of the N,Fe-CDs decreased slightly but still removed over 92% of the S^{2-}. The efficiency of catalysis with the N,Fe-CDs decreased considerably in the fifth cycle, giving a S^{2-} removal of only 66% (Figure 5). The loss of activity of the N,Fe-CDs is possibly because of the S^{2-} oxidation products, e.g., elemental S, adsorbed on surface, reducing the accessibility of the catalyst to free S^{2-}. Therefore, dialysis against water for the desorption of S is necessary for the multiple recovery of the catalyst. Similar to conventional photocatalysts such as TiO_2, N,Fe-CDs can be recycled more than four times, suggesting its good practical performance as a photocatalyst for industrial applications [34]. Good recyclability of CDs has been reported earlier by Hu et al. In their paper, the CDs catalysts also showed outstanding recyclability, being used for six cycles without a substantial loss in catalytic activity [35].

3.3. Remediation of KWE by N,Fe-CDs

Soluble S-containing inorganic compounds in the KWE were mainly S^{2-} (including HS^-), SO_4^{2-} and $S_2O_3^{2-}$, according to the measurement data (Figure 6). The dissociation of S^{2-} into HS^- is majorly because of the strong alkaline pH (12.7) in the KWE. No redox reaction occurs, only dissociation. Both S^{2-} and HS^- could be fairly oxidized with our catalyst and quantified by titration analysis. Thereafter, they were taken into calculation together to simplify stoichiometry. After treatment with the N,Fe-CDs for 12 h, the concentration of S^{2-} in the KWE (pH 12.7) decreased from 1.19 to 0.04 mmol/L. The concentration of $S_2O_3^{2-}$ increased from 2.09 to 2.63 mmol/L (Figure 6a). In other words, 93.9% of S^{2-} was oxidized to become $S_2O_3^{2-}$. Interestingly, the concentration of SO_4^{2-}

increased from 0.56 to 0.84 mmol/L. In terms of total sulfur concentration, there was no difference before and after the catalytic oxidation with the N,Fe-CDs. This suggests no other form of S, such as elemental sulfur, was considerably generated in the catalytic treatment.

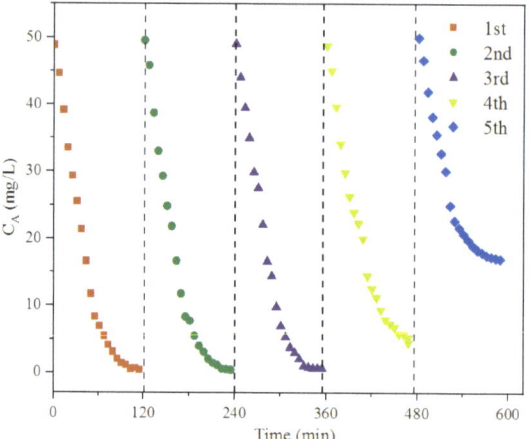

Figure 5. Recyclability of N,Fe-CDs for remediation of S^{2-}.

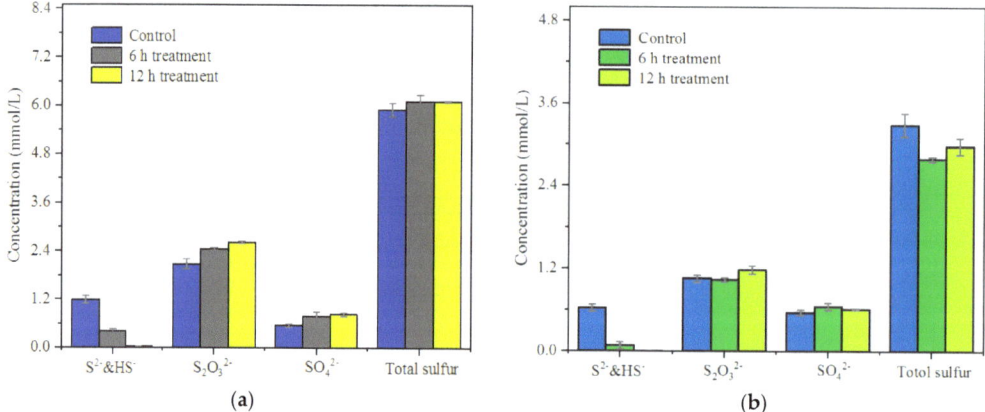

Figure 6. Remediation of S^{2-} in KWE by N,Fe-CDs, (**a**) Original KWE, pH 12.7 and (**b**) Neutralized KWE, pH 7.

The removal of S^{2-} in the neutralized KWE by the N,Fe-CDs is demonstrated in Figure 6b. About half of the S^{2-} in original the KWE was lost after neutralization and the rest was still dissolved in the liquor. Photocatalytic treatment with the N,Fe-CDs removed almost all the free S^{2-} in the neutralized KWE which was decreased from 0.62 to 0.004 mmol/L. The content of $S_2O_3^{2-}$ was increased from 1.05 to 1.18 mmol/L. The concentration of SO_4^{2-} kept relatively stable and total sulfur was slightly reduced by 0.31 mmol/L. There was 42% of converted S^{2-} not detected. A major reason is that the H_2S generated from the neutralization of the KWE slowly volatilized into the air phase. Another possible explanation is that a part of the S^{2-} was converted into elemental sulfur but not detected in the work. In a pulp mill, there is no pH adjustment in the pulp washing process. It is suggested to perform the catalytic oxidation in the KWE storing tank. The catalyst is also applicable in ETS. The risks of releasing H_2S or S due to neutralization still need further investigation.

The effect of treatment with N,Fe-CDs on the COD of KWE is illustrated in Figure 7. The addition of the N,Fe-CDs up to 500 mg/L did not contribute to the COD. It suggests that citric acid and L-glutamic acid have undergone sufficient oxidation during the hydrothermal fabrication process. After treatment with 500 mg/L N,Fe-CDs for 12 h, the total COD of the KWE was decreased by about 50%. Obviously, there are other reductive substances in the KWE, in addition to S^{2-} and $S_2O_3^{2-}$, that could be oxidized by the N,Fe-CDs.

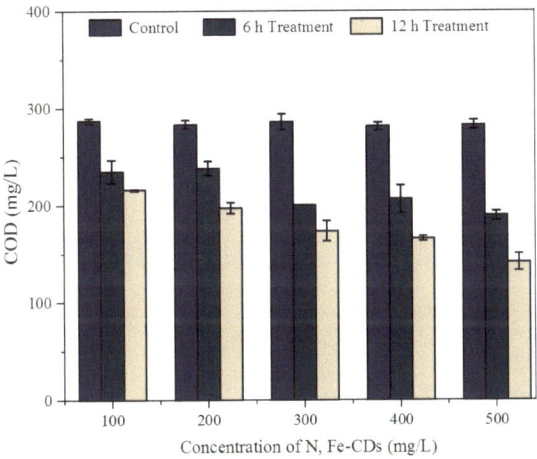

Figure 7. Effects of treatment with N,Fe-CDs on COD of KWE.

Previously, Selvaraj et al. [36] have reported that electro oxidation process aided with UV-light remediated 100% of sulfide, 92% of COD and 70% of the total organic carbon in tannery effluents. Since N,Fe-CDs are charged nanoparticles, coupling with electro oxidation may further enhance the catalytic capacity of N,Fe-CDs. We and Ye [37] fabricated cyclometalated Ir–Zr metal–organic frameworks as a recyclable visible-light photocatalyst. This novel catalyst could convert sulfide into sulfoxide in water. It inspires us that well-designed catalysts enable direct conversion of sulfide into value-added products through photocatalysis.

It is worth noting that advanced analytical tools such as liquid chromatography-mass spectrometry (LC-MS) are powerful tools for examining the intermediates generated from the photodegradation process [38]. The characterization of S-containing intermediates may help better understand the mechanism of sulfide oxidation by N,Fe-CDs.

3.4. Safety Assessment

AchE is a key enzyme in the animal nervous system. The inhibition of AchE is often measured for the determination of the potential neurotoxicity of drinking water after being contaminated with pesticides, detergents and other bioactive chemicals [39]. The discharging of N,Fe-CDs may impact the safety of a water ecosystem. Therefore, we detected the inhibitory effects of N,Fe-CDs on AchE to evaluate their toxicity. Data in Figure 8 show, in presence of less than 300 mg/L of the N,Fe-CDs, over 90% of AchE activity was preserved. Even when the concentration of the N,Fe-CDs reached 500 mg/L, the loss of AchE activity was around 22%, much lower than 50%, a toxic threshold level. Obviously, our N,Fe-CDs catalyst is safe according to the above assessment. Previously, Suner et al. reported that CDs based on citric acid and arginine had no effect on the activity of AchE at concentrations below 1000 mg/L [40]. However, doping of Ag and Cu in CDs presented strong inhibition. The activity of AchE was decreased by over 60% in the presence of 500 mg/L of Arg-Ag-CDs [40]. By comparison with our data in Figure 8, it can be concluded that Fe-doped CDs have better biosafety.

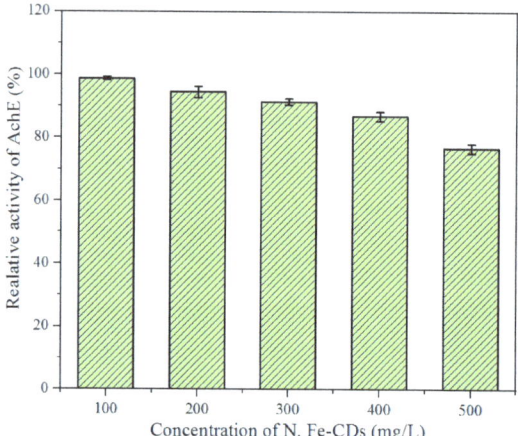

Figure 8. Effect of N,Fe-CDs on the activity of AchE.

4. Conclusions

The N,Fe-CDs presented in this work were fluorescent nanoparticles (average size 3.18 nm) and had a maximum emission wavelength of 410 nm when excited at 340 nm. The catalytic oxidation of S^{2-} by the N,Fe-CDs followed the first-order kinetic model with an activation energy of 33.77 kJ/mol. The N,Fe-CDs had a good recyclability. A high photocatalytic activity (removal ratio > 90%) towards S^{2-} was maintained after being recycled four times. N,Fe-CDs could efficiently remove S^{2-} in kraft washing effluent (KWE), either in original alkaline or neutralized solutions. After the catalytic remediation of S^{2-}, the COD of the KWE was also decreased. The N,Fe-CDs were safe below a content of 300 mg/L according to acetylcholinesterase inhibition tests. The use of N,Fe-CDs may help develop cleaner production processes for kraft brownstock washing. Since odorous S-containing organic compounds such as CH_3SCH_3 or CH_3SSCH_3 are generated during kraft pulping, the N,Fe-CDs could be further studied to catalytically remove these harmful compounds.

Supplementary Materials: The following supporting information can be downloaded at: https://www.mdpi.com/article/10.3390/polym15030679/s1, Figure S1: Automatic sampling and measurement device for photocatalytic oxidation of S^{2-}; Figure S2: Effects of S^{2-} on fluorescence intensity of N,Fe-CDs; Figure S3: High-resolution XPS spectra of N,Fe-CDs.

Author Contributions: Conceptualization, H.L. (Hao Liu); methodology, H.L. (Hao Luo); software, H.L. (Hao Luo); validation, H.L. (Hao Liu) and C.S.; formal analysis, H.L. (Hao Luo), H.L. (Hao Liu) and C.S.; investigation, H.L. (Hao Luo); resources, H.L. (Hao Luo) and C.S.; data curation, H.L. (Hao Luo) and C.S.; writing—original draft preparation, H.L. (Hao Luo); writing—review and editing, H.L. (Hao Liu); visualization, H.L. (Hao Luo) and C.S.; supervision, H.L. (Hao Liu) and C.S.; project administration, H.L. (Hao Liu); funding acquisition, H.L. (Hao Liu). All authors have read and agreed to the published version of the manuscript.

Funding: This work was supported by grants from Fundamental Research Funds for the Central Universities (No. 2022ZYGXZR0073), the Opening Project of Guangxi Key Laboratory of Green Processing of Sugar Resources (No. GXTZYKF202209), Guangdong Natural Science Foundation (No. 2017A030310341) and Funding for Construction of Bengbu-SCUT Research Center for Advanced Manufacturing of Biomaterials (No. 20210190).

Institutional Review Board Statement: Not applicable.

Data Availability Statement: Not applicable.

Conflicts of Interest: The authors declare no conflict of interest.

References

1. Laftah, W.A.; Rahaman, W.A. Chemical pulping of waste pineapple leaves fiber for kraft paper production. *J. Mater. Res. Technol.* **2015**, *4*, 254–261. [CrossRef]
2. Rodríguez, A.; Moral, A.; Serrano, L.; Labidi, J.; Jiménez, L. Rice straw pulp obtained by using various methods. *Bioresour. Technol.* **2008**, *99*, 2881–2886. [CrossRef] [PubMed]
3. Maček, A. Research on combustion of black-liquor drops. *Prog. Energy. Combust. Sci.* **1999**, *25*, 275–304. [CrossRef]
4. Pola, L.; Collado, S.; Oulego, P.; Oulede, P.; Calvo, P.A.; Díaz, M. Characterisation of the wet oxidation of black liquor for its integration in Kraft paper mills. *Chem. Eng. J.* **2021**, *405*, 126610. [CrossRef]
5. Pizzichini, M.; Russo, C.; Di Meo, C. Purification of pulp and paper wastewater, with membrane technology, for water reuse in a closed loop. *Desalination* **2005**, *178*, 351–359. [CrossRef]
6. Wang, R.; Deplazes, R.; Vogel, F.; Baudouin, D. Continuous extraction of black liquor salts under hydrothermal conditions. *Ind. Eng. Chem. Res.* **2021**, *60*, 4072–4085. [CrossRef]
7. Xue, Y.; Ma, L.; Zhang, L.; Zhao, W.; Li, Z. A Green, rapid and efficient dual-sensors for highly selective and sensitive detection of cation (Hg^{2+}) and anion (S^{2-}) ions based on CMS/AgNPs composites. *Polymers* **2020**, *12*, 113. [CrossRef] [PubMed]
8. Puhakka, J.A.; Ferguson, J.F.; Benjamin, M.M.; Salkinoja-Salonen, M. Sulfur reduction and inhibition in anaerobic treatment of simulated forest industry wastewater. *Syst. Appl. Microbiol.* **1989**, *11*, 202–206. [CrossRef]
9. Haile, A.; Gelebo, G.G.; Tesfaye, T.; Mengie, W.; Mebrate, M.A.; Abuhay, A.; Limeneh, D.Y.P. Pulp and paper mill wastes: Utilizations and prospects for high value-added biomaterials. *Bioresour. Bioprocess.* **2021**, *8*, 35. [CrossRef]
10. Csekö, G.; Pan, C.; Gao, Q.; Horváth, A.K. Kinetics of the two-stage oxidation of sulfide by chlorine dioxide. *Inorg. Chem.* **2018**, *57*, 10189–10198. [CrossRef]
11. Meng, X.; Khoso, S.A.; Wu, J.; Tian, M.; Kang, J.; Liu, H.; Zhang, Q.; Sun, W.; Hu, Y. Efficient COD reduction from sulfide minerals processing wastewater using Fenton process. *Miner. Eng.* **2019**, *132*, 110–112. [CrossRef]
12. Hutton, G.A.M.; Martindale, B.C.M.; Reisner, E. Carbon dots as photosensitisers for solar-driven catalysis. *Chem. Soc. Rev.* **2017**, *46*, 6111–6123. [CrossRef] [PubMed]
13. Xu, X.; Ray, R.; Gu, Y.; Ploehn, H.J.; Gearheart, L.; Raker, K.; Scrivens, W.A. Electrophoretic analysis and purification of fluorescent single-walled carbon nanotube fragments. *J. Am. Chem. Soc.* **2004**, *126*, 12736–12737. [CrossRef] [PubMed]
14. Khayal, A.; Dawane, V.; Amin, M.A.; Tirth, V.; Yadav, V.K.; Algahtani, A.; Khan, S.H.; Islam, S.; Yadav, K.K.; Jeon, B. Advances in the methods for the synthesis of carbon dots and their emerging applications. *Polymers* **2021**, *13*, 3190. [CrossRef] [PubMed]
15. Jiang, X.; Shi, Y.; Liu, X.; Wang, M.; Song, P.; Xu, F.; Zhang, X. Synthesis of nitrogen-doped lignin/DES carbon quantum dots as a fluorescent probe for the detection of Fe^{3+} ions. *Polymers* **2018**, *10*, 1282. [CrossRef]
16. Zeng, Y.; Xu, Z.; Liu, X.; Cai, H.; Zhang, M.; Song, J.; Zeng, P.; Qu, D.; Guo, J.; Li, H. Novel N, F co-doped carbon dots to detect sulfide and cadmium ions with high selectivity and sensitivity based on a "turn-off-on" mechanism. *Dyes Pigm.* **2022**, *203*, 110379. [CrossRef]
17. Barati, A.; Shamsipur, M.; Abdollahi, H. Metal-ion-mediated fluorescent carbon dots for indirect detection of sulfide ions. *Sens. Actuators B Chem.* **2016**, *230*, 289–297. [CrossRef]
18. Ge, J.; Shen, Y.; Wang, W.; Li, Y.; Yang, Y. N-doped carbon dots for highly sensitive and selective sensing of copper ion and sulfide anion in lake water. *J. Environ. Chem. Eng.* **2021**, *9*, 105081. [CrossRef]
19. Abi, A.; Kazemi, G.; Safavi, A. A carbon dot-based fluorescence method for selective quantification of sulfide in environmental samples. *Sens. Actuators B Chem.* **2018**, *277*, 1–7. [CrossRef]
20. Wen, R.; Li, Q.H.; Sun, Y.; Li, Y.; Luo, X.; Zhao, X.; Gao, X.F. Synthesis optimization of rich-urea carbon-dots and application in the determination of H_2S in rich- and barren-liquids of desulphurizing solutions. *Analyst* **2021**, *146*, 7635–7644. [CrossRef]
21. Liu, J.; Liu, C.; Zhou, Z.; Wu, S.; Zhou, K.; Wei, W. Enzymatic determination of D-alanine using a cationic poly(fluorenylenephenylene) as the fluorescent probe and MnO_2 nanosheets as quenchers. *Mikrochim. Acta* **2019**, *186*, 1–7. [CrossRef] [PubMed]
22. Rafiq, A.; Imran, M.; Aqeel, M.; Naz, M.; Lkram, M.; Ali, S. Study of transition metal ion doped CdS nanoparticles for removal of dye from textile wastewater. *J. Inorg. Organomet. Polym. Mater.* **2020**, *30*, 1915–1923. [CrossRef]
23. Bhattacharyya, S.; Ehrat, F.; Urban, P.; Teves, R.; Wyrwich, R.; Döblinger, M.; Feldmann, J.; Urban, A.S.; Stolarczyk, J.K. Effect of nitrogen atom positioning on the trade-off between emissive and photocatalytic properties of carbon dots. *Nat. Commun.* **2017**, *8*, 1–9. [CrossRef] [PubMed]
24. Pawlak, Z.; Pawlak, A.S. Modification of iodometric determination of total and reactive sulfide in environmental samples. *Talanta* **1999**, *48*, 347–353. [CrossRef]
25. Dedkov, Y.M.; Elizarova, O.V.; Kelina, S.Y. Dichromate method for the determination of chemical oxygen demand. *J. Anal. Chem.* **2000**, *55*, 777–781. [CrossRef]
26. Ellman, G.L.; Courtney, K.D.; Andres, V.; Featherstone, R.M. A new and rapid colorimetric determination of acetylcholinesterase activity. *Biochem. Pharmacol.* **1961**, *7*, 88–95. [CrossRef] [PubMed]
27. Wu, Y.; Cao, L.; Zan, M.; Hou, Z.; Ge, M.; Dong, W.; Li, L. Iron and nitrogen-co-doped carbon quantum dots for the sensitive and selective detection of hematin and ferric ions and cell imaging. *Analyst* **2021**, *146*, 4954–4963. [CrossRef] [PubMed]
28. Ahmadian-Fard-Fini, S.; Ghanbari, D.; Amiri, O.; Salavati-Niasari, M. Green sonochemistry assisted synthesis of hollow magnetic and photoluminescent $MgFe_2O_4$-carbon dot nanocomposite as a sensor for toxic Ni (II), Cd (II) and Hg (II) ions and bacteria. *RSC Adv.* **2021**, *11*, 22805–22811. [CrossRef]

29. Li, K.; Sun, C.; Chen, Z.; Qu, X.; Xie, H.; Zhong, Q. Boron doped C_3N_5 for photocatalytic nitrogen fixation to ammonia: The key role of boron in nitrogen activation and mechanism. *Chem. Eng. J.* **2022**, *429*, 132440. [CrossRef]
30. Pandit, S.; Behera, P.; Sahoo, J.; De, M. In situ synthesis of amino acid functionalized carbon dots with tunable properties and their biological applications. *ACS Appl. Bio Mater.* **2019**, *2*, 3393–3403. [CrossRef]
31. Wang, L.; Liu, Y.; Yang, Z.; Wang, Y.; Rao, H.; Yue, G.; Wu, C.; Lu, C.; Wang, X. A ratiometric fluorescence and colorimetric dual-mode assay for H_2O_2 and xanthine based on Fe, N co-doped carbon dots. *Dyes Pigm.* **2020**, *180*, 108486. [CrossRef]
32. Cui, X.; Wang, Y.; Liu, J.; Yang, Q.; Zhang, B.; Cao, Y.; Wang, Y.; Lu, G. Dual functional N- and S-co-doped carbon dots as the sensor for temperature and Fe^{3+} ions. *Sens. Actuators B Chem.* **2017**, *242*, 1272–1280. [CrossRef]
33. Santos, R.B.; Hart, P.W. Brownstock washing—A review of the literature. *Tappi J.* **2014**, *13*, 9–19. [CrossRef]
34. Monda, K.; Sharm, A. Recent advances in electrospun metal-oxide nanofiber based interfaces for electrochemical biosensing. *RSC Adv.* **2016**, *6*, 83589–83612. [CrossRef]
35. Hu, Y.; Li, M.; Gao, Z.; Wang, L.; Zhang, J. Waste polyethylene terephthalate derived carbon dots for separable production of 5-hydroxymethylfurfural at low temperature. *Catal. Lett.* **2021**, *151*, 2436–2444. [CrossRef]
36. Ossola, R.; Tolu, J.; Clerc, B.; Erickson, P.R.; Winkel, L.H.E.; McNeill, K. Photochemical production of sulfate and methanesulfonic acid from dissolved organic sulfur. *Environ. Sci. Technol.* **2019**, *53*, 13191–13200. [CrossRef]
37. Wei, L.Q.; Ye, B.H. Cyclometalated Ir–Zr metal–organic frameworks as recyclable visible-light photocatalysts for sulfide oxidation into sulfoxide in water. *ACS Appl. Mater. Interfaces* **2019**, *11*, 41448–41457. [CrossRef]
38. Selvaraj, H.; Aravind, P.; George, H.S.; Sundaram, M. Removal of sulfide and recycling of recovered product from tannery lime wastewater using photoassisted-electrochemical oxidation process. *J. Ind. Eng. Chem.* **2020**, *83*, 164–172. [CrossRef]
39. Holth, T.F.; Tollefsen, K.E. Acetylcholine esterase inhibitors in effluents from oil production platforms in the North Sea. *Aquat. Toxicol.* **2012**, *112*, 92–98. [CrossRef]
40. Suner, S.S.; Sahiner, M.; Ayyala, R.S.; Bhethanabotla, V.R.; Sahiner, N. Versatile fluorescent carbon dots from citric acid and cysteine with antimicrobial, anti-biofilm, antioxidant, and AChE enzyme inhibition capabilities. *J. Fluoresc.* **2021**, *31*, 1705–1717. [CrossRef]

Disclaimer/Publisher's Note: The statements, opinions and data contained in all publications are solely those of the individual author(s) and contributor(s) and not of MDPI and/or the editor(s). MDPI and/or the editor(s) disclaim responsibility for any injury to people or property resulting from any ideas, methods, instructions or products referred to in the content.

Article

Palladium Nanoparticle-Loaded Mesostructural Natural Woods for Efficient Water Treatment

Zirun Wang, Chao Jia *, Hengxue Xiang * and Meifang Zhu

State Key Laboratory for Modification of Chemical Fibers and Polymer Materials, College of Materials Science and Engineering, Donghua University, Shanghai 201620, China
* Correspondence: jiachao0806@dhu.edu.cn (C.J.); hengxuexiang@dhu.edu.cn (H.X.)

Abstract: Natural wood with oriented microchannels and unique multi-level structures is an ideal candidate for making water treatment membranes. Here, palladium nanoparticles are loaded into different kinds of natural woods and the degradation property of the wood membranes for organic pollutants are investigated. The water flux of hardwoods is significantly higher than that of softwood due to the existence of large vessel elements. For the single pollutant, both hardwood and softwood show high degradation efficiency for methylene blue and methylene orange, while the degradation efficiency of the softwoods for 4-nitrophenol is significantly higher than that of the hardwoods due to their lower water flux. For the mixed pollutants, all the wood membranes have a good degradation property for different concentrations of methylene blue in polluted water, while the degradation efficiency of high concentration methylene orange and 4-nitrophenol is low. Our work will provide some guidance for the degradation of organic pollutants in actual polluted water.

Keywords: wood; Pd nanoparticles; water treatment; catalytic degradation

Citation: Wang, Z.; Jia, C.; Xiang, H.; Zhu, M. Palladium Nanoparticle-Loaded Mesostructural Natural Woods for Efficient Water Treatment. Polymers 2023, 15, 658. https://doi.org/10.3390/polym15030658

Academic Editors: Marta Otero and Ricardo N. Coimbra

Received: 10 November 2022
Revised: 17 January 2023
Accepted: 25 January 2023
Published: 27 January 2023

Copyright: © 2023 by the authors. Licensee MDPI, Basel, Switzerland. This article is an open access article distributed under the terms and conditions of the Creative Commons Attribution (CC BY) license (https://creativecommons.org/licenses/by/4.0/).

1. Introduction

With the development of industry, more and more wastewater has been produced and discharged into the environment. The wastewater, especially the wastewater containing organic pollutants, will cause serious pollution to the environment and harm to the human body [1,2]. The removal of organic pollutants from wastewater is essential to the protection of the environment and human health. Several methods have been developed to remove organic pollutants, including physical adsorption [3–6], chemical oxidation [7,8], membrane filtration [9–12], etc. However, these treatment methods still have some problems, including low efficiency, poor durability, secondary pollution and so on. Therefore, it is urgent to develop water treatment materials with high efficiency, durability and no secondary pollution.

Wood, as a kind of natural renewable material, has attracted more and more attention in recent years [13–16] and has been applied in many fields, including transparent papers [17,18], solar steam generation [19,20], etc. Natural wood has oriented microchannels, unique multi-level structures and a large number of active groups, which can be physically or chemically modified to obtain various functional materials. For example, the porous structure of natural wood can be used to obtain highly efficient oil adsorption materials through delignification treatment [21]. Studies have also shown that efficient water purification can be achieved by loading metal nanoparticles into the microchannels within natural wood [22,23]. However, the treatment effect of natural wood loaded with metal nanoparticles on the mixture of organic pollutants needs to be further studied.

Natural wood can be divided into broadleaf wood and coniferous wood, which are called hardwood and softwood, respectively [24,25]. Hardwoods are composed of fiber tracheids and vessel elements, which account for approximately 20% and 50% of wood pore volume, respectively. The vessel elements have a larger diameter than the fiber tracheids. Generally, the vessel elements are connected by perforated plates with micron holes at the cell ends, which allow water to continuously travel along the wood microchannels. Unlike hardwoods, softwoods have only fiber tracheids for transporting water and ions. The significantly different structures of hardwoods and softwoods result in different physical properties [26,27] and functional materials made from them. Therefore, it is of great significance to study the effect of pore structures of natural woods on their water treatment performance.

In this work, we loaded palladium nanoparticles into different kinds of natural woods and prepared wood membranes (palladium nanoparticle-loaded natural wood membranes, PNNW membranes) with high efficiency for water treatment. Three hardwoods, including basswood, beech wood and balsa wood, and two softwoods, including pine wood and fir wood, were selected as the base materials. The effect of the microstructures of the different natural woods on the catalytic degradation of organic pollutants by the wood membranes was investigated. In addition, we also studied the degradation performance of wood membranes on mixtures of different organic pollutants to evaluate their practical application effect. This work will provide some guidance for the application of natural wood in practical water treatment.

2. Experiment

2.1. Materials and Chemicals

Palladium chloride ($PdCl_2$), methylene bromide (MB), methylene orange (MO), 4-nitrophenol (4-NP), sodium borohydride ($NaBH_4$), hydrochloric acid (HCl) and sodium hydroxide (NaOH) were purchased from National Pharmaceutical Group Corporation (Shanghai, China). Basswood (*Tilia*), beech wood (*Fagus sylvatica* L.), balsa wood (*O. pyramidale*), pine wood (*Pelargonium hortorum*) and fir wood (*Pseudotsuga sinensis* Dode) were purchased from Anhui Meiru Company (Hefei, China). These wood samples were taken from healthy parts of mature wood.

2.2. Preparation of PNNW Membranes

$PdCl_2$ solution (0.5–2 mg mL^{-1}) was prepared by adding $PdCl_2$ powders (50–200 mg) and HCl (1 mL, 2 mol L^{-1}) to deionized water (100 mL), and heating at 60 °C for 1 h. Natural wood membranes with a dimension of 8 mm × 50 mm × 50 mm (thickness × length × width) were immersed in $PdCl_2$ solution and heated at 40–100 °C for 4–14 h to obtain PNNW membranes. Unless otherwise stated, the PNNW membranes for dye degradation property test were prepared by heating natural wood in 2 mg mL^{-1} $PdCl_2$ solution at 80 °C for 12 h.

2.3. Water Flux Test

The water flux of the PNNW membranes was calculated by testing the time it took 1 L of water to pass through a certain area of wood membranes at a pressure of 0.08 MPa. The water flux (F, L m^{-2} h^{-1}) was obtained using the following equation:

$$F = V/(A \times T)$$

where V represents the volume of filtered water (1 L), A (m^2) represents the effective area of PNNW membranes and T (h) is the time for water to pass through the membranes.

2.4. Dye Degradation Property Test

The degradation of methylene blue, methyl orange and 4-nitrophenol in the presence of $NaBH_4$ was used to evaluate the water treatment performance of PNNW membranes. 50 mg $NaBH_4$ and different amounts of dyes were mixed with deionized water to obtain 500 mL aqueous solution. For mixed solutions containing two or three dyes, the concentration of each dye is the same. The solutions then flowed through the PNNW membrane to degrade the organic dyes. The pH value of the solutions was adjusted using HCl and

NaOH. The absorbance measurement was performed for the solutions before and after filtration. The degradation efficiency (η, %) was calculated using the following equation:

$$\eta\ (\%) = (C_0 - C)/C_0 \times 100\%$$

where C_0 is the initial concentration of the solution and C is the concentration of the filtered solution.

3. Characterization

The microstructures of wood were characterized using a scanning electron microscope (JSM-5600LV, JEOL, Tokyo, Japan). The absorbance of dye solution before and after filtration was measured using a UV-vis spectrophotometer in the wavelength of 200–800 nm (UV1750 lambda 35, Mettler Toledo, Columbus, OH, USA). Wood density was calculated based on the measured mass and volume.

4. Results and Discussion

4.1. Preparation of Pd Nanoparticle-Loaded Wood Membranes and Degradation Mechanism

Natural wood is mainly composed of cellulose, hemicellulose and lignin, which account for more than 90% of the total weight of wood. Cellulose is formed by dehydration and condensation of glucoses, and mainly acts as a skeleton in wood cells. Hemicellulose is polymerized by glucoses and other kinds of monosaccharides, and its molecular weight is generally lower than that of cellulose. Hemicellulose contains a large number of branched chains, and mainly plays a filling role in wood cells. Lignin is a three-dimensional network polymer composed of p-hydroxyphenyl (H), guaiacyl (G) and syringyl (S) units. Lignin serves primarily as a binder in wood, binding cellulose and hemicellulose together.

Natural wood has excellent physical and mechanical properties. Due to the oriented microchannels, natural wood exhibits anisotropic physical and mechanical properties, including anisotropic mass transport characteristics [14] and anisotropic mechanical properties [16]. An interconnected pore network can be formed in natural wood by oriented microchannels and pits on cell walls, which is particularly suitable for liquid transport. Therefore, natural wood has been widely used for liquid-based transport applications, such as solar steam generation [16,28] and cleanup of viscous crude oil spill [29]. In addition, by loading catalysts into natural wood and treating wastewater containing organic dyes in the natural wood's inherent microchannels, the efficient catalytic degradation of organic pollutants can be achieved, and the consumption of non-renewable resources and energy can be effectively reduced.

In order to load Pd nanoparticles into wood, wood was immersed in a palladium nanoparticle precursor solution. The lignin distributed uniformly in wood microchannels can effectively reduce Pd ions to Pd nanoparticles. Cellulose and hemicellulose contain large amounts of hydroxyl groups, which can firmly immobilize Pd nanoparticles to wood microchannels. The oriented microchannels in natural wood are conducive to water transport, and water can effectively contact Pd nanoparticles on the microchannels during transport, so as to achieve efficient degradation of organic pollutants in wastewater.

Natural wood has a light yellow color due to the presence of lignin components (Figure S1). When Pd nanoparticles were loaded into wood microchannels, the wood turned black (Figure S2). Softwoods have smaller diameter microchannels, which result in lower water flux. Because of the slower flow rate of water and the longer contact time between organic pollutants in water and catalysts, softwoods are generally more efficient in degradation (Figure 1a). Hardwoods have larger vessels with larger pores than softwoods, so water flux is relatively higher. However, the degradation efficiency is low due to the short contact time between organic pollutants and catalyst (Figure 1b).

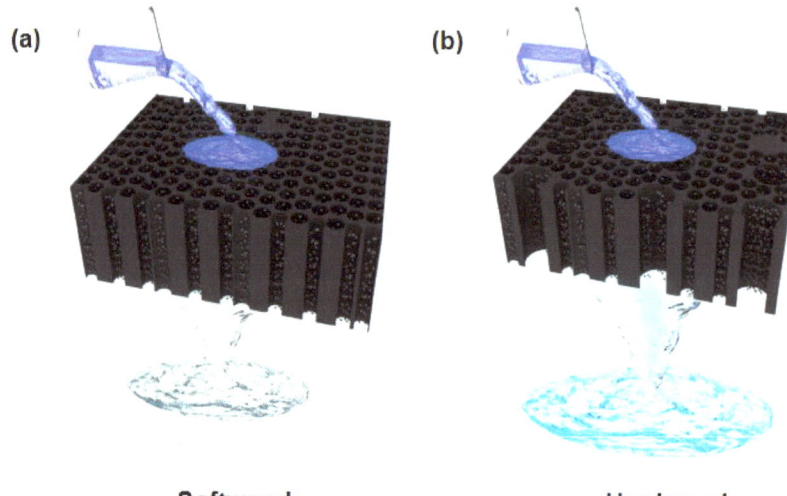

Figure 1. Schematic showing the catalytic degradation of pollutants in wastewater by (**a**) softwood and (**b**) hardwood loaded with Pd nanoparticles.

4.2. Microstructures of Natural Woods

Trees rely on internally oriented microchannels to transport water and nutrients from the bottom up to support their growth. Both hardwoods and softwoods from trees have anisotropic aligned microchannels. Figure 2a,d show a typical hardwood, basswood, and a typical softwood, pine wood, respectively. It can be seen from the SEM image that basswood includes vessels with a diameter of about 100 microns and tracheids with a diameter of dozens of microns (Figure 2b). If we zoom in on the cell walls of basswood, we can see that they have a large number of pits with a diameter of several microns on their surface (Figure 2c).

In contrast to basswood, pine wood contains only tracheids, which are about 20 microns in diameter and are very uniform (Figure 2e). There are also pits on the cell walls of pine wood, but the number is significantly less than that of basswood (Figure 2f). The orientated microchannels and pits on the cell walls in natural wood form an interconnected pore network. Water can be transported not only along the microchannels but also horizontally through the pits.

Pore structure and porosity have an important influence on the water treatment performance of natural wood. Figure 3a–c shows the SEM images of the upper surfaces of three different hardwoods, beech wood, balsa wood and basswood, and it can be observed that they have significantly different lumen diameters. In addition, the difference in vessel and tracheid diameter can be clearly seen from the top view of natural wood. In contrast to the complex structure of hardwoods, softwoods have only tracheids with uniform lumen diameter (Figure 3d,e), and the pore diameter of different softwoods does not differ much.

Different natural woods have different pore structures, resulting in their different densities and porosities (Figure 3f). In the case of hardwoods, the difference between pore structures of different hardwoods is very large, resulting in markedly different densities and porosity. Beech wood has a density of 731 kg m^{-3} and 51% porosity, while balsa wood, also a hardwood, has a density of 121 kg m^{-3} and a porosity of 92%. For softwoods, their density and porosity may also differ quite a lot, for example, the densities of pine wood and fir wood are 503 kg m^{-3} and 358 kg m^{-3}, respectively, and their porosities are 67% and 76%, respectively. However, the difference in density and porosity between softwoods is not as great as that of hardwoods [16].

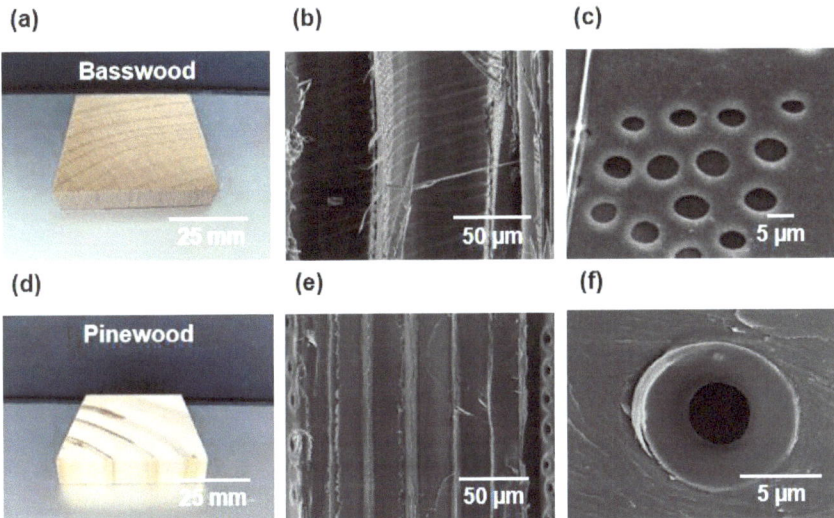

Figure 2. Microstructures of hardwood and softwood. (**a**) Digital image of a typical hardwood, basswood. (**b**) SEM image showing the vessels and tracheids of basswood. (**c**) SEM image of the pits on the cell walls of basswood. (**d**) Digital image of a typical softwood, pine wood. (**e**) SEM image showing the tracheids of pine wood. (**f**) SEM image of the pits on the cell walls of pine wood.

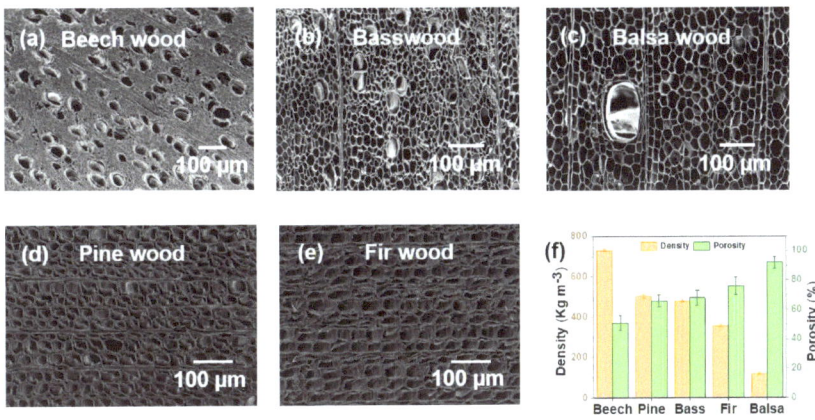

Figure 3. Comparison of pore structure and porosity of hardwood and softwood. SEM images of different hardwoods, (**a**) beech wood, (**b**) basswood and (**c**) balsa wood. SEM images of different softwoods, (**d**) pine wood and (**e**) fir wood. (**f**) Density and porosity of different wood species.

4.3. Degradation Property of PNNW Membranes for Single Organic Pollutant

The PNNW membranes were prepared by loading Pd nanoparticles into the wood microchannels. The Pd nanoparticles with a size of 30–100 nm can be clearly visualized in the wood microchannels (Figure 4). A home-made filter device was used to test the degradation property of PNNW membranes for different organic pollutants (Figure 5a). We first determined the water flux of different woods to evaluate their water treatment efficiency. The water fluxes of beech wood, basswood and balsa wood are 6541, 31,177 and 25,459 L m^{-2} h^{-1}, respectively, which are much higher than those of pine wood and fir wood (Figure 5b). Note that the water fluxes of pine wood and fir wood are only 487 and

561 L m^{-2} h^{-1}, respectively. The significantly higher water fluxes of hardwoods can be attributed to their large vessel elements, which enable faster water transport.

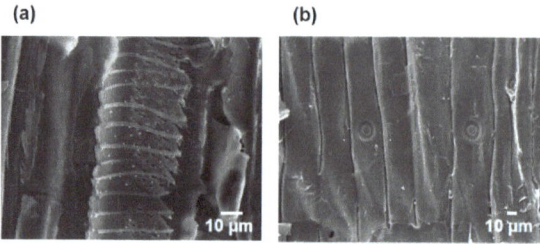

Figure 4. SEM images of the microchannels in the (**a**) basswood and (**b**) pine wood to show the Pd nanoparticles.

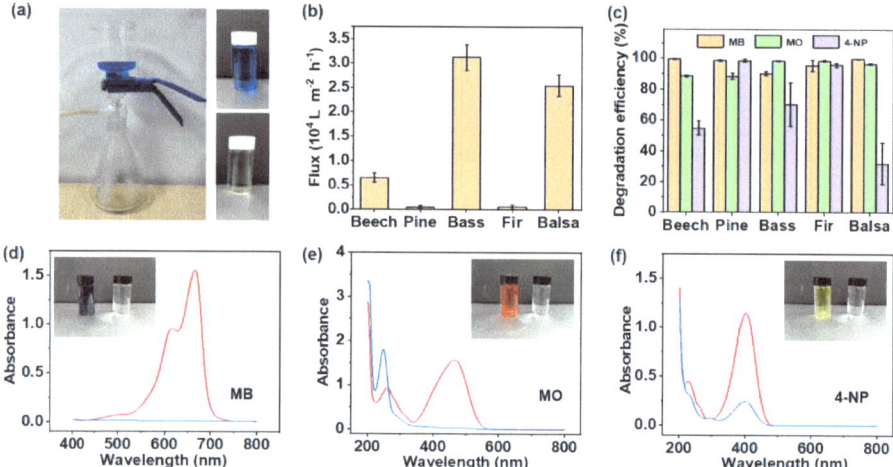

Figure 5. Degradation property of PNNW membranes for single organic pollutant. (**a**) Digital images showing the home-made filter device and solution before and after filtration. (**b**) Water fluxes of different natural woods. (**c**) Degradation efficiency of PNNW membranes to methylene blue (MB), methylene orange (MO) and 4-nitrophenol (4-NP) solutions with concentration of 20 mg L^{-1}. Absorbance curves of (**d**) MB, (**e**) MO and (**f**) 4-NP solutions before (red curve) and after (blue curve) filtration.

In addition, different hardwoods also exhibit significantly different water fluxes, which may be caused by the different diameters of the microchannels in different woods. Although beech wood contains a large number of vessel elements, the average diameter of its fiber tracheids is about 15 µm, significantly smaller than that of basswood and balsa wood (Figures 3a–c and S3), resulting in a low water flux. Although the average diameter of fiber tracheids of balsa wood is larger than that of basswood, the number of vessel elements with a larger diameter is significantly less than that of basswood (Figures 3b,c and S3), so the water flux of balsa wood is slightly lower than that of basswood.

In the textile industry, azo dyes are widely used, so we chose azo dyes, methylene blue (MB), methylene orange (MO) and 4-nitrophenol (4-NP), as model dyes to evaluate the degradation property of PNNW membranes. The absorbance of MB, MO and 4-NP solutions before and after filtration was measured to calculate the degradation efficiency of these solutions by PNNW membranes (Figure 5c–f). The characteristic absorption peaks of MB, MO and 4-NP are located at 640, 460, and 400 nm, respectively. After filtration, the characteristic absorption intensity of various organic dye solutions decreased significantly. The color of MB, MO and 4-NP solution changed from blue, orange and light yellow, respectively, to colorless, which directly reflected the degradation effect of the PNNW membranes.

Different PNNW membranes have different degradation properties for different organic dyes. All the PNNW membranes showed excellent degradation performance for MB and MO without significant difference. In particular, the degradation efficiency of beech wood, pine wood and balsa wood for MB and the degradation efficiency of basswood and fir wood for MO reached more than 99% in single dye solution. These results indicate that Pd nanoparticles have very high degradation efficiency for MB and MO. Even under the condition of the high water flux of hardwood, they can achieve a similar degradation effect to softwood with low water flux.

The degradation efficiency of 4-NP by the five PNNW membranes was quite different. The 4-NP degradation efficiencies of beech wood, pine wood, basswood, fir wood and balsa wood were 55%, 99%, 67%, 96% and 32%, respectively. The degradation efficiency of the two softwood species, pine wood and fir wood, for 4-NP was significantly higher than that of the hardwoods. These results indicate that the degradation of 4-NP by Pd nanoparticles takes a longer time than MB and MO. Since there are many large-diameter vessel elements in hardwoods, the 4-NP solution can pass through the filter membranes quickly, and the 4-NP cannot be effectively degraded by the Pd nanoparticles. However, softwoods only contain small diameter tracheids, so the water flux is low and the solution passes through the filtration membranes for a longer time. Under such a condition, 4-NP has enough time to contact Pd nanoparticles and be degraded, so the degradation efficiency is higher.

The effects of pH value, $PdCl_2$ solution concentration, heat treatment temperature and time on the degradation efficiency of the basswood-based PNNW membrane for the MB solutions were investigated. The results showed that the property of PNNW membrane under alkaline conditions is better than that under acidic conditions. $NaBH_4$ is weakly alkaline and easy to decompose under acidic conditions, so the degradation efficiency is low under acidic conditions (Figure S4). With the increase in $PdCl_2$ solution concentration, the color of the wood membranes obtained in the same treatment conditions gradually deepened (Figure S5). The degradation efficiency remained high even with low-concentration $PdCl_2$ treatment (Figure S6).

We prepared different basswood-based PNNW membranes by changing the heat treatment time (4–14 h) and temperature (40–100 °C), and investigated the effect of heating time and temperature on the degradation efficiency (Figures S7–S9). It can be seen from Figure S8 that the degradation efficiency of PNNW membranes for 20 mg L^{-1} MB solution gradually increases with the increasing heating time. When the heating time increases from 4 h to 8 h, the degradation efficiency increases rapidly from 37% to 88%. When the heating time is increased to more than 10 h, the degradation efficiency is essentially unchanged with the increase in heating time, which is about 95%. With the increase in heating time, more Pd^{2+} can be reduced to Pd nanoparticles, thus improving the catalytic property. The effect of heat treatment temperature on the catalytic degradation efficiency of PNNW membranes was small. Compared with that of PNNW membranes treated at 40 °C, the degradation efficiency of PNNW membranes treated at 100 °C was only slightly improved.

In order to verify that organic dyes are degraded by Pd nanoparticles and not by some components in the wood, we performed filtration experiments using natural wood. When the organic dye solution was filtered through natural wood, no obvious color change was observed. The characteristic absorption peaks of various dyes before and after filtration did not change clearly, as observed through infrared absorption spectrum analysis (Figure S10). These results indicate that natural wood has no degradation effect on organic dyes.

4.4. Degradation Property of PNNW Membranes for Mixed Organic Pollutant

Industrial wastewater usually contains a variety of organic pollutants, so it is necessary to study the degradation property of the prepared PNNW membranes for mixed organic pollutants. Therefore, we studied the degradation property of PNNW membranes for pollutants in mixed solutions containing two dyes and three dyes. For the mixed solution containing MB and MO, the PNNW membranes had high degradation efficiency for both organic pollutants in the mixed solution. The degradation efficiency of MO by fir wood-based PNNW membrane was about 80%, and the degradation efficiency of both dyes by other PNNW membranes was more than 99% (Figure 6a).

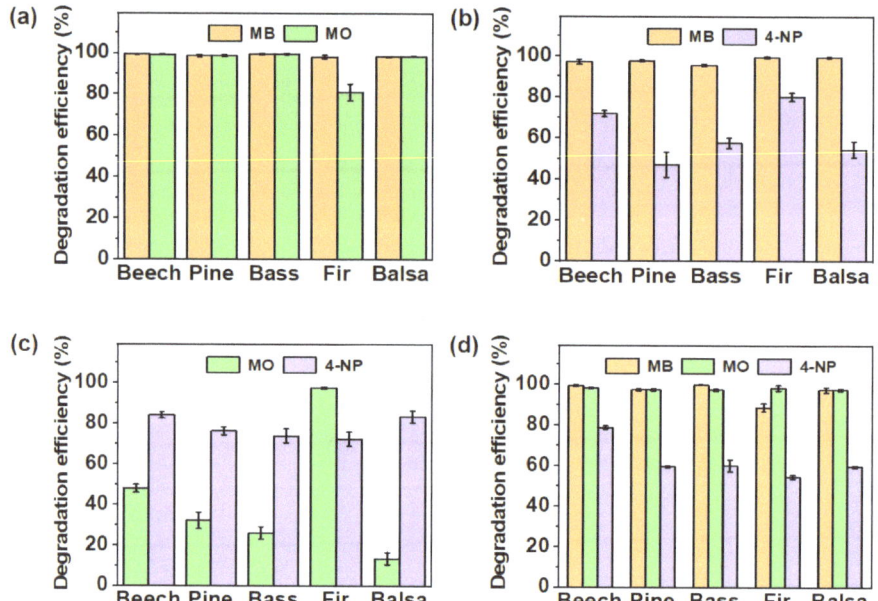

Figure 6. Degradation property of PNNW membranes for mixed organic pollutant. Degradation property of PNNW membranes for (**a**) MB and MO mixed solution, (**b**) MB and 4-NP mixed solution, (**c**) MO and 4-NP mixed solution and (**d**) MB, MO and 4-NP mixed solution. The concentrations of MB, MO and 4-NP in the mixed solution are 20 mg L^{-1}.

For the mixed solution containing MB and 4-NP, the degradation efficiency of the five PNNW membranes for MB in the mixed solution was more than 95%, but their degradation property for 4-NP was poor (Figure 6b). We found that all PNNW membranes had higher degradation efficiency for 4-NP in the mixed solution of MO and 4-NP. The degradation efficiency of the fir wood-based PNNW membrane for MO in the MO and 4-NP mixed solution was about 98%, and the high degradation efficiency may be caused by the longer catalytic degradation time. However, the degradation efficiency of other PNNW membranes for MO in the MO and 4-NP mixed solution was lower than 50%, and tends to decrease with the increase in wood porosity (Figure 6c).

The degradation property of PNNW membranes for the MB, MO and 4-NP mixture was further investigated. All PNNW membranes showed a degradation efficiency of more than 90% for MB and MO in the mixed solution. In particular, the beech wood-, pine wood-, basswood- and balsa wood-based PNNW membranes had a degradation efficiency of more than 98% for both pollutants. However, the degradation efficiency of all PNNW membranes for 4-NP in the mixed solution was less than 80%, similar to the situation of the two pollutants (Figure 6d). Through the above analysis, it can be concluded that the degradation property of PNNW membranes for different organic pollutants in the mixed solutions is affected by the types of pollutants.

When the dye concentration in the mixed solutions was increased from 20 mg L^{-1} to 30 mg L^{-1} and 40 mg L^{-1}, the degradation property of all the PNNW membranes for MB in the mixed solutions basically did not change significantly. However, the degradation efficiencies of the MO component in the MB and MO mixture solution, 4-NP component in the MB and 4-NP mixture solution, and MO component in the MB, MO and 4-NP mixture solution were significantly reduced (Figures S11 and S12). The above results show that all PNNW membranes have a good degradation property for different concentrations of MB in polluted water, but when the concentration of MO and 4-NP in polluted water is too high, the catalysts in the wood membranes are too late to degrade these dyes, resulting in low degradation efficiency.

5. Conclusions

Five kinds of natural wood membranes loaded with Pd nanoparticles were prepared and the catalytic degradation efficiency of these wood membranes for single and mixed dyes was studied. Softwoods have a low water flux because they contain only small-diameter fiber tracheids. Compared with hardwood, softwood has a significantly higher degradation efficiency for single 4-NP due to sufficient contact time between the dye and catalyst. For the single MB and MO, both hardwood and softwood show high degradation efficiency. For the mixed pollutants, there are some interferences between different pollutants, resulting in different catalytic degradation efficiency. All the wood membranes have a good degradation property for different concentrations of MB, while the degradation efficiency of high-concentration MO and 4-NP is low. Therefore, the appropriate natural wood-based water treatment membrane should be selected according to the kinds of pollutants contained in the polluted water in practical applications.

Supplementary Materials: The following supporting information can be downloaded at: https://www.mdpi.com/article/10.3390/polym15030658/s1, Figure S1: Digital image of a natural wood with light yellow. Figure S2: Digital image of a Pd nanoparticle loaded natural wood. Figure S3: Distribution of pore diameter of different woods. Figure S4: Catalytic degradation prop-erty of basswood based PNNW membrane to MB solution under different pH conditions. Figure S5: PNNW membranes prepared using PdCl2 solutions with different concentrations. Figure S6: Catalytic degradation efficiency of the basswood based PNNW membranes treated with different PdCl2 concentra-tions to 20 mg L^{-1} MB solution. Figure S7: PNNW membranes prepared with different heating time. Figure S8: Catalytic degradation efficiency of the basswood based PNNW membranes treated with different heat-ing time for 20 mg L^{-1} MB solution. Figure S9: Catalytic degradation efficiency of basswood based PNNW membrane treated with different heating temperatures for 20 mg L^{-1} MB solution. Figure S10: Absorbance curves of MB, MO and 4-NP solutions before and after filtration using natural wood. Figure S11: Degrada-tion property of PNNW membranes to mixed organic pollutant (The concentrations of MB, MO and 4-NP in the mixed solution are 30 mg L^{-1}). Figure S12: Degradation property of PNNW membranes to mixed organic pollutant (The concentrations of MB, MO and 4-NP in the mixed solution are 40 mg L^{-1}).

Author Contributions: C.J., H.X. and M.Z. conceived the idea and supervised the research. Z.W. contributed to the material preparation and characterization. Z.W. conducted filtration performance measurement. Z.W. and C.J. contributed to the writing of the manuscript. All authors have read and agreed to the published version of the manuscript.

Funding: This research was funded by the National Natural Science Foundation of China, grant number 52073047 and 52102090, the Science and Technology Commission of Shanghai Municipality, grant number 20JC1414900, the Program of Shanghai Academic/Technology Research Leader, grant number 20XD1433700, the Joint Funds of the National Natural Science Foundation of China, grant number U20A20257, and the Fundamental Research Funds for the Central Universities, grant number 2232022D-04.

Institutional Review Board Statement: Not applicable.

Informed Consent Statement: Not applicable.

Data Availability Statement: The data are available from the corresponding authors upon reasonable request.

Conflicts of Interest: The authors declare no conflict of interest.

References

1. Grant, S.B.; Saphores, J.D.; Feldman, D.L.; Hamilton, A.J.; Fletcher, T.D.; Cook, P.L.M.; Stewardson, M.; Sanders, B.F.; Levin, L.A.; Ambrose, R.F.; et al. Taking the "waste" out of "wastewater" for human water security and ecosystem sustainability. *Science* **2012**, *337*, 681–686. [CrossRef]
2. Zhang, Y.X.; Huo, J.Z.; Zheng, X.J. Wastewater: China's next water source. *Science* **2021**, *374*, 1332. [CrossRef] [PubMed]
3. Ren, Z.J.; Umble, A.K. Recover wastewater resources locally. *Nature* **2016**, *529*, 25. [CrossRef]
4. Ersan, G.; Apul, O.G.; Perreault, F.; Karanfil, T. Adsorption of organic contaminants by graphene nanosheets: A review. *Water Res.* **2017**, *126*, 385–398. [CrossRef]
5. Joseph, L.; Flora, J.R.V.; Park, Y.G.; Badawy, M.; Saleh, H.; Yoon, Y. Removal of natural organic matter from potential drinking water sources by combined coagulation and adsorption using carbon nanomaterials. *Sep. Purif. Technol.* **2012**, *95*, 64–72. [CrossRef]
6. Li, N.; Sheng, G.P.; Lu, Y.Z.; Zeng, R.J.; Yu, H.Q. Removal of antibiotic resistance genes from wastewater treatment plant effluent by coagulation. *Water Res.* **2017**, *111*, 204–212. [CrossRef] [PubMed]
7. Benidris, E.; Ghezzar, M.R.; Ma, A.; Ouddane, B.; Addou, A. Water purification by a new hybrid plasma-sensitization-coagulation process. *Sep. Purif. Technol.* **2017**, *178*, 253–260. [CrossRef]
8. Ma, J.T.; Chen, Y.L.; Nie, J.X.; Ma, L.M.; Huang, Y.X.; Li, L.; Liu, Y.; Guo, Z.G. Pilot-scale study on catalytic ozonation of bio-treated dyeing and finishing wastewater using recycled waste iron shavings as a catalyst. *Sci. Rep.* **2018**, *8*, 7555. [CrossRef] [PubMed]
9. Pi, Y.H.; Li, X.Y.; Xia, Q.B.; Wu, J.L.; Li, Y.W.; Xiao, J.; Li, Z. Adsorptive and photocatalytic removal of persistent organic pollutants (POPs) in water by metal-organic frameworks (MOFs). *Chem. Eng. J.* **2018**, *337*, 351–371. [CrossRef]
10. Ghadiri, M.; Mohammadi, M.; Asadollahzadeh, M.; Shirazian, S. Molecular separation in liquid phase: Development of mechanistic model in membrane separation of organic compounds. *J. Mol. Liq.* **2018**, *262*, 336–344. [CrossRef]
11. Ling, S.J.; Qin, Z.; Huang, W.W.; Cao, S.F.; Kaplan, D.L.; Buehler, M.J. Design and function of biomimetic multilayer water purification membranes. *Sci. Adv.* **2017**, *3*, e1601939. [CrossRef]
12. Cao, C.X.; Yuan, J.Y.; Cheng, J.P.; Han, B.H. Synthesis of porous polymer/tissue paper hybrid membranes for switchable oil/water separation. *Sci. Rep.* **2017**, *7*, 3101. [CrossRef]
13. Jia, C.; Chen, C.J.; Kuang, Y.D.; Fu, K.; Wang, Y.L.; Yao, Y.G.; Kronthal, S.; Hitz, E.; Song, J.W.; Xu, F.J.; et al. From wood to textiles: Top-down assembly of aligned cellulose nanofibers. *Adv. Mater.* **2018**, *30*, e1801347. [CrossRef] [PubMed]
14. Jia, C.; Jiang, F.; Hu, P.; Kuang, Y.D.; He, S.M.; Li, T.; Chen, C.J.; Murphy, A.; Yang, C.P.; Yao, Y.G.; et al. Anisotropic, mesoporous microfluidic frameworks with scalable, aligned cellulose nanofibers. *ACS Appl. Mater. Interfaces* **2018**, *10*, 7362–7370. [CrossRef] [PubMed]
15. Jia, C.; Li, T.; Chen, C.J.; Dai, J.Q.; Kierzewski, I.M.; Song, J.W.; Li, Y.J.; Yang, C.P.; Wang, C.W.; Hu, L.B. Scalable, anisotropic transparent paper directly from wood for light management in solar cells. *Nano Energy* **2017**, *36*, 366–373. [CrossRef]
16. Jia, C.; Li, Y.J.; Yang, Z.; Chen, G.; Yao, Y.G.; Jiang, F.; Kuang, Y.D.; Pastel, G.; Xie, H.; Yang, B.; et al. Rich mesostructures derived from natural woods for solar steam generation. *Joule* **2017**, *1*, 588–599. [CrossRef]
17. Hu, Y.C.; Hu, F.Q.; Gan, M.X.; Xie, Y.M.; Feng, Q.H. Facile one-step fabrication of all cellulose composites with unique optical performance from wood and bamboo pulp. *Carbohydr. Polym.* **2021**, *274*, 118630. [CrossRef] [PubMed]
18. Fujisaki, Y.; Koga, H.; Nakajima, Y.; Nakata, M.; Tsuji, H.; Yamamoto, T.; Kurita, T.; Nogi, M.; Shimizu, N. Transparent nanopaper-based flexible organic thin-film transistor array. *Adv. Funct. Mater.* **2014**, *24*, 1657–1663. [CrossRef]
19. Chen, C.J.; Li, Y.J.; Song, J.W.; Yang, Z.; Kuang, Y.; Hitz, E.; Jia, C.; Gong, A.; Jiang, F.; Zhu, J.Y.; et al. Highly flexible and efficient solar steam generation device. *Adv. Mater.* **2017**, *29*, 1701756. [CrossRef]
20. Hai, L.V.; Muthoka, R.M.; Panicker, P.S.; Agumba, D.O.; Pham, H.D.; Kim, J. All-biobased transparent-wood: A new approach and its environmental-friendly packaging application. *Carbohydr. Polym.* **2021**, *264*, 118012. [CrossRef]
21. Guan, H.; Cheng, Z.Y.; Wang, X.Q. Highly compressible wood sponges with a spring-like lamellar structure as effective and reusable oil absorbents. *ACS Nano* **2018**, *12*, 10365–10373. [CrossRef] [PubMed]
22. Che, W.B.; Xiao, Z.F.; Wang, Z.; Li, J.; Wang, H.G.; Wang, Y.G.; Xie, Y.J. Wood-based mesoporous filter decorated with silver nanoparticles for water purification. *ACS Sustain. Chem. Eng.* **2019**, *7*, 5134–5141. [CrossRef]
23. Chen, F.J.; Gong, A.S.; Zhu, M.W.; Chen, G.; Lacey, S.D.; Jiang, F.; Li, Y.F.; Wang, Y.B.; Dai, J.Q.; Yao, Y.G.; et al. Mesoporous, three-dimensional wood membrane decorated with nanoparticles for highly efficient water treatment. *ACS Nano* **2017**, *11*, 4275–4282. [CrossRef]
24. Kirbas, I.; Cifci, A. An effective and fast solution for classification of wood species: A deep transfer learning approach. *Ecol. Inform.* **2022**, *69*, 101633. [CrossRef]
25. Wang, C.K.; Zhao, P. Study on simultaneous classification of hardwood and softwood species based on spectral and image characteristics. *Spectrosc. Spectr. Anal.* **2021**, *41*, 1713–1721.

26. Roman, K.; Barwicki, J.; Rzodkiewicz, W.; Dawidowski, M. Evaluation of mechanical and energetic properties of the forest residues shredded chips during briquetting process. *Energies* **2021**, *14*, 3270. [CrossRef]
27. Liu, H.; Li, Z.; Zhang, X.; Tang, B.; Wan, C.; Wang, K. The effect of different moderate thermal modification durations on the wood properties of american alder. *Materials* **2022**, *15*, 8839. [CrossRef]
28. He, S.M.; Chen, C.J.; Kuang, Y.D.; Mi, R.Y.; Liu, Y.; Pei, Y.; Kong, W.Q.; Gan, W.T.; Xie, H.; Hitz, E.; et al. Nature-inspired salt resistant bimodal porous solar evaporator for efficient and stable water desalination. *Energy Environ. Sci.* **2019**, *12*, 1558–1567. [CrossRef]
29. Kuang, Y.D.; Chen, C.J.; Chen, G.; Pei, Y.; Pastel, G.; Jia, C.; Song, J.W.; Mi, R.Y.; Yang, B.; Das, S.; et al. Bioinspired solar-heated carbon absorbent for efficient cleanup of highly viscous crude oil. *Adv. Funct. Mater.* **2019**, *29*, 1900162. [CrossRef]

Disclaimer/Publisher's Note: The statements, opinions and data contained in all publications are solely those of the individual author(s) and contributor(s) and not of MDPI and/or the editor(s). MDPI and/or the editor(s) disclaim responsibility for any injury to people or property resulting from any ideas, methods, instructions or products referred to in the content.

Article

Highly Efficient Biosorption of Cationic Dyes via Biopolymeric Adsorbent-Material-Based Pectin Extract Polysaccharide and Carrageenan Grafted to Cellulosic Nonwoven Textile

Yassine EL-Ghoul [1,2,*] and Salman Alsamani [1]

1. Department of Chemistry, College of Science, Qassim University, Buraidah 51452, Saudi Arabia; 421100253@qu.edu.sa
2. Textile Engineering Laboratory, University of Monastir, Monastir 5019, Tunisia
* Correspondence: y.elghoul@qu.edu.sa; Tel.: +96-65-9551-9071

Citation: EL-Ghoul, Y.; Alsamani, S. Highly Efficient Biosorption of Cationic Dyes via Biopolymeric Adsorbent-Material-Based Pectin Extract Polysaccharide and Carrageenan Grafted to Cellulosic Nonwoven Textile. *Polymers* **2024**, *16*, 585. https://doi.org/10.3390/polym16050585

Academic Editors: Marta Otero and Ricardo N. Coimbra

Received: 2 January 2024
Revised: 30 January 2024
Accepted: 7 February 2024
Published: 21 February 2024

Copyright: © 2024 by the authors. Licensee MDPI, Basel, Switzerland. This article is an open access article distributed under the terms and conditions of the Creative Commons Attribution (CC BY) license (https://creativecommons.org/licenses/by/4.0/).

Abstract: Water scarcity and contamination have emerged as critical global challenges, requiring the development of effective and sustainable solutions for the treatment of contaminated water. Recently, functionalized polymer biomaterials have garnered significant interest because of their potential for a wide range of water treatment applications. Accordingly, this paper highlights the design of a new adsorbent material based on a cellulosic nonwoven textile grafted with two extracted biopolymers. The layer-by-layer grafting technique was used for the polyelectrolyte multi-layer (PEM) biosorbent production. Firstly, we extracted a Suaeda fruticosa polysaccharide (SFP) and confirmed its pectin-like polysaccharide structure via SEC, NMR spectroscopy, and chemical composition analyses. Afterward, the grafting was designed via an alternating multi-deposition of layers of SFP polymer and carrageenan crosslinked with 1,2,3,4-butanetetracarboxylic acid (BTCA). FT-IR and SEM were used to characterize the chemical and morphological characteristics of the designed material. Chemical grafting via polyesterification reactions of the PEM biosorbent was confirmed through FT-IR analysis. SEM revealed the total filling of material microspaces with layers of grafted biopolymers and a rougher surface morphology. The assessment of the swelling behavior revealed a significant increase in the hydrophilicity of the produced adsorbent system, a required property for efficient sorption potential. The evaluation of the adsorption capabilities using the methylene blue (MB) as cationic dye was conducted in various experimental settings, changing factors such as the pH, time, temperature, and initial concentration of dye. For the untreated and grafted materials, the greatest adsorbed amounts of MB were 130.6 mg/g and 802.6 mg/g, respectively (pH = 4, T = 22 C, duration = 120 min, and dye concentration = 600 mg/L). The high adsorption performance, compared to other reported materials, was due to the presence of a large number of hydroxyl, sulfonate, and carboxylic functional groups in the biosorbent polymeric system. The adsorption process fitted well with the pseudo-first-order kinetic model and Langmuir/Temkin adsorption isotherms. This newly developed multi-layered biosorbent shows promise as an excellent adsorption resultant and cheap-cost/easy preparation alternative for treating industrial wastewater.

Keywords: polyelectrolyte multi-layers; suaeda fruticosa polysaccharide; extraction; carrageenan; cellulosic nonwoven; grafting; NMR; SEM; adsorption; isotherms

1. Introduction

The issue of organic-matter-induced water pollution is worldwide, with varying aspects and consequences based on the developed statuses of individual nations [1,2]. The concentrations of pollutants in products must be as low as possible. Prevention is therefore essential and is based on the following three aspects: the regulatory aspect involves setting standards; the health aspect includes, in particular, the technical control of installations; and finally, the scientific and technological aspect corresponds to the improvement of the pollution control processes [3,4].

Nevertheless, there is still a lot of work to be done, particularly with regard to textile effluents. These discharges are part of the worst-treated wastewater and are recognized for their physically powerful staining, increased pH variations, high chemical oxygen demand, and higher bio-toxicity to microorganisms [5]. Because it is less noticeable than direct pollution (odor, cloudiness), pollution from toxic organic waste is more subtle [6]. Aquatic fauna are not the only species concerned as the health of living things is gradually declining, their lives are being shortened, their progeny may be impacted by deformities, and their likelihood of developing cancer will rise [7–9]. By eating the meat of these living things, as well as fruits, vegetables, and other foods, we consume these same harmful pollutants without realizing it because we are part of the food chain.

Among the most harmful organic pollutants are dyes [10–12]. Large volumes of polluting dye discharges into wastewater are reportedly caused by the pulp, food coloring, paper, textile, and cosmetic industries. Living aquatic organisms are seriously endangered by these toxic and even carcinogenic colored wastewater pollutants [13–15].

This imposes the necessity and the vital urgency to find an adequate and effective solution to treat these highly toxic and harmful discharges. Different techniques have been proposed in this regard. The various biological treatments, filtration combined with coagulation, precipitation, coagulation/flocculation, ozonation, reverse osmosis, adsorption, ion exchange, and advanced electrochemical oxidation processes have been investigated for the treatment of textile dye discharges [16–21]. These technologies have revealed deficiencies and limitations given their high investment and operating costs. The processes based on reverse osmosis and ion exchange have proved to be more interesting given their ability to recover pollutant products during their elimination from the effluents [22–24]. However, these processes as well as those of advanced oxidation are not economically feasible due to their investment and relatively high operating cost.

Alternatively, the adsorption process using solid adsorbents has gained specific attention from researchers, revealing great effective potential to treat and remove organic pollutants in wastewater treatment. This technique is shown to be the most advantageous thanks to its straightforward design and cheap cost of investment. In this perspective, different adsorbents based on synthetic and natural polymers have been investigated for the elimination of dyes from contaminated waste [25–28]. Moreover, adsorbents based on textile materials are not often reported in the literature. A few studies have been conducted on the removal of cationic dyes via the use of certain pretreated synthetic textiles. In these studies, their chemical modification did not improve their low adsorption capacities due to their inert and hydrophobic character [29–31]. The functionalization of cellulosic textiles by some commercialized natural polymers such as k-carrageenan, alginate, and chitosan during our previous studies allowed effective biosorbents with excellent adsorption capacities [32].

In line with this emerging topic, the present research work suggests a new biosorbent material based on nonwoven cellulose grafted with a pectin polysaccharide extract and carrageenan polymer as two proposed potent natural adsorbents. The grafting is provided by the intermediate of a BTCA polycarboxylic acid as a crosslinking agent. The performance of the made adsorbent in eliminating cationic dyes (methylene blue in the current study) will be warranted by the different carboxylic acid, sulfate, and ester groups of the two grafted polyanions and the different hydroxyl functions presented in both the cellulosic support and the two biopolymers. The textile support material used is based on cellulose polymer, which is well known for its favorable hydrophilic properties [33–36]. The first polymer is a carrageenan biopolymer from edible red algae, a natural sulfated polysaccharide recognized for its gelling and biological properties [37,38]. The investigated pectin-like polysaccharide is extracted from Suaeda fruticosa leaves. The Suaeda fruticosa polysaccharide (SFP) has revealed, in the literature, various biological and bacteriological properties [39]. Thus, in the current paper, we will take advantage of both the different properties of the two crosslinked natural polysaccharides as well as the effectiveness of the new multi-layer grafting method to design a natural adsorbent that is bioactive and particularly capable of revealing excellent sorption capacity. In addition, such a textile material treated with robust and durable grafting offers the possibility and the potential for multiple reuses.

In our method, after the extraction and characterization of the SFP biopolymer, the latest natural polysaccharide and the carrageenan biopolymer are grafted on the cellulosic material in the presence of the BTCA polycarboxylic acid as a crosslinking agent. The different parameters of grafting, such as time and curing temperature, will be optimized. Then, different characterization analyses (FT-IR, SEM, TGA/DTA, and swelling capacity) will be carried out to evaluate the effectiveness of the designed PEM material and the pad–dry–cure grafting process. After that, the performance of the biosorbent will be assessed in various experimental settings, such as pH, temperature, time, and methylene blue concentration. The experimental data will finally be analyzed and modeled via the exploration of theoretical kinetic and isotherm equations.

2. Materials and Methods

2.1. Materials

A low-cost nonwoven textile based on cellulosic fibers was used as the base material for the prepared adsorbent. The material had a surface weight of 240 g/m^2 and a thickness of 0.6 mm. The nonwoven material was produced via a calendaring thermal consolidation. 1,2,3,4-butanetetracarboxylic acid (BTCA), used as a crosslinking agent, and the iota-carrageenan (a tiny white powder) were acquired from Sigma-Aldrich (Taufkirchen, Germany). All chemicals and reagents employed for the extraction of the pectin-like polysaccharide or the grafting procedures were used without further purification. Methylene blue (MB, M.W = 319.85 g/mol, λ_{max} in water = 665 nm, chemical formula: $C_{16}H_{18}ClN_3S$) as a reference of cationic dyes used as adsorbate was provided by the Central Drug House (India). Distilled water was used to prepare the MB solutions for the batch adsorption experiments.

2.2. Extraction of the SFP Polymer

A productive extraction process that our research team had previously published was examined [40] with slight adjustments. Three kilograms of Fresh Suaeda fruticosa leaves were chopped and repeatedly cleaned in distilled water. They were next ground and stored for three days at 40 °C in an oven after being dried. Following a depigmentation process in a soxhlet containing 95% ethanol, the resultant powder (2.4 kg) was extracted using citric acid using an ultrasonic bath (pH = 3.5, temperature = 70 °C, and duration = 3 days). The acidic extract was filtered, neutralized with NaOH solution, and then precipitated with 95% ethanol after the residue was removed. Before deproteination, the precipitate was dissolved in distilled water. Ultimately, using a Biobase Vacuum Freeze Dreyer in Shandong, China, the aqueous phase was dialyzed in water and lyophilized to provide 17% of SFP polymer.

2.3. SEC Analysis of the SFP Polymer

By using size exclusion chromatography analysis, the macromolecular properties of SFP were obtained. The viscometer detector, differential refractive index (RI), and multiangle light scattering were all included in the SEC device. For the analysis, DMF was used as a solvent. We leveraged an OmniSEC program for data analysis and extrapolation.

2.4. NMR Analysis

The ^{1}H and ^{13}C NMR spectra of the extracted SFP polymer were obtained via a Bruker Avance DRX 500 spectrometer (Bruker Instruments, Inc., Rheinstetten, Germany). Spectra were recorded in D_2O (99.8 Atom% D). Chemical shifts in different spectra were expressed in ppm using tetramethylsilane as an internal standard. In each experiment, the number of scans was fixed according to the sample concentration.

2.5. Analysis of Carbohydrate Content in the SFP Extract

Considering galactose as a standard, the phenol-sulfuric acid technique was used to determine the carbohydrate content of SFP [41]. To assess the amount of uronic acid, the carbazole pathway was adopted, with galacturonic acid serving as a reference [42] (Bitter and Muir 1962). According to Lowry et al., the protein amount was identified using the Lowry method [43].

2.6. Designing of the Biosorbent Material

A pad–dry–cure grafting process was performed for the functionalization of different cellulosic materials. The samples were washed beforehand with distilled water at 40 °C. Then, they were passed into an impregnation solution containing 60 g/L of SFP, 40 g/L of BTCA (crosslinking agent), 60 g/L of carrageenan, 10 g/L of sodium hypophosphite (catalyst for the polyesterifiacation reaction), and 3 g/L of ammonium hydrogen phosphate. Distilled water was used as a solvent for the impregnation bath. Nonwoven samples were padded in the impregnation bath then roll-squeezed and dried at 100 ° C for 30 min. After that, the samples were thermofixed at variable times and curing temperatures. Finally, to eliminate the nonfixed polymers, the functionalized samples were washed with distilled water and dried for 40 min at 100 °C. To evaluate the grafting rate, the samples were weighed before and after functionalization. Equation (1) was used to calculate the grafting rate of the functionalized samples (expressed as % − Wt). For each value recorded, 10 replicates were investigated.

$$\% - \text{Wt} = \frac{[mf - mi]}{mi} \times 100 \tag{1}$$

Here, % − Wt is the grafting rate, and m_i and m_f represent the weights of samples taken before and after the grafting procedure.

2.7. Grafting Characterization Procedures

The analysis of the chemical grafting was evaluated via an infrared spectroscopy study. An FTIR spectrometer from Agilent Technologies equipped with an ATR system (attenuated total reflection) was used for the different assessments of the untreated and functionalized samples. The different spectra were measured from 4000 to 400 cm^{-1}.

For an efficient adsorption behavior, the swelling capacity of the adsorbent is usually considered a crucial characteristic. Swelling measurements of virgin and functionalized cellulosic material were assessed via a gravimetric technique. Samples were first dried to determine the initial weight (mi). Samples were then impregnated for 2 days in distilled water. The time of impregnation was varied, and after wiping, the weight was measured for each duration of impregnation. Equation (2) was utilized to ascertain the rate of swelling:

$$\% - SR = \frac{(mf - mi)}{mi} \times 100 \tag{2}$$

Here, the weights of the dried and swollen samples are, respectively, mi and mf.

An FEI Quanta SEM microscope was explored for the assessment of the surface morphology of virgin and grafted PEM adsorbents. For the various measurements, the accelerating voltage was fixed at 5 KV. Varying magnifications were selected to identify the surface morphological appearance of different samples. To increase the samples' conductivity, a thin layer of carbon was applied on their surfaces before SEM analysis.

2.8. Adsorption Batch Experiments

The different adsorption experiments were performed using a batch reactor where the MB dye was set in contact with the designed adsorbent and stirred vigorously at 140 rpm. Different condition parameters affecting the adsorption performance were studied, including the pH (varied from 3 to 9), time (ranged from 0 to 120 min.), concentration of the MB dye (varying from 25 to 1000 mg/L), and contact temperature (22, 40, and 60 °C). Following each experiment, the residual concentration was measured using a Shimadzu (UV-2600) UV-visible spectrophotometer.

The following Equation (3) was used to evaluate the amount, expressed as q (mg/g), of the MB dye adsorbed on the different samples grafted with the PEM polymeric system.

$$q\ (mg/g) = \frac{C_0 - C_e}{m} \times v \qquad (3)$$

Here, v is the MB dye volume (L), m is the adsorbent weight (g), and C_0 and C_e are the original and residual concentrations (mg/L), respectively.

3. Results and Discussion

3.1. SEC Evaluation and Carbohydrate Content in the SFP Extract

Physicochemical evaluations of the naturally extracted polysaccharide isolated from the Suaeda fruticosa plant were determined using steric exclusion chromatography. The macromolecular properties and the carbohydrate amount of the polysaccharide are summarized in Table 1.

Table 1. Extracted SFP's macromolecular properties and carbohydrate contents.

Extracted Product	M_n (g/mol)	M_w (g/mol)	$Đ$ (M_w/M_n)	$[\eta]$ (mL/g)	Carbohydrate Content (%)		
					Neutral sugar	Proteins	Gal Acid
SFP polysaccharide	131×10^3	267×10^3	2.03	289	55.61	-	44.39

The polymer's average molecular weight per number was around 131,000 g/mol. The isolated polysaccharide had a polydispersity value of 2.03. It was regarded as an extremely acceptable value, even for an extract of natural polymers. This value demonstrated the extracted polymer's satisfactory homogeneity and the effectiveness of the extraction technique used.

The optimized natural extract's primary components, as shown by the findings of the carbohydrate composition analysis, were 44.39% galacturonic acid and 55.61% neutral monosaccharides. This measured rate of galacturonic acid is somewhat lower than that from the same fruit grown in a different place (47.5%) [39] and greater than that found in other reported fruit pectins, such as in grapefruit peel (27.34%) and citrus peel (33.20%) [44]. These variable results are usually due to distinct fruit pectin origins, different cultivation regions, different plant parts investigated, or different extraction methods applied. Additionally, the isolated SFP did not contain any proteins according to the results of the Lowry technique. According to the literature, the natural polymer extract may be regarded as comprising pectin-like polysaccharides due to the higher determined quantity of Gal acid [39].

NMR Spectroscopy Analysis

To improve NMR spectra quality and to obtain more structure information, the SFP polymer was partially hydrolyzed before NMR analysis. Figure 1 shows the ^{1}H spectrum of the extracted SFP. The different SFP signals were assigned according to the literature values. Two signals that appeared close to 5.18 and 4.94 were ascribed to the anomeric H-1 and H-5 protons from non-esterified GlA [45,46]. In addition, Tow overlapped signals appeared around d 5.16 and were assigned to the anomeric H-1 and H-5 from the esterified carboxyl groups in the galacturonic acid residues. The signal at 4.02 ppm showed the characteristics of protons of methyl ester that existed in (1,4)-linked GalA [47]. The signals at 3.66, 3.88, and 4.22 ppm were attributed to the H-2, H-3, and H-4 protons of (1,4)-linked GalA, respectively.

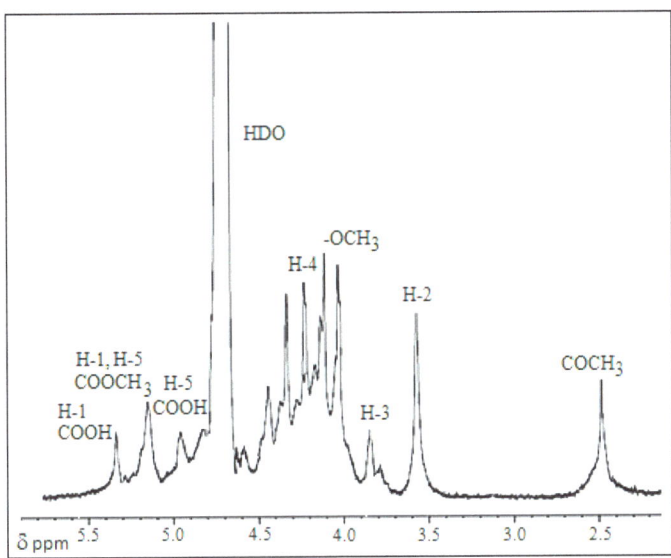

Figure 1. SFP extracted polymer—^{1}H NMR spectrum.

The ^{13}C NMR spectrum in Figure 2 showed two major signals that were assigned to the C-6 of the GalA carboxyl group at d 174.9 ppm (esterified) and d 172.3 ppm (non-esterified). Two anomeric signals appeared around d 102.1 and 101.0 ppm and were ascribed to the esterified and non-esterified GalA carboxyl units. Signals at 80.1 (C-4), 73.7 (C-5), 71.4 (C-3), and 70.1 (C-2) were attributed to the GalA carboxyl moieties [47]. The signal around 55.3 ppm was assigned to the O-CH3 ester group (O-methyl) linked to the C-6 [45]. Overall, ^{1}H and ^{13}C NMR analysis in turn confirmed the polysaccharide-like pectin structure of the SFP extracted polymer.

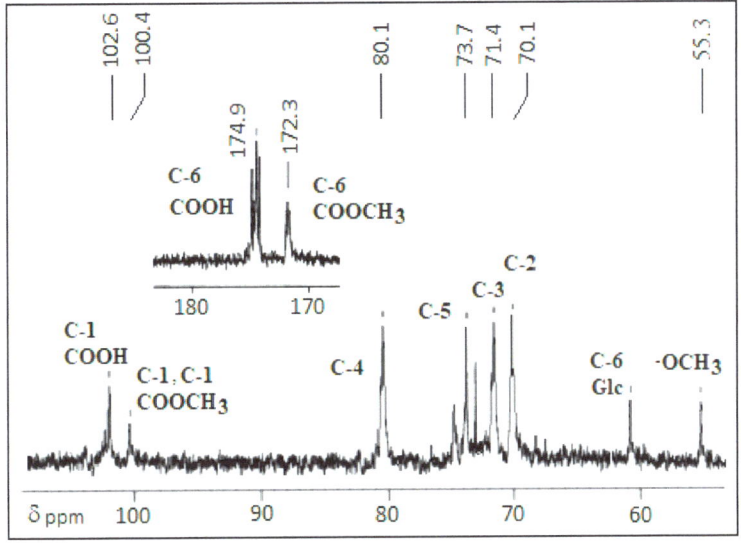

Figure 2. ^{13}C NMR spectrum of SFP extracted polymer.

3.2. Preparation of the PEM Biopolymer Adsorbent

The different biosorbent materials were designed via the layer-by-layer deposition technique. In this method, the materials obtained in the form of alternating layers of polymers contain one to six pairs of layers. The grafting rate variation is depicted in Figure 3 according to the number of pairings of successive layers of applied biopolymers. The weight gain increases progressively with the number of PEM layers grafted onto the cellulose biosorbent. We notice, from a functionalization of three pairs of polymers, the presence of a jump in the rate of grafting. Samples with three grafted pairs of layers will be selected for the characterization and adsorption studies.

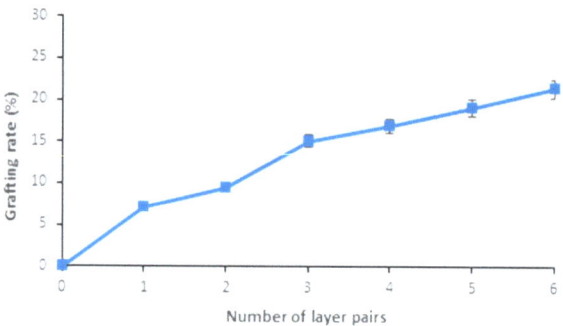

Figure 3. Grafting rate variation according to the number of layers grafted onto the biosorbent material.

3.3. Swelling Behavior

At varying periods of impregnation, swelling tests were conducted on functionalized PEM samples with varied pair layers and untreated cellulosic material. Results for the virgin cellulose material are shown in Figure 4, where a pseudo-plateau was reached upon 5 h, showing a maximum saturation of 184%. The swelling ratio increased gradually with time impregnation. Both of the functionalized PEM samples displayed a progressive rising trend. The sample that functionalized with three pairings had a swelling ratio more than twice as high as the untreated sample. The PEM material with five layers displayed a nearly three-fold swelling ratio, indicating a more hydrophilic capacitance. This resulted from the carrageenan biopolymer's hydrophilic nature and the SFP extracted polysaccharide functionalizing the cellulosic material, both of which are renowned for their high hydrophilicity [48,49].

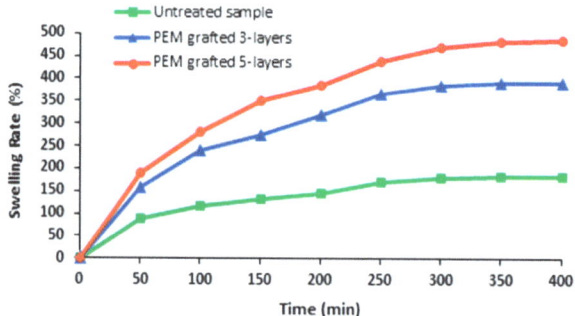

Figure 4. Variation in virgin and PEM biosorbent swelling capacities according to the number of grafted layers.

A superior hydrophilic composition exhibiting an increased water penetration capacity was attained through the functionalization process, employing natural hydrophilic biopolymers during the layer-by-layer deposition procedure. This is a necessary property to enhance adsorption capability.

3.4. FT-IR Analysis

FT-IR–ATR analysis was performed to identify the different functional groups that emerged during functionalization and to characterize the chemical grafting of the biosorbent material. This characterization was used to assess the PEM-grafted material as well as the virgin cellulose sample. Figure 5 displays the spectra of functionalized material with three polymeric layers and a cellulosic sample that had not been treated. The chemical grafting was demonstrated by various new peaks that developed in the PEM-grafted cellulosic material. The emergence of a peak with a center at 1724 cm^{-1} confirmed the polyesterification reaction between the hydroxyl functionalities of the SFP and carrageenan biological polymers and the carboxylic functions of the BTCA polycarboxylic acid. This is consistent with earlier research studies where FT-IR was employed to support the evidence of an esterification polymerization process using crosslinking compounds such as polycarboxylic acids to link cellulosic material to different polymers [25,50,51]. The grafted biosorbent material displays the various absorbances associated with the carrageenan sulfate groups, such as sulfate at 1214 cm^{-1}, galactose-4-sulfate at 920 cm^{-1}, and galactose-2-sulfate appearing at 834 cm^{-1} [40]. The SFP polymer's symmetric vibration of the COO$^-$ groups of the galacturonic acid was represented by the peak that was closed to 1310 cm^{-1} [52]. Additionally, a wider, more significant peak around 3290 cm^{-1} that was observed with treated material was assigned to the OH functions of the cellulose material, the SFP, and carrageenan biopolymers. In a nutshell, we were able to confirm the PEM biosorbent system's chemical interconnection using FT-IR analysis, and the results revealed the effectiveness of the functionalizing chemical procedure using the layer-by-layer grafting technique.

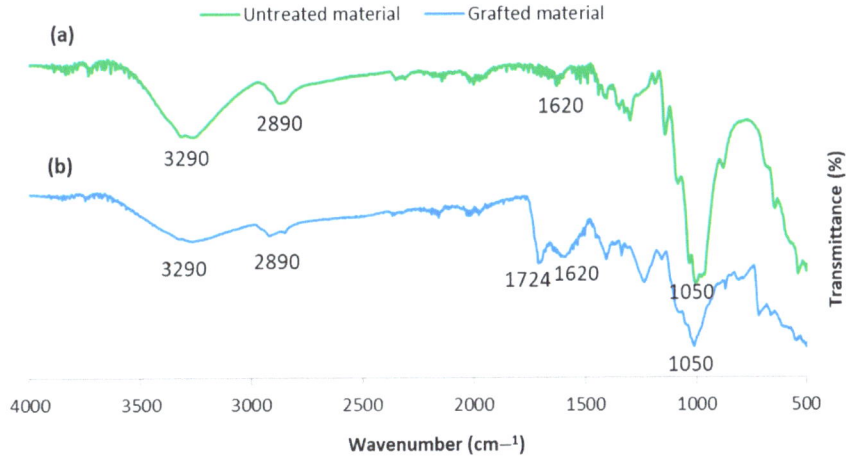

Figure 5. Spectra via FT-IR analysis of (**a**) untreated cellulose material and (**b**) grafted PEM adsorbent material.

3.5. SEM Analysis

Figure 6 displayed micrographs of virgin and functionalized biosorbent materials with three- and five-layer pairs. After grafting, we observed a sizable surface change. The microspaces across the fibers were filled with the layers of the two grafted polymers and did not affect the surface porosity of the samples. When comparing the untreated sample to the grafted PEM, there was an increase in surface roughness. The significant change in surface morphology proved the permanence and efficacy of the grafting technique used.

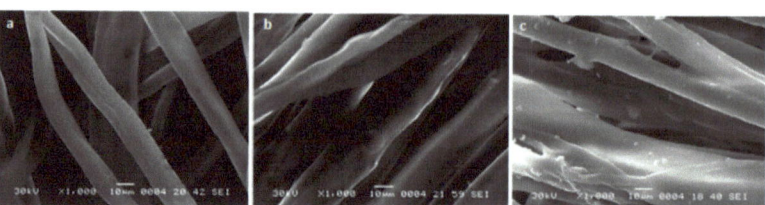

Figure 6. SEM analysis of virgin cellulose material (**a**) and grafted adsorbent designed at different layers: 3 layers (**b**) and 5 layers (**c**).

3.6. Application to MB Dye Adsorption

In this study, the produced materials, both untreated and grafted, were used as MB dye adsorbents by modifying the pH, time, temperature, and dye concentration. The progress of the amount of adsorbed MB dye according to the pH variation is shown in Figure 7a. At pH = 6, the highest adsorbed amount was achieved. In actuality, poor adsorbed rates were caused by the positively charged methylene blue ions' opposition to the adsorbent surface that is positively charged under highly acidic conditions. The adsorbent surface becomes negative at higher pH levels (pH ≈ 6), promoting an electrostatic interaction with the MB dye. The repulsive forces between dye molecules and the interface of the biosorbent could be the cause of the decreased amount of adsorbed dye under basic conditions.

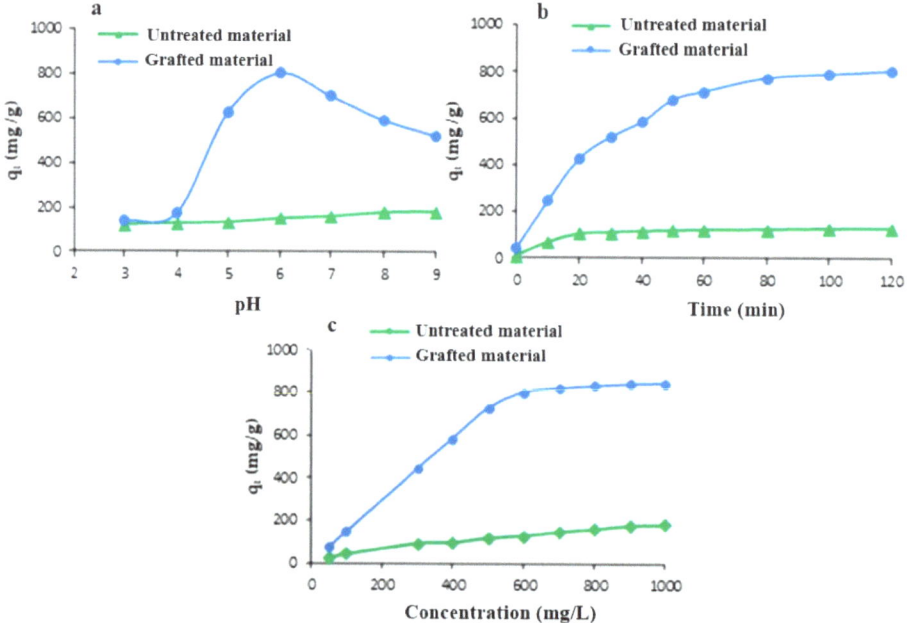

Figure 7. (**a**) Effect of pH (C_0 = 600 mg/L, t = 120 min), (**b**) time (pH = 6, C_0 = 600 mg/L, T = 22 °C), and (**c**) dye concentration (pH = 6, T = 22 °C, t = 120 min) on the adsorption of MB dye (samples are with 3 grafted pairs of layers).

The amount of time needed to reach equilibrium was found to be 120 min (Figure 7b). In the first 50 min, when more than 80% of the target was achieved, it can be seen that adsorption was quick. The fact that there are numerous adsorption sites available at the surface of the adsorbent during this initial stage may help explain this trend. Adsorption reached a constant state after this amount of time; it could be accounted for by the adsorption sites' saturation. The highest amounts of methylene blue that could be adsorbed for untreated and grafted materials were 130.4 mg/g and 802.4 mg/g, respectively. The grafting of additional reactive functions (sulfonate and carboxylate groups) onto cellulose surfaces may account for this variation in sorption capacities.

It was found that when the initial MB concentration increased, correspondingly, the amount of dye that was adsorbed increased. For the untreated samples, it reached 181.4 mg/g while, for the grafted materials it reached 802.6 mg/g (Figure 7c). Concerning functionalized samples, for instance, this adsorbed amount reduced from 844.5 mg/g at 22 °C to 794.5 mg/g at 60 °C as a result of the temperature variation (Figure 8). This demonstrated that the methylene blue adsorption in this instance was exothermic. Indeed, at higher temperatures, the MB dye may desorb from the samples.

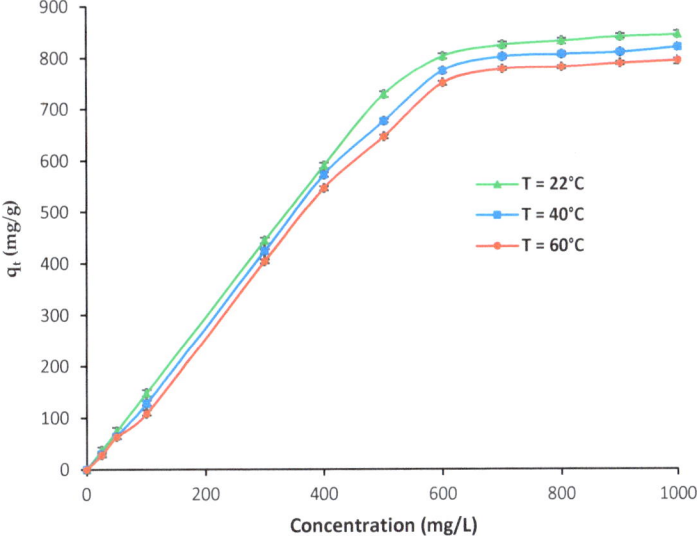

Figure 8. Effect of temperature on the MB dye adsorption (pH = 6, t = 120 min) (samples were with 3 grafted pairs of layers, and for each measurement, 3 replicates were made).

Table 2 summarizes the adsorption performance of some recent natural polymeric adsorbents reported in the literature. The comparison with our studied adsorbent reveals the importance of the currently designed material as a potent biosorbent. The highest adsorption capacity reflects the adsorbent efficiency achieved and the effectiveness of the multi-layer process.

Table 2. An overview of the most recent studies on the adsorption of MB dye from industrial wastewater using natural polymeric absorbents.

Adsorbent	Adsorbate	q_t (mg/g)	Adsorption Efficiency (%)	Reference
k-carrageenan/alginate/cellulose	MB	522.4	98.6	[53]
Activated carbon from Date Press Cake	MB	613.8	83.3	[54]
Chitosan-epichlorohydrin/zeolite composite	MB	44.2	90	[55]
PAM-chitosan magnetic nanoparticles	MB	104.06	76.1	[56]
Chitosan/carboxymethyl cellulose capsules	MB	64.6	4.4	[57]
Chitosan/Zeolite composite	MB	19.23	84.85	[58]
Alginate/gelatin/graphene oxide composite	MB	322.6	-	[59]
Maize silk powder	MB	132.1	--	[60]
Prickly (peel) bark of cactus fruit	MB	222	-	[61]
Dead *Typha angustifolia* (L.) leaves	MB	106.75	89.83	[62]
watermelon rind	MB	200	99	[63]
Fe_3O_4–κ-carrageenan/chitosan	MB	123	-	[64]
$Fe_3O_4@SiO_2$–κ-carrageenan	MB	530	-	[65]
Gelatin-based magnetic beads	MB	465	-	[66]
Carboxymethyl chitosan-modified magnetic-cored dendrimers	MB	20.85	-	[67]
Cactus pear seed cake	MB	260	56.48	[68]
Rattan (*Lacosperma secundiflorum*)	MB	359	96	[69]
Karanj fruit hulls	MB	239.4	94.4	[70]
Lignocellulosic Raspberry Leaves	MB	244.6	-	[71]
PAni@CTAB-Mt	MB	108.82	91.22	[72]
Chitosan/PAAM	MB	62.5	75	[73]
Alginate/PAA	MB	120	98	[73]
Cellulose Populus tremula seed fibers	MB	140.4	98	[74]
Alginate/polyethyleneimine	MB	400	99	[75]
Chitosan/carboxymethyl cellulose-GO	MB	405	82	[76]
SFP/carrageenan/cellulose PEM	MB	802.6	99	Current study

A schematic illustration of hydrogen bonds and ionic interactions between the PEM crosslinked to the surface of a cellulose textile material and MB dye molecules is shown in Scheme 1. Indeed, through the establishment of a hydrogen bond with the nitrogen atom of methylene blue, the free hydroxyl groups of cellulose nonwoven material might interact. However, through ionic interaction, the carboxylate groups of the SFP polymer (COO^-) and BTCA crosslinking agent, as well as the sulfonate groups (SO_3^- of carrageenan), may react with the N^+ cations of the dye.

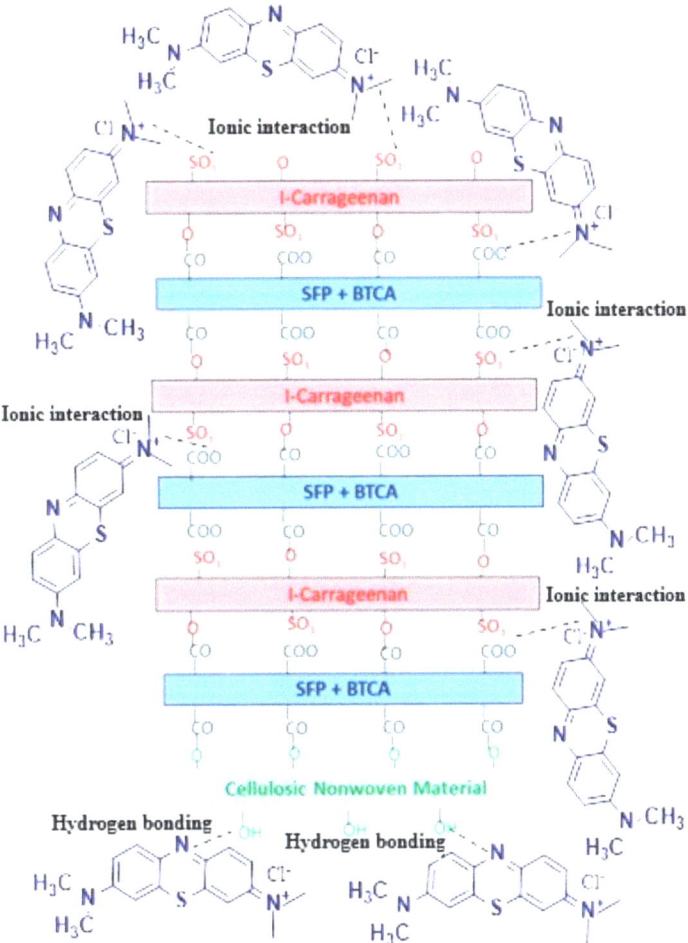

Scheme 1. A schematic illustration of the different interactions between methylene blue dye and the textile nonwoven material grafted with the crosslinked polymers.

3.7. Kinetic Modeling

To understand the equilibrium attraction between adsorbates and adsorbents, kinetic data were required. These might indicate whether the examined mechanism is physical, chemical, or involved in mass transfer. Here, the effectiveness of modeling kinetic data via Elovich and pseudo-first order, pseudo-second order, and intra-particular diffusion was evaluated. The outcomes are shown in Figure 9. In Table 3, the calculated kinetic parameters for the various equations are compiled. The calculated kinetic parameters and acquired curves suggested that the kinetic data might be well described by a pseudo-first-order equation ($0.96 < R^2$). High correlation coefficients ($0.99 < R^2$) were also achieved within the pseudo-second order. These findings demonstrated the complexity of the adsorption process, which could be classified as comprising both physical and chemical sorption modes [77]. The plots of the intraparticle diffusion model diverged from the starting point, indicating that it was not the only rate-controlling mechanism [78].

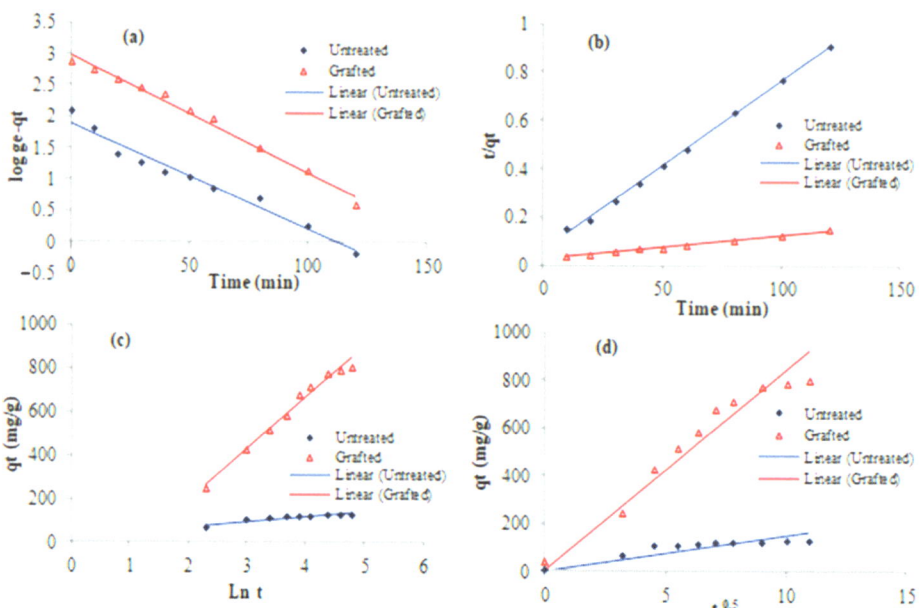

Figure 9. Adsorption kinetic data, mapped onto untreated and grafted biosorbent materials including (**a**) pseudo-first order, (**b**) pseudo-second order, (**c**) Elovich,, and (**d**) intraparticular diffusion models (samples are with 3 grafted pairs of layers).

Table 3. Synopsis of the ungrafted and grafted samples' kinetic data models.

Equations	Parameters	Untreated Material	Grafted Biosorbent
Pseudo-first order	K_1 (1/min)	0.033	0.025
	q_e (mg/g)	79.305	962.948
	R^2	0.969	0.988
Pseudo-second order	K_2	0.000845	0.000036
	q	142.86	1000
	h	17.24	35.71
	R^2	0.999	0.995
Elovich	α (mg/g/min)	72.447	78.046
	β (mg/g/min)	0.0042	0.044
	R^2	0.982	0.857
Intra-particular-diffusion	K_1 (mgg^{-1}·min$^{1/2}$)	15.168	84.507
	R^2	0.956	0.989

3.8. Isotherms and Thermodynamic Study

The equations [79] that follow are used to obtain the different thermodynamic parameters of the adsorption, which are the Gibbs free energy change ($\Delta G°$), entropy change ($\Delta S°$), and enthalpy change ($\Delta H°$):

$$\Delta G° = RT \times lnkd \qquad (4)$$

$$kd = qe \times Ce \qquad (5)$$

$$lnkd = \frac{\Delta S°}{R} - \frac{\Delta H°}{RT} \qquad (6)$$

$\Delta H°$ and $\Delta S°$ were obtained by plotting $lnkd$ against $1/T$, with the slope and intercept denoting the respective values.

The equations of Langmuir, Freundlich, Temkin, and Dubinin were used to assess the interaction between the grafted materials and the investigated adsorbate (Figure 10). Table 4 displays the derived parameters. In contrast to the other examined equations, the Langmuir equation showed a better fit with the experimental data ($0.99 \leq R^2$). This pattern implied a monolayer adsorption process and homogeneous sorption sites with comparable adsorption capabilities [80]. On the other hand, we observe that at 60 °C, the Temkin equation better fitted the adsorption data ($0.91 \leq R^2$) compared to the other models. This indicates that at relatively high temperatures (60 °C), the adsorption binding energy decreases with the increase in surface coverage [81]. Furthermore, at this temperature, we obtained a high value of the heat of sorption (the Temkin constant bt = 374.1 J mol^{-1}), suggesting a chemical adsorption process [82].

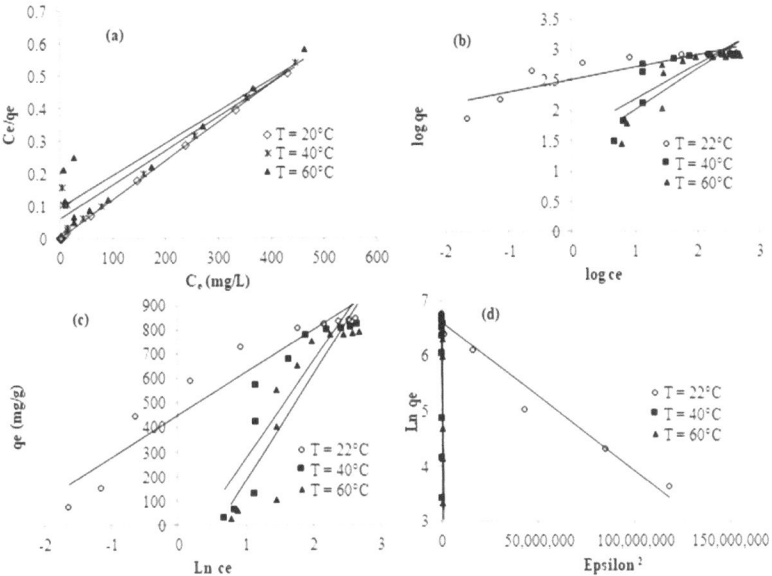

Figure 10. Models for isotherms data include (**a**) Langmuir, (**b**) Freundlich, (**c**) Temkin, and (**d**) Dubinin models.

Table 4. Different thermodynamic parameters and isotherm constants of adsorption.

T (°C)	Langmuir			Freundlich			Temkin			Dubinin			Thermodynamic		
	K_L	q_L	R^2	K_F	n_F	R^2	B_T	A_T	R^2	q_m	E	R^2	$\Delta G°$ (KJ/Mol)	$\Delta H°$ (KJ/Mol)	$\Delta S°$ (J/mol)
22	0.025	1000	0.999	321.07	2.044	0.772	438.3	0.07	0.821	740.40	2672.6	0.972	13.74		
40	0.022	1000	0.932	40.07	2.411	0.642	400.8	0.73	0.80	773.29	707.1	0.851	31.50	−2.604	−56.103
60	0.0093	1000	0.834	21.28	1.724	0.712	374.1	0.55	0.917	541.31	158.11	0.797	50.18		

The values of (1/n) exhibited surface heterogeneity or adsorption intensity. In fact, good adsorption may be displayed when the value of $1/n_F$ falls between 0.1 and 1.0 [83]. In our work, $1.72 \leq n_F \leq 2.41$, which reveals that the methylene blue molecules adsorb so effectively to the surface of the grafted samples. The exothermic nature of the mechanism of adsorption was demonstrated by the drop in the adsorption energy constant values (B_T), calculated using the Temkin model and a temperature rise. This is in line with the patterns shown in the temperature effect we mentioned previously.

The negative enthalpy value ($\Delta H° = -2.604$ KJ/mol) suggests that the reaction between the RR198 reactive dye and the produced adsorbent is exothermic. The increase in adsorption capacity with temperature was supported by these data. The entropy change's negative result ($\Delta S° = -56.103$ J/mol) indicated that there was less disorder and randomness at the interface between the MB dye's solid solution and the produced adsorbent. The computed free energy ($\Delta G° = 13.74 - 50.18$ KJ/mol) showed positive values, indicating a non-spontaneous sorption mechanism.

4. Conclusions

In the current study, a method of extraction applied to a Suaeda fruticosa plant yielded a naturally occurring polysaccharide, and various physicochemical characterizations (SEC, carbohydrate content, and NMR study) validated its pectin-like polysaccharide structure. Then, a layer-by-layer grafting method was investigated to design a novel multi-layered polymeric biosorbent. Chemical and morphological characteristics were obtained for the produced cellulosic biosorbent material that was grafted with extracted pectin polysaccharide and carrageenan biopolymers. The stability of the multi-layer grafting and the chemical functionalization during the polyesterification process were confirmed through FT-IR/ATR and SEM analysis. Methylene blue dye was the adsorbate selected as the reference cationic dye for biosorption assessment on PEM grafting material. The impact of the various process parameters, including pH, temperature, time, and MB concentration, on the sorption equilibrium was studied, revealing their significant effect. The biosorbent material created revealed exceptional sorption capacities of MB, reaching 802.6 mg/g higher than several polymeric materials reported in the literature. The enhanced sorption of the dye was caused by the incorporation of various carboxylate and sulfonate moieties into the crosslinked carrageenan/SFP polymer grafted onto the cellulose biosorbent system. The relationship between the equations from the theory and the data from experiments indicated that the data on kinetics could be accounted for in both pseudo-first order and pseudo-second order models, suggesting that the adsorption process involved both chemical and physical interactions. In heterogeneous adsorption sites, the adsorption phenomena took place, revealing an exothermic and spontaneous process. Adsorption data were better suited for the Langmuir isotherm. They demonstrated that all adsorption sites on the biosorbent material were homogeneous and had the same adsorption efficiency and that the adsorption process was confined to a monolayer (due to the excellent fit with the Langmuir model), which was further confirmed by the non-adequacy obtained with the Frundlich model. We developed a straightforward and novel low-cost polymeric biosorbent material to remove one of the most harmful pollutants from the water discharges of the textile industry. It might be extended to investigate how this substance might be used to remove metals and pesticides among other contaminants.

Author Contributions: Data curation, Y.E.-G. and S.A.; formal analysis, Y.E.-G. and S.A.; investigation, Y.E.-G. and S.A.; methodology, Y.E.-G.; project administration, Y.E.-G.; software, Y.E.-G. and S.A.; supervision, Y.E.-G.; validation, Y.E.-G.; writing—original draft, Y.E.-G. and S.A.; writing—review & editing, Y.E.-G. All authors have read and agreed to the published version of the manuscript.

Funding: The authors gratefully acknowledge Qassim University, represented by the Deanship of Scientific Research, on the financial support for this research under the number COS-2022-1-1-J-26650 during the academic year 1444 AH/2022 AD.

Institutional Review Board Statement: Not applicable.

Informed Consent Statement: Not applicable.

Data Availability Statement: The data are contained within the article.

Conflicts of Interest: The authors declare no conflicts of interest.

References

1. Inyinbor Adejumoke, A.; Adebesin Babatunde, O.; Oluyori Abimbola, P.; Adelani Akande Tabitha, A.; Dada Adewumi, O.; Oreofe Toyin, A. Water pollution: Effects, prevention, and climatic impact. *Water Chall. Urban. World* **2018**, *33*, 33–47.
2. Lin, L.; Yang, H.; Xu, X. Effects of water pollution on human health and disease heterogeneity: A review. *Front. Environ. Sci.* **2022**, *10*, 880246. [CrossRef]
3. Boviatsis, M.; Alexopoulos, A.B.; Vlachos, G.P. Evaluation of the response to emerging environmental threats, focusing on carbon dioxide (CO_2), volatile organic compounds (VOCs), and scrubber wash water (SOx). *Euro-Mediterr. J. Environ. Integr.* **2022**, *7*, 391–398. [CrossRef]
4. Sun, Y.; Li, H.; Lei, S.; Semple, K.T.; Coulon, F.; Hu, Q.; Gao, J.; Guo, G.; Gu, Q.; Jones, K.C. Redevelopment of urban brownfield sites in China: Motivation, history, policies and improved management. *Eco-Environ. Health* **2022**, *1*, 63–72. [CrossRef]
5. Arslan, I.; Balcioglu, I.A. Advanced oxidation of raw and biotreated textile industry wastewater with O_3, H_2O_2/UV-C and their sequential application. *J. Chem. Technol. Biotechnol. Int. Res. Process Environ. Clean Technol.* **2001**, *76*, 53–60. [CrossRef]
6. Zhang, S.; Li, B.; Wang, X.; Zhao, G.; Hu, B.; Lu, Z.; Wen, T.; Chen, J.; Wang, X. Recent developments of two-dimensional graphene-based composites in visible-light photocatalysis for eliminating persistent organic pollutants from wastewater. *Chem. Eng. J.* **2020**, *390*, 124642. [CrossRef]
7. Al-Tohamy, R.; Ali, S.S.; Li, F.; Okasha, K.M.; Mahmoud, Y.A.G.; Elsamahy, T.; Jiao, H.; Fu, Y.; Sun, J. A critical review on the treatment of dye-containing wastewater: Ecotoxicological and health concerns of textile dyes and possible remediation approaches for environmental safety. *Ecotoxicol. Environ. Saf.* **2022**, *231*, 113160. [CrossRef]
8. Kishor, R.; Purchase, D.; Saratale, G.D.; Saratale, R.G.; Ferreira, L.F.R.; Bilal, M.; Chandra, R.; Bharagava, R.N. Ecotoxicological and health concerns of persistent coloring pollutants of textile industry wastewater and treatment approaches for environmental safety. *J. Environ. Chem. Eng.* **2022**, *9*, 105012. [CrossRef]
9. Islam, T.; Repon, M.; Islam, T.; Sarwar, Z.; Rahman, M.M. Impact of textile dyes on health and ecosystem: A review of structure, causes, and potential solutions. *Environ. Sci. Pollut. Res. Int.* **2023**, *30*, 9207–9242. [CrossRef]
10. Tkaczyk, A.; Mitrowska, K.; Posyniak, A. Synthetic organic dyes as contaminants of the aquatic environment and their implications for ecosystems: A review. *Sci. Total Environ.* **2020**, *717*, 137222. [CrossRef]
11. Ismail, M.; Akhtar, K.; Khan, M.I.; Kamal, T.; Khan, M.A.; M Asiri, A.; Seo, J.; Khan, S.B. Pollution, toxicity and carcinogenicity of organic dyes and their catalytic bio-remediation. *Curr. Pharm. Des.* **2019**, *25*, 3645–3663. [CrossRef]
12. Khan, M.D.; Singh, A.; Khan, M.Z.; Tabraiz, S.; Sheikh, J. Current perspectives, recent advancements, and efficiencies of various dye-containing wastewater treatment technologies. *J. Water Process Eng.* **2023**, *53*, 103579. [CrossRef]
13. Kumar, S.; Yadav, S.; Kataria, N.; Chauhan, A.K.; Joshi, S.; Gupta, R.; Kumar, P.; Chong, J.W.R.; Khoo, K.S.; Show, P.L. Recent Advancement in Nanotechnology for the Treatment of Pharmaceutical Wastewater: Sources, Toxicity, and Remediation Technology. *Curr. Pollut. Rep.* **2023**, *9*, 110–142. [CrossRef]
14. Ramesh, B.; Saravanan, A.; Kumar, P.S.; Yaashikaa, P.R.; Thamarai, P.; Shaji, A.; Rangasamy, G.A. Review on algae biosorption for the removal of hazardous pollutants from wastewater: Limiting factors, prospects and recommendations. *Environ. Pollut.* **2023**, *327*, 121572. [CrossRef] [PubMed]
15. Karri, R.R.; Ravindran, G.; Dehghani, M.H. Wastewater—Sources, toxicity, and their consequences to human health. In *Soft Computing Techniques in Solid Waste and Wastewater Management*; Elsevier: Amsterdam, The Netherlands, 2021; pp. 3–33.
16. Wang, B.; Liu, Y.; Zhang, H.; Shi, W.; Xiong, M.; Gao, C.; Cui, M. Hydrodynamic cavitation and its application in water treatment combined with ozonation: A review. *J. Ind. Eng. Chem.* **2022**, *114*, 33–51. [CrossRef]
17. Ahmad, M.; Yousaf, M.; Nasir, A.; Bhatti, I.A.; Mahmood, A.; Fang, X.; Jian, X.; Kalantar-Zadeh, K.; Mahmood, N. Porous eleocharis@ MnPE layered hybrid for synergistic adsorption and catalytic biodegradation of toxic Azo dyes from industrial wastewater. *Environ. Sci. Technol.* **2019**, *53*, 2161–2170. [CrossRef]
18. Altowayti, W.A.H.; Shahir, S.; Othman, N.; Eisa, T.A.E.; Yafooz, W.M.S.; Al-Dhaqm, A.; Soon, C.Y.; Yahya, I.B.; Che Rahim, N.A.N.b.; Abaker, M.; et al. The Role of Conventional Methods and Artificial Intelligence in the Wastewater Treatment: A Comprehensive Review. *Processes* **2022**, *10*, 1832. [CrossRef]
19. El-taweel, R.M.; Mohamed, N.; Alrefaey, K.A.; Husien, S.; Abdel-Aziz, A.B.; Salim, A.I.; Mostafa, N.G.; Said, L.A.; Fahim, I.S.; Radwan, A.G. A review of coagulation explaining its definition, mechanism, coagulant types, and optimization models; RSM, and ANN. *Curr. Res. Green Sustain. Chem.* **2023**, *6*, 100358. [CrossRef]
20. Anisuzzaman, S.M.; Joseph, C.G.; Pang, C.K.; Affandi, N.A.; Maruja, S.N.; Vijayan, V. Current Trends in the Utilization of Photolysis and Photocatalysis Treatment Processes for the Remediation of Dye Wastewater: A Short Review. *Chem. Eng.* **2022**, *6*, 58. [CrossRef]
21. Zakaria, N.; Rohani, R.; Wan Mohtar, W.H.M.; Purwadi, R.; Sumampouw, G.A.; Indarto, A. Batik Effluent Treatment and Decolorization—A Review. *Water* **2023**, *15*, 1339. [CrossRef]
22. Fernández-Medrano, V.; Cuartas-Uribe, B.; Bes-Piá, M.-A.; Mendoza-Roca, J.-A. Application of Nanofiltration and Reverse Osmosis Membranes for Tannery Wastewater Reuse. *Water* **2022**, *14*, 2035. [CrossRef]
23. Hamad, H.N.; Idrus, S. Recent Developments in the Application of Bio-Waste-Derived Adsorbents for the Removal of Methylene Blue from Wastewater: A Review. *Polymers* **2022**, *14*, 783. [CrossRef]
24. Devaisy, S.; Kandasamy, J.; Aryal, R.; Johir, M.A.H.; Ratnaweera, H.; Vigneswaran, S. Removal of Organics with Ion-Exchange Resins (IEX) from Reverse Osmosis Concentrate. *Membranes* **2023**, *13*, 136. [CrossRef]

25. Borpatra Gohain, M.; Karki, S.; Yadav, D.; Yadav, A.; Thakare, N.R.; Hazarika, S.; Lee, H.K.; Ingole, P.G. Development of Antifouling Thin-Film Composite/Nanocomposite Membranes for Removal of Phosphate and Malachite Green Dye. *Membranes* **2022**, *12*, 768. [CrossRef] [PubMed]
26. Jabli, M.; Sebeia, N.; El-Ghoul, Y.; Soury, R.; Al-Ghamdi, Y.O.; Saleh, T.A. Chemical modification of microcrystalline cellulose with polyethyleneimine and hydrazine: Characterization and evaluation of its adsorption power toward anionic dyes. *Int. J. Biol. Macromol.* **2023**, *229*, 210–223. [CrossRef]
27. Akartasse, N.; Azzaoui, K.; Mejdoubi, E.; Hammouti, B.; Elansari, L.L.; Abou-salama, M.; Aaddouz, M.; Sabbahi, R.; Rhazi, L.; Siaj, M. Environmental-Friendly Adsorbent Composite Based on Hydroxyapatite/Hydroxypropyl Methyl-Cellulose for Removal of Cationic Dyes from an Aqueous Solution. *Polymers* **2022**, *14*, 2147. [CrossRef] [PubMed]
28. Ferkous, H.; Rouibah, K.; Hammoudi, N.-E.-H.; Alam, M.; Djilani, C.; Delimi, A.; Laraba, O.; Yadav, K.K.; Ahn, H.-J.; Jeon, B.-H.; et al. The Removal of a Textile Dye from an Aqueous Solution Using a Biocomposite Adsorbent. *Polymers* **2022**, *14*, 2396. [CrossRef]
29. Sharma, M.; Singh, G.; Vaish, R. Diesel soot-coated non-woven fabric for oil-water separation and adsorption applications. *Sci. Rep.* **2019**, *9*, 8503. [CrossRef]
30. Ren, Y.; Guo, J.; Lu, Q.; Xu, D.; Qin, J.; Yan, F. Polypropylene Nonwoven Fabric@Poly(ionic liquid)s for Switchable Oil/Water Separation, Dye Absorption, and Antibacterial Applications. *ChemSusChem* **2018**, *11*, 1092–1098. [CrossRef] [PubMed]
31. Haji, A.; Mousavi Shoushtari, A.; Abdouss, M. Plasma activation and acrylic acid grafting on polypropylene nonwoven surface for the removal of cationic dye from aqueous media. *Desalination Water Treat.* **2013**, *53*, 3632–3640. [CrossRef]
32. EL-Ghoul, Y.; Ammar, C.; Alminderej, F.M.; Shafiquzzaman, M. Design and Evaluation of a New Natural Multi-Layered Biopolymeric Adsorbent System-Based Chitosan/Cellulosic Nonwoven Material for the Biosorption of Industrial Textile Effluents. *Polymers* **2021**, *13*, 322. [CrossRef] [PubMed]
33. Costa, C.; Viana, A.; Oliveira, I.S.; Marques, E.F. Interactions between Ionic Cellulose Derivatives Recycled from Textile Wastes and Surfactants: Interfacial, Aggregation and Wettability Studies. *Molecules* **2023**, *28*, 3454. [CrossRef] [PubMed]
34. El-Ghoul, Y.; Alminderej, F.M. Bioactive and superabsorbent cellulosic dressing grafted alginate and Carthamus tinctorius polysaccharide extract for the treatment of chronic wounds. *Text. Res. J.* **2020**, *91*, 235–248. [CrossRef]
35. Somogyi Škoc, M.; Stevelić, N.; Rezić, I. Development and Characterization of Sustainable Coatings on Cellulose Fabric and Nonwoven for Medical Applications. *Sustainability* **2024**, *16*, 857. [CrossRef]
36. El-Ghoul, Y.; Ammar, C.; El-Achari, A. New polymer based modified cyclodextrins grafted to textile fibers; characterization and application to cotton wound dressings. *Int. J. Appl. Res. Text.* **2014**, *2*, 11–21.
37. Davydova, V.N.; Krylova, N.V.; Iunikhina, O.V.; Volod'ko, A.V.; Pimenova, E.A.; Shchelkanov, M.Y.; Yermak, I.M. Physicochemical Properties and Antiherpetic Activity of κ-Carrageenan Complex with Chitosan. *Mar. Drugs* **2023**, *21*, 238. [CrossRef]
38. EL-Ghoul, Y.; Al-Fakeh, M.S.; Al-Subaie, N.S. Synthesis and Characterization of a New Alginate/Carrageenan Crosslinked Biopolymer and Study of the Antibacterial, Antioxidant, and Anticancer Performance of Its Mn(II), Fe(III), Ni(II), and Cu(II) Polymeric Complexes. *Polymers* **2023**, *15*, 2511. [CrossRef]
39. Mzoughi, Z.; Abdelhamid, A.; Rihouey, C.; Le Cerf, D.; Bouraoui, A.; Majdoub, H. Optimized extraction of pectin-like polysaccharide from *Suaeda fruticosa* leaves: Characterization, antioxidant, anti-inflammatory and analgesic activities. *Carbohydr. Polym.* **2018**, *185*, 127–137. [CrossRef]
40. Alminderej, F.M.; Ammar, C.; El-Ghoul, Y. Functionalization, characterization and microbiological performance of new biocompatible cellulosic dressing grafted chitosan and Suaeda fruticosa polysaccharide extract. *Cellulose* **2021**, *28*, 9821–9835. [CrossRef]
41. You, L.; Gao, Q.; Feng, M.; Yang, B.; Ren, J.; Gu, L.; Zhao, M. Structural characterization of polysaccharides from Tricholoma matsutake and their antioxidant and antitumor activities. *Food Chem.* **2013**, *138*, 2242–2249. [CrossRef]
42. Bitter, T.; Muir, H.M. A modified uronic acid carbazole reaction. *Anal. Biochem.* **1962**, *4*, 330–334. [CrossRef]
43. Lowry, O.H.; Rosebrough, N.J.; Farr, A.L.; Randall, R.J. Protein measurement with the Folin phenol reagent. *J. Biol. Chem.* **1951**, *193*, 265–275. [CrossRef]
44. Wang, W.; Ma, X.; Xu, Y.; Cao, Y.; Jiang, Z.; Ding, T.; Ye, X.; Liu, D. Ultrasound-assisted heating extraction of pectin from grapefruit peel: Optimization and comparison with the conventional method. *Food Chem.* **2015**, *178*, 106–114. [CrossRef]
45. Marcon, M.V.; Carneiro, P.I.B.; Wosiacki, G.; Beleski-Carneiro, E.; Petkowicz, C.L.O. Pectins from apple pomace -Characterization by 13C and 1H NMR spectroscopy. *Annu. Magn. Reson.* **2005**, *4*, 56–63.
46. Rosenbohm, C.; Lundt, I.; Christensen, T.I.; Young, N.G. Chemically methylated and reduced pectins: Preparation, characterisation by 1H NMR spectroscopy, enzymatic degradation, and gelling properties. *Carbohydr. Res.* **2003**, *338*, 637–649. [CrossRef] [PubMed]
47. Wang, H.; Wei, G.; Liu, F.; Banerjee, G.; Joshi, M.; Bligh, S.; Wang, S. Characterization of Two Homogalacturonan Pectins with Immunomodulatory Activity from Green Tea. *Int. J. Mol. Sci.* **2014**, *15*, 9963–9978. [CrossRef] [PubMed]
48. Chen, Y.; Sheng, Q.; Hong, Y.; Lan, M. Hydrophilic nanocomposite functionalized by carrageenan for the specific enrichment of glycopeptides. *Anal. Chem.* **2019**, *91*, 4047–4054. [CrossRef]
49. Uranga, J.; Nguyen, B.T.; Si, T.T.; Guerrero, P.; de la Caba, K. The Effect of Cross-Linking with Citric Acid on the Properties of Agar/Fish Gelatin Films. *Polymers* **2020**, *12*, 291. [CrossRef]
50. El-Ghoul, Y. Biological and microbiological performance of new polymer-based chitosan and synthesized amino-cyclodextrin finished polypropylene abdominal wall prosthesis biomaterial. *Text. Res. J.* **2020**, *90*, 2690–2702. [CrossRef]

51. Rochas, C.; Lahaye, M.; Yaphe, W. Sulfate Content of Carrageenan and Agar Determined by Infrared Spectroscopy. *Bot. Mar.* **1986**, *29*, 335–340. [CrossRef]
52. Zhao, M.; Yang, N.; Yang, B.; Jiang, Y.; Zhang, G. Structural characterization of water-soluble polysaccharides from Opuntia monacantha cladodes in relation to their anti-glycated activities. *Food Chem.* **2007**, *105*, 1480–1486. [CrossRef]
53. Ammar, C.; Alminderej, F.M.; EL-Ghoul, Y.; Jabli, M.; Shafiquzzaman, M. Preparation and Characterization of a New Polymeric Multi-Layered Material Based K-Carrageenan and Alginate for Efficient Bio-Sorption of Methylene Blue Dye. *Polymers* **2021**, *13*, 411. [CrossRef]
54. Heidarinejad, Z.; Rahmanian, O.; Fazlzadeh, M.; Heidari, M. Enhancement of methylene blue adsorption onto activated carbon prepared from Date Press Cake by low frequency ultrasound. *J. Mol. Liq.* **2018**, *264*, 591–599. [CrossRef]
55. Jawad, A.H.; Abdulhameed, A.S.; Abdallah, R.; Yaseen, Z.M. Zwitterion composite chitosan-epichlorohydrin/zeolite for adsorption of methylene blue and reactive red 120 dyes. *Int. J. Biol. Macromol.* **2020**, *163*, 756–765. [CrossRef] [PubMed]
56. Zheng, X.; Zheng, H.; Xiong, Z.; Zhao, R.; Liu, Y.; Zhao, C.; Zheng, C. Novel anionic polyacrylamide-modify-chitosan magnetic composite nanoparticles with excellent adsorption capacity for cationic dyes and pH-independent adsorption capability for metal ions. *Chem. Eng. J.* **2020**, *392*, 123706. [CrossRef]
57. Kong, Q.; Wang, X.; Lou, T. Preparation of millimeter-sized chitosan/carboxymethyl cellulose hollow capsule and its dye adsorption properties. *Carbohydr. Polym.* **2020**, *244*, 116481. [CrossRef]
58. Azari, A.; Noorisepehr, M.; Dehganifard, E.; Karimyan, K.; Hashemi, S.Y.; Kalhori, E.M.; Norouzi, R.; Agarwal, S.; Gupta, V.K. Experimental Design, Modeling and Mechanism of Cationic Dyes Biosorption on to Magnetic Chitosan-lutaraldehyde Composite. *Int. J. Biol. Macromol.* **2019**, *131*, 633–645. [CrossRef]
59. Jiao, C.; Li, T.; Wang, J.; Wang, H.; Zhang, X.; Han, X.; Zhaofang, D.; Yali, S.; Yuyue, C. Efficient removal of dyes from aqueous solution by a porous sodium alginate/gelatin/graphene oxide triple-network composite aerogel. *J. Polym. Environ.* **2020**, *28*, 1492–1502. [CrossRef]
60. Miraboutalebi, S.M.; Nikouzad, S.K.; Peydayesh, M.; Allahgholi, N.; Vafajoo, L.; McKay, G. Methylene blue adsorption via maize silk powder: Kinetic, equilibrium, thermodynamic studies and residual error analysis. *Process. Saf. Environ. Prot.* **2017**, *106*, 191–202. [CrossRef]
61. Abdelkarim, S.; Mohammed, H.; Nouredine, B. Sorption of Methylene Blue Dye from Aqueous Solution Using an Agricultural Waste. *Trends Green Chem.* **2017**, *3*, 1–7. [CrossRef]
62. Boumaza, S.; Yenounne, A.; Hachi, W.; Kaouah, F.; Bouhamidi, Y.; Trari, M. Application of *Typha angustifolia* (L.) Dead Leaves Waste as Biomaterial for the Removal of Cationic Dye from Aqueous Solution. *Int. J. Environ. Res.* **2018**, *12*, 561–573. [CrossRef]
63. Jawad, A.H.; Razuan, R.; Appaturi, J.N.; Wilson, L.D. Adsorption and mechanism study for methylene blue dye removal with carbonized watermelon (Citrullus lanatus) rind prepared via one-step liquid phase H_2SO_4 activation. *Surf. Interfaces* **2019**, *16*, 76–84. [CrossRef]
64. Mahdavinia, G.R.; Mosallanezhad, A. Facile and green route to prepare magnetic and chitosan-crosslinked κ-carrageenan bionanocomposites for removal of methylene blue. *J. Water Process. Eng.* **2016**, *10*, 143–155. [CrossRef]
65. Soares, S.F.; Simões, T.R.; Trindade, T.; Daniel-da-Silva, A.L. Highly efficient removal of dye from water using magnetic carrageenan/silica hybrid nano-adsorbents. *Water Air Soil Pollut.* **2017**, *228*, 87. [CrossRef]
66. Saber-Samandari, S.; Joneidi-Yekta, H.; Mohseni, M. Adsorption of anionic and cationic dyes from aqueous solution using gelatin-based magnetic nanocomposite beads comprising carboxylic acid functionalized carbon nanotube. *Chem. Eng. J.* **2017**, *308*, 1133–1144. [CrossRef]
67. Kim, H.R.; Jang, J.W.; Park, J.W. Carboxymethyl chitosan-modified magnetic-cored dendrimer as an amphoteric adsorbent. *J. Hazard. Mater.* **2016**, *317*, 608–616. [CrossRef] [PubMed]
68. El Maguana, Y.; Elhadiri, N.; Bouchdoug, M.; Benchanaa, M.; Jaouad, A. Activated carbon from prickly pear seed cake: Optimization of preparation conditions using experimental design and its application in dye removal. *Int. J. Chem. Eng.* **2019**, *2019*, 8621951. [CrossRef]
69. Islam, M.A.; Ahmed, M.J.; Khanday, W.A.; Asif, M.; Hameed, B.H. Mesoporous activated carbon prepared from NaOH activation of rattan (*Lacosperma secundiflorum*) hydrochar for methylene blue removal. *Ecotoxicol. Environ. Saf.* **2017**, *138*, 279–285. [CrossRef]
70. Islam, M.A.; Sabar, S.; Benhouria, A.; Khanday, W.A.; Asif, M.; Hameed, B.H. Nanoporous activated carbon prepared from karanj (*Pongamia pinnata*) fruit hulls for methylene blue adsorption. *J. Taiwan Inst. Chem. Eng.* **2017**, *74*, 96–104. [CrossRef]
71. Mosoarca, G.; Popa, S.; Vancea, C.; Dan, M.; Boran, S. Removal of Methylene Blue from Aqueous Solutions Using a New Natural Lignocellulosic Adsorbent—Raspberry (*Rubus idaeus*) Leaves Powder. *Polymers* **2022**, *14*, 1966. [CrossRef]
72. Mennas, N.; Lahreche, S.; Chouli, F.; Sabantina, L.; Benyoucef, A. Adsorption of Methylene Blue Dye by Cetyltrimethylammonium Bromide Intercalated Polyaniline-Functionalized Montmorillonite Clay Nanocomposite: Kinetics, Isotherms, and Mechanism Study. *Polymers* **2023**, *15*, 3518. [CrossRef]
73. ALSamman, M.T.; Sánchez, J. Chitosan- and Alginate-Based Hydrogels for the Adsorption of Anionic and Cationic Dyes from Water. *Polymers* **2022**, *14*, 1498. [CrossRef]
74. Almutairi, F.M.; El-Ghoul, Y.; Jabli, M. Extraction of Cellulose Polymeric Material from Populus tremula Fibers: Characterization and Application to the Adsorption of Methylene Blue and Crystal Violet. *Polymers* **2021**, *13*, 3334. [CrossRef] [PubMed]
75. Godiya, C.B.; Xiao, Y.; Lu, X. Amine functionalized sodium alginate hydrogel for efficient and rapid removal of methyl blue in water. *Int. J. Biol. Macromol.* **2020**, *144*, 671–681. [CrossRef] [PubMed]

76. Mittal, H.; Al Alili, A.; Morajkar, P.P.; Alhassan, S.M. GO crosslinked hydrogel nanocomposites of chitosan/carboxymethyl cellulose–A versatile adsorbent for the treatment of dyes contaminated wastewater. *Int. J. Biol. Macromol.* **2021**, *167*, 1248–1261. [CrossRef]
77. Gucek, A.; Sener, S.; Bilgen, S.; Mazmanci, A. Adsorption and kinetic studies of cationic and anionic dyes on pyrophyllite from aqueous solutions. *J. Coll. Interf. Sci.* **2005**, *286*, 53–60. [CrossRef]
78. Ho, Y.S.; McKay, G. The kinetics of sorption of basic dyes from aqueous solution by sphagnum moss peat. *Can. J. Chem. Eng.* **1998**, *76*, 822–827. [CrossRef]
79. Abdulhameed, A.S.; Jawad, A.H.; Mohammad, A.T. Synthesis of chitosan-ethylene glycol diglycidyl ether/TiO_2 nanoparticles for adsorption of reactive orange 16 dye using a response surface methodology approach. *Bioresour. Technol.* **2019**, *293*, 122071. [CrossRef] [PubMed]
80. Mall, I.D.; Srivastava, V.C.; Agarwal, N.K. Removal of Orange-G and Methyl Violet dyes by adsorption onto bagasse fly ash-kinetic study and equilibrium isotherm analyses. *Dye. Pigment.* **2006**, *69*, 210–223. [CrossRef]
81. Temkin, M.I.; Pyzhev, V. Kinetic of ammonia synthesis on promoted iron catalyst. *Acta Physiochim.* **1940**, *12*, 327–356.
82. Ngulube, T.; Gumbo, J.R.; Masindi, V.; Maity, A. Calcined magnesite as an adsorbent for cationic and anionic dyes: Characterization, adsorption parameters, isotherms and kinetics study. *Heliyon* **2018**, *4*, 838. [CrossRef] [PubMed]
83. Kuang, Y.; Zhang, X.; Zhou, S. Adsorption of Methylene Blue in Water onto Activated Carbon by Surfactant Modification. *Water* **2020**, *12*, 587. [CrossRef]

Disclaimer/Publisher's Note: The statements, opinions and data contained in all publications are solely those of the individual author(s) and contributor(s) and not of MDPI and/or the editor(s). MDPI and/or the editor(s) disclaim responsibility for any injury to people or property resulting from any ideas, methods, instructions or products referred to in the content.

Article

A Novel Chitosan/Nano-Hydroxyapatite Composite for the Adsorptive Removal of Cd(II) from Aqueous Solution

Rachid El Kaim Billah [1,†], Ikrame Ayouch [2,3,†], Youness Abdellaoui [4,5], Zineb Kassab [6], Moonis Ali Khan [7,*], Mahfoud Agunaou [1], Abdessadik Soufiane [1], Marta Otero [8,*] and Byong-Hun Jeon [9]

1. Laboratory of Coordination and Analytical Chemistry, Department of Chemistry, Faculty of Sciences, University of Chouaib Doukkali, El Jadida 24000, Morocco
2. Laboratory of Materials and Interfacial Systems, Faculty of Sciences Tétouan, University Abdelmalek Essaadi (UAE), P.O. Box 2121, Tétouan 93000, Morocco
3. MASCIR Foundation, Rabat Design, Rue Mohamed EL Jazouli, Madinat EL Irfane, Rabat 10100, Morocco
4. Faculty of Engineering, Autonomous University of Yucatan, Mérida 97000, Mexico
5. Department of Sustainability of Natural Resources and Energy, Center for Research and Advanced Studies of the National Polytechnic Institute, Saltillo 25900, Mexico
6. Materials Science Energy and Nanoengineering Department (MSN), Mohammed VI Polytechnic University (UM6P), Ben Guerir 43150, Morocco
7. Chemistry Department, College of Science, King Saud University, Riyadh 11451, Saudi Arabia
8. Departmento de Química y Física Aplicadas, Universidad de Leon, Campus de Vegazana s/n, 24071 Leon, Spain
9. Department of Earth Resources and Environmental Engineering, Hanyang University, Seoul 04763, Republic of Korea
* Correspondence: mokhan@ksu.edu.sa (M.A.K.); marta.otero@unileon.es (M.O.)
† These authors contributed equally to this work.

Citation: El Kaim Billah, R.; Ayouch, I.; Abdellaoui, Y.; Kassab, Z.; Khan, M.A.; Agunaou, M.; Soufiane, A.; Otero, M.; Jeon, B.-H. A Novel Chitosan/Nano-Hydroxyapatite Composite for the Adsorptive Removal of Cd(II) from Aqueous Solution. *Polymers* **2023**, *15*, 1524. https://doi.org/10.3390/polym15061524

Academic Editors: Alexander Böker and Daniela Predoi

Received: 13 November 2022
Revised: 27 February 2023
Accepted: 14 March 2023
Published: 19 March 2023

Copyright: © 2023 by the authors. Licensee MDPI, Basel, Switzerland. This article is an open access article distributed under the terms and conditions of the Creative Commons Attribution (CC BY) license (https://creativecommons.org/licenses/by/4.0/).

Abstract: A novel polymer bio-composite based on nano-hydroxyapatite (n-Hap) and chitosan (CS) (CS/n-Hap) was synthesized to effectively address toxic cadmium ions removal from water. The composition and structure of CS/n-Hap bio-composite were analyzed through different characterization techniques. XRD patterns affirmed that the crystalline structure of n-Hap remained unaltered during CS/n-Hap synthesis, while FT-IR spectrum sustained all the characteristic peaks of both CS and n-Hap, affirming the successful synthesis of CS/n-Hap. Adsorption studies, including pH, adsorbent dosage, contact time, initial Cd(II) concentration, and temperature, were carried out to explain and understand the adsorption mechanism. Comparatively, CS/n-Hap bio-composite exhibited better Cd(II) adsorption capacity than pristine CS, with an experimental maximum uptake of 126.65 mg/g under optimized conditions. In addition, the kinetic data were well fitted to the pseudo-second-order model, indicating the formation of chemical bonds between Cd(II) and CS/n-Hap during adsorption. Furthermore, the thermodynamic study suggested that Cd(II) adsorption onto CS/n-Hap was endothermic and spontaneous. The regeneration study showed only about a 3% loss in Cd(II) uptake by CS/n-Hap after five consecutive cycles. Thus, a simple and facile approach was here developed to synthesize an eco-friendly and cost-effective material that can be successfully employed for the removal of toxic heavy metal ions from water.

Keywords: nano-hydroxyapatite; chitosan; bio-composite; cadmium; adsorption mechanism

1. Introduction

Environmental pollution induced through rapid technological development is a matter of critical ecological concern. In particular, the contamination by heavy metal ions poses a serious risk to the ecosystem [1]. Among them, cadmium [Cd(II)] is the most concerning heavy metal as it is a cornerstone of recent widespread diseases in developing countries [2].

Cadmium is a rare but naturally occurring element in the earth's crust, where it is not present in pure state but in combination with other elements (e.g., oxygen (cadmium oxide), chlorine (cadmium chloride), or sulfur (cadmium sulfide)) [3], possibly leaching into ground and surface waters. In addition, significant Cd(II) emissions can be anthropogenically caused by various industrial activities, including metal plating, cadmium–nickel batteries, phosphate fertilizer, mining, pigments, stabilizers, and alloys [4]. Cd(II) is a highly toxic element; excessive human exposure to it can lead to bone degeneration, liver damage, lung failure, hypertension, kidney dysfunction and, eventually, cancer [5].

Adhering to environmental limits and eventually safeguarding the ecosystem reduction in Cd(II) concentration is essential. Various treatment methods, such as chemical precipitation, ultrafiltration, membrane separation, electrochemical deposition, and adsorption, have been engineered in this regard. Although these methods have shown promising results in removing Cd(II) from wastewater, they may present some significant drawbacks, such as high sludge yield, high energy requirements, and secondary pollution generation [6,7]. These shortcomings are avoided by adsorptive methods, which have been proven to be practical in terms of cost, simplicity, versatility, and flexibility. Therefore, adsorption is actually the commercially preferred option for the removal of Cd(II) from water. However, this is conditioned by the choice of a suitable adsorbent based on its specific surface area (SSA), adsorption potential, high availability, and mechanical and chemical stabilities [8,9]. Hence, it is necessary to develop appropriate, efficient, environmentally-friendly and cost-effective adsorbents for Cd(II) removal.

Activated carbon is the most extensively used adsorbent for removing inorganic and organic contaminants from water. However, producing and regenerating activated carbon involves superimposable costs, counting on the precursor raw material and synthesis methodology, which usually involves specific physicochemical processes to refine the adsorbent's performance [10]. Alternately, synthetic polymers are appealing materials for developing adsorbents by different methods. In addition, their characteristics, such as SSA and chemical functionalities, can be controlled and are highly effective. However, the synthesis approaches are usually complicated and involve specific reagents, which increases net costs [11,12]. In this regard, natural polymers (biopolymers), especially polysaccharides, have received particular attention. Cellulose, chitosan (CS), and alginate are three major natural biopolymers that have been actively employed for the synthesis of materials aimed at the adsorption of heavy metals from contaminated aqueous solutions due to their compelling properties [13,14]. After cellulose, CS is the second most abundant natural polymer. It is widely found in nature and is usually produced by chemical (alkaline) or enzymatic deacetylation of chitin. It has potential applications in biotechnological, agricultural, packaging, pharmaceutical, textile, cosmetic, and many other industries [15–20]. As CS is biodegradable, non-toxic, sustainable, and is composed of many amino and hydroxyl groups that can interact electrostatically with heavy metals, it is considered as a valuable material for water treatment [21,22]. However, due to its crystallized structure, CS has poor acid resistance, low porosity, and low mechanical strength [23]. Therefore, CS cannot be used without adequate preparatory modification or combination so to get a chemically stable material in the aqueous phase over a wide pH range, to improve its adsorption performance towards heavy metals by creating more active sites and to enhance porosity. Several research works have been conducted using modified CS as adsorbent for heavy metals. For instance, Sutirman et al. [23] studied divalent heavy metals adsorption on modified CS beads. Wu et al. [24] fabricated phosphorylated magnetic $CS/CoFe_2O_4$ for divalent heavy metal ions uptake. CS-silica hybrid aerogel was developed for divalent heavy metal removal with a maximum uptake of 64.74 mg/g [25]. In addition, Rathinam et al. [26] synthesized a chitosan–lysozyme bio-composite using glutaraldehyde as a crosslinker and investigated its adsorption ability towards Cr(VI). Furthermore, to eliminate Cd(II) and Pb(II) from water, Chen et al. [27] used chitosan/vermiculite bio-composite with an epichlorohydrin (ECH) cross-linking agent, achieving maximum adsorption capacities of 58.5 mg/g and 166.7 mg/g for Cd(II) and Pb(II), respectively, at pH 4.

Nano-crystalline materials generally have a higher SSA and adsorption capacity than conventional materials. In this sense, one of the most promising materials to be combined with CS is nano-hydroxyapatite (n-Hap). N-Hap ($Ca_{10}(PO_4)_6(OH)_2$) is regarded as an environmentally friendly adsorbent due to its biosecurity, low-cost, and excellent storage ability [27,28]. In addition, it displays outstanding biocompatibility and high removal ability towards heavy metals through mechanisms such as ion-exchange, surface complex formation, or dissolution–precipitation. Moreover, the crystalline and porous structure of n-Hap favors the adsorption of metal ions onto n-Hap [27,29]. However, n-Hap nanoparticles tend to aggregate due to the extremely high surface energy induced by van der Waals force, which causes a significant decrease of their SSA and eventually affects their removal capacity [30]. Therefore, the compositing of n-Hap with CS can synergically evolve desired adsorption properties that cannot be attained individually. In the composite, CS serves as a carrier to prevent n-Hap aggregation and increases its adsorption capacities by improving the number of active sites as well as generating functional groups; meanwhile, n-Hap can impart vital chemical and mechanical stability, the potential to immobilize metal cations on its P-OH group sites, and also a large SSA.

In the described context, this work aimed at enhancing the adsorptive performance of the abundant natural polymer CS (extracted from shrimp chitin) by combining it with n-Hap. A new green and innovative CS/n-Hap composite was synthesized, fully characterized, and compared with CS in the removal of Cd(II) from water. A comprehensive assessment of the adsorptive performance under varying experimental conditions such as pH, adsorbent dosage, presence of competing species, contact time, initial Cd(II) concentration, temperature, and competition by counter ions and organic contaminants was accomplished. Kinetic, equilibrium, and thermodynamic parameters were evaluated, and the adsorption mechanism was discussed. In addition, and in view of assessing the sustainable application of the CS/n-Hap composite for wastewater treatment, its reusability was investigated through a cyclic regeneration study.

2. Materials and Methods

2.1. Chemicals and Reagents

Crude chitin was extracted from shrimp shells waste collected from El-Jadida region in Morocco; calcium nitrate ($Ca(NO_3)_2, 4H_2O$, N° CAS: 13477-34-4), acetic acid ($C_2H_4O_2$, N° CAS: 64-19-7), diammonium hydrogen phosphate (($NH_4)_2HPO_4$, N° CAS: 7783-28-0), nitric acid (HNO_3, N° CAS: 7697-37-2), cadmium nitrate ($Cd(NO_3)_2$, N° CAS: 10022-68-1), sodium hydroxide (NaOH, N° CAS: 1310-73-2), hydrochloric acid (HCl) magnesium nitrate ($Mg(NO_3)_2, 6H_2O$, CAS N°: 13446-18-9), sodium nitrate ($NaNO_3$, CAS N°: 7631-99-4), potassium nitrate (KNO_3, CAS N°: 7757-79-1), 2-nitrophenol ($O_2NC_6H_4OH$, N° CAS: 88-75-5), paracetamol ($CH_3CONHC_6H_4OH$, N° CAS: 103-90-2), and amoxicillin ($CH_3CONHC_6H_4OH$, N° CAS: 61336-70-7) were purchased from Sigma-Aldrich, St. Louis, MO 63103, USA.

2.2. Adsorbent Synthesis

2.2.1. Preparation of Chitosan (CS)

The chitin was treated with a solution of NaOH (48 wt.%) at 100 °C to remove acetyl groups from the chitin, the solid–liquid ratio being 1:20. The so-treated chitin was separated by filtration, washed with deionized (D.I.) water until the solution reached pH ~7, and then oven-dried at 50 °C [31] to obtain the CS to be used in this work.

The CS deacetylation degree was determined by conductometric analysis; 150 mg of CS was completely solubilized in 10 mL of 0.1 N HCl. Thereafter, the volume of the solution was adjusted to 200 mL with D.I. water. The CS solution was titrated with 0.1 N NaOH solution and the solution conductivity was measured after each addition of NaOH.

The degree of deacetylation (DDA) was calculated as:

$$\text{DDA} = \frac{203 \, (V2 - V1)N}{m + 42(V2 - V1)N} \times 100 \qquad (1)$$

where N is the normality of the NaOH solution (N), V1 and V2 are the equivalent volumes of NaOH (L), and m is the mass of CS (g). An inoLab™ Cond 7310 Conductometer (Xylem Analytics Germany GmbH, Weilheim, Germany) was used during the analysis.

Additionally, the molecular weight of CS was as well determined by viscometry; one of the most used methods for the determination of the molecular weight of CS is from the intrinsic viscosity η by applying the Marque–Houwink equation as:

$$[\eta] = K \times M \times a \qquad (2)$$

where K and a are constants that depend on the polymer–solvent system at a given temperature, M is the molecular weight in dalton and η is the intrinsic viscosity. The values of K and a are respectively 0.078 mL/g and 0.76 for material solutions prepared in 0.1 M acetic acid/0.2 M sodium acetate at 25 °C.

To measure intrinsic viscosity, a capillary Ubbelohde-type viscometer (0.53 μm) was used. Measurements consisted of the determination (at a given temperature) of the flow time in a vertical capillary tube for the solvent and for the same volume of a CS solution. The photoelectric cells were connected to a digital chronometer giving the flow time in seconds with a precision of 10^{-2} s. Then, the reduction in viscosity was measured for CS solutions at different concentrations. The plot of viscosity versus CS concentration of the solutions results in a line whose intercept is equal to the intrinsic viscosity. From the so-determined value, the M of the prepared CS was calculated. The extracted CS had a M = 345 and a DDA = 93%.

2.2.2. Preparation of Nano-Hydroxyapatite (n-Hap)

The n-Hap was obtained by wet chemical synthesis, as described by Abidi et al. [32]. Briefly, 0.1 M of $Ca(NO_3)_2 \cdot 4H_2O$ and 0.06 M $(NH_4)_2HPO_4$ were prepared separately. Subsequently, 100 mL of previously prepared $Ca(NO_3)_2$ solution was transferred into a 500 mL three-neck round-bottom flask and heated at 90 °C for an hour in an oil bath with magnetic stirrer. 100 mL of aqueous $(NH_4)_2HPO_4$ solution followed by 200 mL of NH_4OH solution were added dropwise through the left neck of the flask by continuously monitoring the pH of the solution. The solution was continuously stirred for 3 h to obtain the n-Hap powder. The chemical precipitation reaction for the synthesis of n-Hap is presented as:

$$10Ca(NO_3)_2 \cdot 4H_2O + 6(NH_4)_2HPO_4 + 8NH_4OH \rightarrow Ca_{10}(PO_4)_6(OH)_2 + 20NH_4NO_3 + 46H_2O \qquad (3)$$

2.2.3. Preparation of CS/n-Hap Composite

To prepare the CS/n-Hap composite, CS powder (2 g) and n-Hap (1 g) were suspended in 80 mL of a 5% (v/v) acetic acid solution for 24 h. Then, the solid material was separated by filtration and washed with D.I. water until the solution reached pH ~7. The composite obtained was oven-dried for 12 h at 65 °C. Finally, the composite was manually ground. Figure S1 (Supplementary Materials) schematic presents CS/n-Hap bio-composite synthesis.

2.3. Characterization

A Bruker D8 diffractometer was used for X-ray diffraction (XRD) analysis. The sample's morphology was analyzed by a XL 30 ESEM scanning electron microscopy (SEM) acquired from Philips (Amsterdam, The Netherlands). Active functional groups over the material's surface were determined by Fourier transform infra-red (FT-IR) spectroscopy using a Nicolet 6700 Spectrometer (Thermo-Fisher Scientific, Massachusetts, USA). Thermal analysis was performed under N_2 atmosphere using a thermogravimetric analyzer Discovery TGA from TA instruments (Waters Corporation, New Castle, USA). The determination of the point of zero charge (pH_{PZC}) was done by the method described by Zaini et al. [33]. Briefly, 0.10 g of material was mixed with 50 mL of 0.01 M NaCl at different initial pHs (pH = 1, 2, 3, 4, 5, 6, 7, and 8). These pH values were adjusted by adding a small amount of 0.1 M HCl/NaOH solutions. The suspensions were allowed to equilibrate for 72 h under mechanical agitation at 25 °C, then centrifuged at 5000 rpm for 10 min, and the final pH of each supernatant was measured using a pH meter inoLab® pH 7310 from Xylem Analytics (Weilheim, Germany).

2.4. Adsorption Studies

Cd(II) batch adsorption experiments were done using CS or CS/n-Hap bio-composite as adsorbents. The adsorption experiments were carried out in 150 mL glass beakers containing 50 mL of Cd(II) solutions under stirring at 400 rpm. At equilibrium, residual Cd(II) concentration was analyzed by inductively coupled plasma atomic emission spectrometry (ICP-AES; Thermo Jarrell Ash Corporation Atom Scan 16, Williamston, SC, USA). The limit of detection (LOD) and limit of quantification (LOQ) were respectively determined as 5 and 15 µg/L. The effect of experimental parameters, namely pH, presence of competing species, adsorbent dose, contact time, initial Cd(II) concentration, and temperature on Cd(II) adsorption onto CS/n-Hap was examined. The Cd(II) solution pH (in range 2–8) was adjusted using 0.1 M HCl/NaOH solution, while the CS and CS/n-Hap bio-composite concentration (m) was ranged between 25 and 250 mg, the temperature (T) was varied between 15 and 45 °C, and the initial Cd(II) concentration (C_o) was varied in range 20–300 mg/L. The contact time (t) during the study was varied in range 5–180 min.

The adsorbed concentration at equilibrium (q_e, mg/g) and removal percentage were respectively calculated as:

$$q_e \ (mg/g) = \frac{(C_o - C_e)V}{m} \qquad (4)$$

$$\text{Removal } (\%) = \frac{(C_o - C_e)V}{C_o} \times 100\% \qquad (5)$$

where C_o and C_e are the initial and residual Cd(II) concentrations in the solution at equilibrium, respectively. V is the solution volume (L), and m is the mass of adsorbent (g). The adsorbed concentration at a time t (q_t, mg/g) was calculated using Equation (4) but replacing C_e by C_t (the residual Cd(II) concentration (mg/L) at time t).

The mechanistic insight during Cd(II) removal was investigated by applying experimental data to adsorption kinetic and isotherm models. In the case of the kinetic study, the used models were pseudo-first order, pseudo-second order and intra-particle diffusion linear models, while for isotherm study, Langmuir and Freundlich models were applied (Tables S1 and S2, Supplementary Materials).

2.5. Regeneration Study

During the study, 100 mg CS/n-Hap was saturated with 50 mL Cd(II) solution with a C_o of 100 mg/L. The Cd(II) saturated CS/n-Hap was separated and washed with D.I. water to remove unadsorbed traces. Thereafter, the CS/n-Hap sample was separately treated with 50 mL of 0.05 N HCl under stirring for 2 h. Then, the regenerated CS/n-Hap was filtered, washed, and reused for the adsorption of Cd(II). The procedure was repeated for ten consecutive regeneration cycles.

3. Results and Discussion

3.1. Characterization

3.1.1. X-ray Diffraction Analysis

The XRD results obtained for CS, n-Hap, and CS/n-Hap composite are illustrated in Figure 1a. The XRD pattern of CS showed two well-defined peaks at 19.4° and 9.6°, corresponding to the characteristic peaks of CS [3,34]. For n-Hap, the peaks at 2θ = 25.88°, 31.88°, 32.28°, 34.08°, 39.78°, and 49.58°, corresponding to the diffraction planes (0 0 2), (2 1 1), (1 1 2), (2 0 2), (1 3 0), and (2 1 3) (JCPDS no. 01-073-8417) respectively, confirm the formation of n-Hap [34]. These diffraction peaks were also found in CS/n-Hap composite, at 20.0°, 29.15°, 31.7°, and 33.1°, corresponding to the diffraction planes of CS and n-Hap, respectively [35]. Compositing with CS did not affect the n-Hap crystalline structure, although shorter and broader signals were observed, in line with previous studies [36,37].

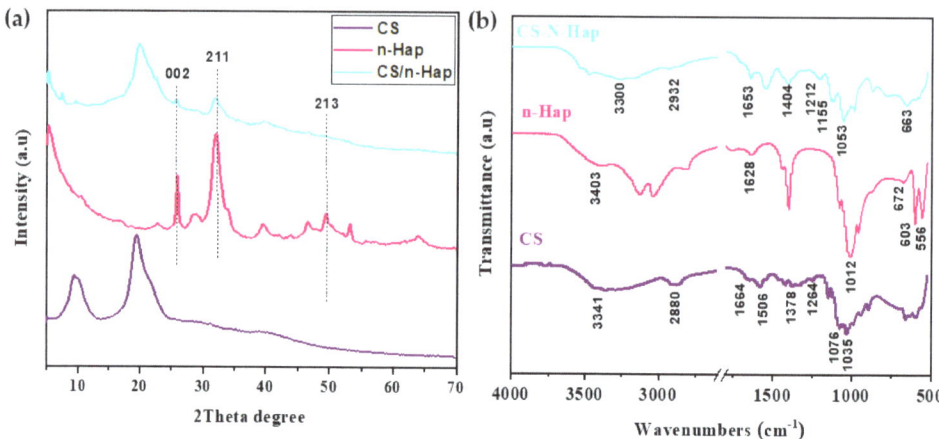

Figure 1. (a) XRD patterns, and (b) FT-IR spectra of CS, n-Hap, and CS/n-Hap.

3.1.2. Fourier Transform Infra-Red Analysis

A characteristic band at 3341 cm^{-1} was observed in the CS spectrum which corresponds to the stretching vibrations of hydroxyl groups, overlapping with the -NH$_2$ stretching vibration peak of CS [35] (Figure 1b). Bands ranging from 1664 to 1506 cm^{-1} can be attributed to the C-O stretching vibrations and the N-H in-plane bending vibrations characteristic of amide I and II structures [35,36], band at 1264 cm^{-1} corresponds to the amide III structure [35]. The band at 2880 cm^{-1} was attributed to -CH vibrations, while at 1378 cm^{-1} it was attributed to -CH$_3$ and -CH$_2$ vibrations. The bands at 1076 and 1035 cm^{-1} correspond to glucosamine stretching vibrations and C–O–C stretching, respectively [35]. n-Hap spectrum displayed bands at 3403, 1628, and 672 cm^{-1} assigned to the stretching and bending of hydroxyl groups [36]. The band observed at 1012 cm^{-1} was assigned to asymmetric stretching, while the bands at 603 and 556 cm^{-1} were assigned to symmetric stretching of phosphate groups [35].

The spectrum of CS/n-Hap composite retained all the characteristic peaks of CS and n-Hap. The band at 3300 cm^{-1} was attributed to -OH groups stretching vibrations; this band decreased in intensity and shifted towards lower wavenumbers in the CS/n-Hap composite due to the intermolecular or intramolecular hydrogen bonds between compounds [38]. The bands at 2932 and 1404 cm^{-1} corresponding to the stretching and bending (–CH) group appeared in the CS/n-Hap composite confirming its formation [35]. The –NH group of CS was included in the 1653 cm^{-1} band [39]. The bands at around 1053 cm^{-1} and 663 cm^{-1} were respectively attributed to the stretching and bending vibrations of the PO$_4^{3-}$ group present in hydroxyapatite [36,38]. The vibration bands of C–O group were overlapped with phosphate bands at 1150–1040 cm^{-1} [36]. The appearance of the ether bond in the pyranose ring at 1155 cm^{-1} and the amide III band at 1212 cm^{-1} further evidenced the adequate compositing of CS/n-Hap. Moreover, the band broadening around 1053 cm^{-1} showed the presence of the CS polymer and its interaction with PO$_4^{3-}$ groups of hydroxyapatite [39]. Thus, physical interactions, namely electronic forces and hydrogen bonds between n-Hap and CS may have probably been established during the composite formation [35].

3.1.3. Morphological and Elemental Analysis

Figure S2a,b (Supplementary Materials) displays the SEM micrographs of n-Hap and the CS/n-Hap composite. The morphological images showed that n-Hap particles exhibit a non-uniform shape and size aggregations. Moreover, the CS/n-Hap bio-composite surface displayed a rough and plastic-like appearance. Furthermore, n-Hap particles appeared to be homogeneously dispersed on the polymer surface, indicating good component incorporation [40] and ability of the polymer to bind the solid n-Hap particles [41]. The energy dispersive spectroscopy (EDX) spectra confirm the presence of calcium and phosphate in n-Hap (Figure S2c, Supplementary Materials). Meanwhile, for the CS/n-Hap composite, significant proportions of nitrogen and carbon resulting from the CS structure are evident (Figure S2d, Supplementary Materials).

3.1.4. Thermogravimetric Analysis

Three main stages of weight loss were observed for both CS and CS/n-Hap samples (Figure 2a,b). The first stage in the range 50–100 °C was attributed to the dehydration of physically retained water, which starts at relative low temperature [41,42]. The thermal decomposition and degradation of the main chain and deacetylation of the CS molecules was responsible for the dominating weight loss at around 228–400 °C in the case of CS [42]. For the CS/n-Hap bio-composite, this weight loss stage occurred at lower temperatures than for CS. This may be due to the improved heat transfer to the CS matrix by the inorganic n-Hap particles present on the surface, leading to rapid degradation of the organic CS [43]. Both CS and CS/n-Hap exhibited other weight loss stage in the range 330–500 °C owing to decomposition of the degraded fragments and slow char oxidation [42]. Finally, residual weight loss at temperatures above 600 °C may be related to the n-Hap dehydroxylation during its thermal decomposition [44].

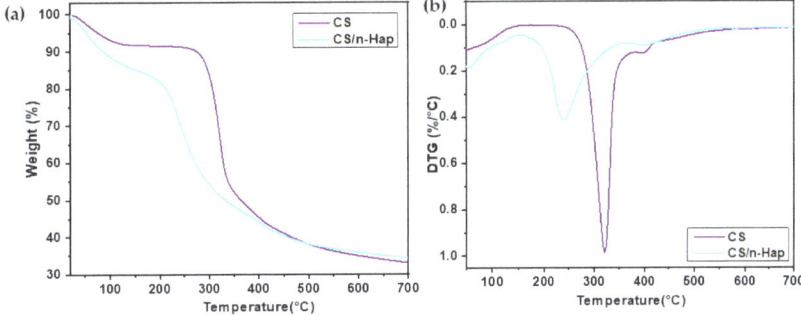

Figure 2. TGA (**a**) and DTG (**b**) plots of CS and CS/n-Hap.

3.1.5. N_2 Adsorption–Desorption Isotherm Analysis

N_2 adsorption/desorption isotherm curves of CS, n-Hap, and CS/n-Hap are depicted in Figure S3 (Supplementary Materials). The isotherms of the three investigated materials were in accordance to type IV isotherms with a H2 hysteresis loop (IUPAC, 1984) [44], which are typical of adsorbents forming aggregates with pores in the shape of plates [45] and suggest that CS, n-Hap, and CS/n-Hap are meso-porous, and that capillary condensation occurs in the pores [46]. Furthermore, compared to CS, which exhibits a SSA of 17 m^2/g, CS/n-Hap has a higher SSA (about five-fold higher, 87.3 m^2/g), thus proving that n-Hap incorporation (with SSA 107.5 m^2/g) enhanced the porous structure of the resulting composite material.

3.2. Adsorption Studies

3.2.1. pH Effect, Counter Ions, and Organic Compounds

The pH of aqueous phase is a parameter that largely affects the adsorption. Figure 3a illustrates that the performance of CS and CS/n-Hap composite towards Cd(II) removal was reliant on the initial solution pH. The maximum Cd(II) uptake on CS/n-Hap and CS occurred at pH 5 and 6, respectively. It was observed that Cd(II) uptake progressively increased from pH 2 to 5 on CS/n-Hap and from pH 2 to 6 on CS, then stabilized with no significant changes at larger pH values. The observed lower Cd(II) uptake under highly acidic conditions must be ascribed to interferences generated by the high concentration of hydronium ions that competed with Cd(II) ions to occupy the active binding sites on the surface of CS/n-Hap and CS, thus inhibiting Cd(II) adsorption [23]. The functional groups involved in binding Cd(II) ions during adsorption were the amine (-NH$_2$) groups of CS and the phosphate (PO$_4^{3-}$) groups of n-Hap. The PO$_4^{3-}$ groups in CS/n-Hap played an important role during Cd(II) adsorption given the fact that they can display negative charges beyond pH 3. Consequently, as pH increases, the deprotonation degree of these functional groups will also increase and there will be more negative binding sites that allow interaction with metal ions [47]. The pH study results were further supported by the point of zero charge (pH$_{PZC}$) of the investigated materials, which was 6.9 for CS and 6.5 for CS/n-Hap (Figure 3b), denoting that CS and CS/n-Hap surface displayed positive charge at pH < pH$_{PZC}$, while the adsorbent surfaces were predominately negatively charged at pH > pH$_{PZC}$ [48].

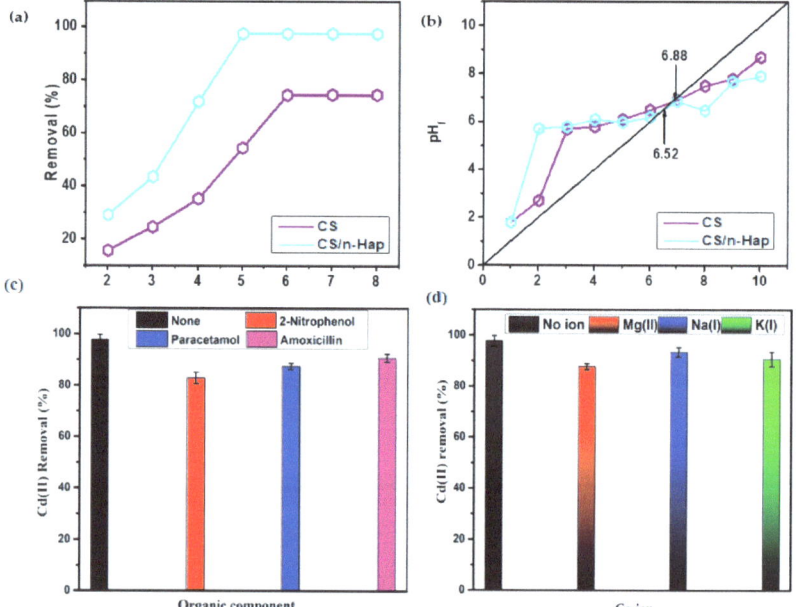

Figure 3. (a) Effect of pH on Cd(II) removal, (b) point of charge zero (pH$_{PZC}$) plot, (c) effect of organic compounds, and (d) effect of counter ions on Cd(II) removal onto CS/n-Hap (t: 120 min; m: 2.0 g/L; C$_0$: 100 mg/L; T: 25 °C, and counter ions/organic C$_0$: 50 mg/L).

The competitive effect of co-existing ions and organic compounds in solution on Cd(II) adsorption was investigated by adding nitrate salts of K$^+$, Na$^+$, and Mg^{2+} and organic compounds such as 2-nitrophenol, paracetamol, and amoxicillin. Removal percentages of Cd(II) by CS/n-Hap in the presence of mineral salts were higher than 90%. This indicates excellent selectivity of CS/n-Hap for Cd(II) removal in their mineral salts presence (Figure 3c). Meanwhile, the removal percentage of Cd(II) in the presence of organic compounds was slightly lower, although >85% of that in the absence of organic compounds (Figure 3d). Overall, CS/n-Hap showed outstanding selectivity toward Cd(II).

3.2.2. Adsorbent Dose Effect

The effect of CS and CS/n-Hap dosage on Cd(II) adsorption was investigated by carrying out experiments with different masses of these adsorbents, namely in the range 25–250 mg (Figure S4, Supplementary Materials). The results revealed that Cd(II) adsorption percentage on both adsorbents initially increased as their doses increased until reaching their maximum adsorption efficiencies (97.65% on CS/n-Hap and 74.7% on CS). However, increasing adsorbents' dose did not lead to a significant increase in Cd(II) removal, which indicates that the total available surface area was not enhanced, probably because the adsorption sites were aggregated and/or the adsorbate surface was not fully accessible for adsorption [49].

3.2.3. Contact Time Effect and Kinetic Modeling

Figure 4a displayed that Cd(II) uptake takes place in two consecutive steps. The first step consisted of a progressive augmentation in the removal percentage, which took about 30 to 50 min for the two investigated adsorbents. The second step corresponded to the surface saturation with Cd(II). However, equilibration time on CS/n-Hap composite was found to be comparatively shorter than on CS, which evidenced the larger efficiency of the composite in the adsorption of Cd(II).

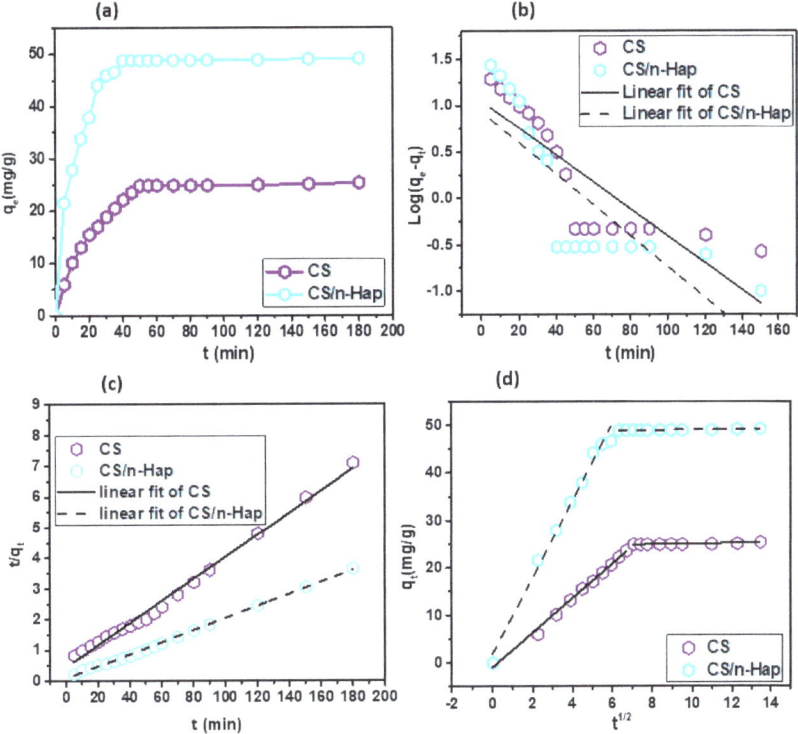

Figure 4. (a) Effect of contact time on Cd(II) adsorption, (b) linearized pseudo-first order kinetic model fitting, (c) linearized pseudo-second order kinetic model fitting, and (d) intra-particle diffusion model fitting. (m: 2.0 g/L; pH: 6.0; C_0: 100 mg/L; and T: 25 °C).

Modeling of kinetic data was done to understand the adsorption dynamics and determine the kinetic rate constant. The pseudo-first order, pseudo-second order, and intra-particle diffusion kinetic models given by equations displayed in Table S1 (Supplementary Materials) were tested to describe the adsorption of Cd(II) onto CS and CS/n-Hap. The linear fittings of the models are illustrated in Figure 4b,c and the kinetic parameters are displayed in Table 1.

Table 1. Kinetics parameters for Cd(II) adsorption on CS and CS/n-Hap.

Kinetic Model	Parameters	CS	CS/n-Hap
Pseudo-first order	k_f (1/min)	0.1457	0.0167
	q_e (mg/g)	11.24	8.59
	R^2	0.7221	0.6463
Pseudo-second order	k_s (g/mg.min)	0.0028	0.0045
	q_e (mg/g)	27.85	50.94
	R^2	0.9923	0.9981
Intra-particle diffusion	k_1 (mg/min$^{1/2}$ g)	3.6242	8.0621
	C_1	0.9386	1.8667
	R_1^2	0.9947	0.9886
	k_2 (mg/min$^{1/2}$ g)	0.0616	0.0382
	C_2	24.394	48.557
	R_2^2	0.7501	0.7589

The experimental q_e values determined for CS (25.36 mg/g) and CS/n-Hap (49.15 mg/g) were nearer to the q_e values inferred by the pseudo-second order than by the first-order kinetic model. In addition, the correlation coefficients (R^2) values for the pseudo-second order model were comparatively higher than those for the pseudo-first order model. Therefore, it can be assumed that the interaction between the two adsorbents and the adsorbate was mainly through chemisorption and that adsorption was mainly controlled by the mass transport rate of the liquid phase or the intra-particle mass transport rate [50]. Intra-particle diffusion rate constants can be obtained from the amount of metal ion adsorbed versus $t^{1/2}$ plots (Figure 4d). The obtained plots presented a double-linearity, indicating that adsorption occurred in two consecutive stages. The first stage corresponds to the external mass transfer (boundary layer diffusion), in which Cd(II) diffuses through the aqueous phase towards the adsorbent. Then, the second stage represents the intra-particle diffusion of metal ions throughout the porous surface of CS and CS/n-Hap.

3.2.4. Initial Adsorbate Concentration Effect and Isotherm Modeling

As depicted in Figure 5a, the effect of initial concentration on the adsorbed concentration at the equilibrium (q_e) was significant for the two studied adsorbents. The Cd(II) uptake was notably improved by increasing the initial concentration of the metal ions and practically no saturation was noticed at higher concentrations. Indeed, at the maximum tested initial Cd(II) concentration, namely 300 mg/L, the q_e on CS and CS/n-Hap were 67.5 and 126.65 mg/g, respectively. The increase of initial Cd(II) concentration yielded a crucial driving force to exceed all resistances of mass transfer of the adsorbate through the aqueous and solid phases, so favoring its adsorption [51].

The equilibrium concentration of Cd(II) (C_e) after adsorption onto CS or CS/n-Hap and the corresponding adsorption equilibrium isotherms are plotted in Figure 5b. Both isotherms were positive, regular, and showed an increase in the q_e with increasing equilibrium Cd(II) concentration (C_e), not showing an apparent saturation under the studied experimental conditions, which indicates that there was no formation of a complete monolayer of Cd(II) over the surface of the adsorbents [52].

Linear fittings of experimental data to Langmuir and Freundlich isotherm models presented by equations given in Table S2 (Supplementary Materials) are displayed in Figure 5c,d, which evidence that Freundlich isotherm model gave comparatively better fittings, especially in the case of Cd(II) adsorption onto CS. This was further confirmed by the higher R^2 obtained for Freundlich than for Langmuir fittings (Table 2). The Freundlich model is associated with adsorbents with heterogeneous surface in which active sites and their corresponding energies are exponentially distributed. Moreover, n > 1 indicates slightly reduced uptake capacity at relatively low equilibrium concentrations. This isotherm model does not predict the adsorbent saturation by the adsorbate; hence, infinite surface coverage is mathematically predicted by the Freundlich model, indicating multilayer adsorption [51]. This agrees well with the observed plot of q_e vs. C_e for the adsorption of Cd(II) onto CS (Figure 5b).

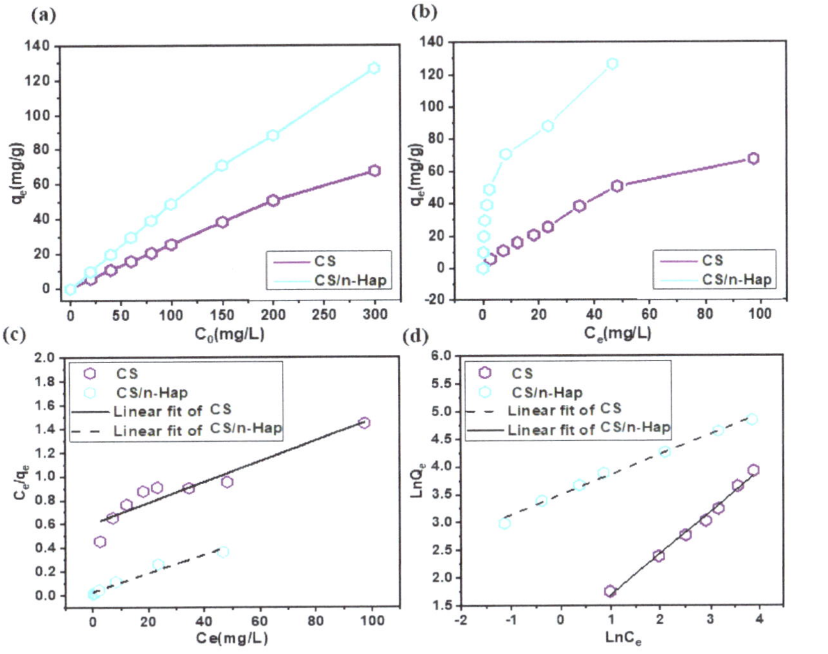

Figure 5. (a) Effect of initial Cd(II) concentration on adsorption, (b) adsorption equilibrium isotherms, (c) Langmuir isotherm model plot, and (d) Freundlich isotherm model plot. (t: 120 min; m: 2.0 g/L; pH: 6.0; C_0: 20–300 mg/L; and T: 25 °C).

Table 2. Isotherm parameters for Cd(II) adsorption on CS and CS/n-Hap.

Adsorbent	Isotherm Models					
	Langmuir			Freundlich		
	q_{max} (mg/g)	K_L (L/mg)	R^2	K_F (mg/g) (L/mg)$^{1/n}$	n	R^2
CS	115.07	0.0143	0.8941	2.5804	1.3401	0.9915
CS/n-Hap	126.58	0.2743	0.9637	5.818	2.7616	0.9923

It should be noted that Cd(II) adsorption performance of CS/n-Hap was better than that of CS, which indicates the favorable effects of compositing n-Hap and CS on Cd(II) adsorption. In the case of Cd(II) adsorption on CS, it was due to electrostatic interaction between the metal ions and the amine groups of CS. Meanwhile, in the case of CS/n-Hap, the amine groups of CS and the phosphate groups of n-Hap synergically provided interaction between these groups and the Cd(II) ions resulting in an improved uptake, as compared with that by CS.

Table S3 [15,25,27,53–58] (Supplementary Materials) outlines an overview of the maximum monolayer adsorption capacities along with experimental conditions of different biopolymers, composites, and/or modified adsorbents used for the adsorption of Cd(II) in the literature. The current study demonstrated that CS/n-Hap composite had a good adsorption capacity for Cd(II) in comparison with that of adsorbents in Table S3 (Supplementary Materials).

3.2.5. Temperature Effect and Thermodynamic Modeling

The Cd(II) adsorption on CS and CS/n-Hap was studied at 298, 308, and 318 K and at initial Cd(II) concentration of 100 mg/L. Obtained results (Figure S5a, Supplementary Materials) showed that the adsorption increased with rise in temperature for studied temperature range, with Cd(II) adsorption percentage onto CS displaying a linear trend with temperature. This adsorption behavior may be due to an increase in the mobility of metal ions with rise in temperature. Moreover, the augmentation in temperature can produce a swelling effect in the internal structure of the adsorbents, leading to further infiltration of Cd(II) ions [59].

Thermodynamic parameters, namely the change of Gibbs free energy ($\Delta G°$), enthalpy ($\Delta H°$), and entropy ($\Delta S°$) for Cd(II) adsorption onto CS and CS/n-Hap were determined by applying the van't Hoff's law (Equation (5)) to adsorption experimental results obtained at different temperatures [32].

$$\text{Ln}\left(K_e^0\right) = \frac{\Delta S°}{R} - \frac{\Delta H°}{RT} = -\frac{\Delta G°}{RT} \quad (6)$$

where the equilibrium constant (K_e^0) was expressed as [39]:

$$K_e^0 = \frac{1000 \times K_L \times M(\text{Adsorbate}) \times [\text{Adsorbate}]°}{\gamma} \quad (7)$$

where R is the universal gas constant (8.314 J/K.mol), T is the temperature in K, M(Adsorbate) is the molar mass of the adsorbate, γ is the activity coefficient, and [Adsorbate]° represents the standard concentration of the adsorbate (1 mol/L). $\Delta H°$ and $\Delta S°$ were calculated from the slope and intercept of the representation of $\ln K_e^0$ versus $1/T$ (Figure S5b, Supplementary Materials). The magnitudes of these parameters are presented in Table 3. As it may be seen by the parameters in Table 3, the adsorption process of Cd(II) was spontaneous for both CS and CS/n-Hap, and the values of $\Delta G°$ were negative at all the studied temperatures. The $\Delta H°$ values for both adsorbents were found to be positive, suggesting endothermic process. Furthermore, the positive $\Delta S°$ values assume an associated increase in the degree of randomness of the solid/liquid interface, with consequent structural changes in the adsorbate/adsorbent system [53,60].

Table 3. Thermodynamics parameters for Cd(II) adsorption on CS and CS/n-Hap.

Adsorbent	Thermodynamic Parameters				
	$\Delta G°$ (kJ/mol)			$\Delta H°$ (kJ/mol)	$\Delta S°$ (J/mol K)
	298 K	308 K	318 K		
CS	−28.77	−30.75	−32.17	124.73	460.7
CS/n-Hap	−36.33	−40.35	−47.18	31.17	122.66

3.2.6. Adsorption Mechanisms

Infra-red spectroscopy and elemental analyses of pristine and Cd(II) saturated CS/n-Hap were performed to depict the mechanism involved during Cd(II) adsorption on CS/n-Hap, illustrated in Figure 6. As presented in FT-IR spectrum (Figure 6a), a band shift corresponding to the hydroxyl groups from 3300 to 3232 cm^{-1} was observed, which could be due to the electrostatic interaction between these functional groups and Cd(II) ions. In addition, adsorption by chelation of Cd(II) ions with the lone electron pair of CS nitrogen occurred due to the shift in the band of –NH group of CS from 1653 to 1610 cm^{-1} [45]. From the EDX analysis (Figure 6b), it was observed that, after adsorption, the Cd(II) percentage on CS/n-Hap was 13.94% and Ca(II) present on CS/n-Hap decreased by 10.39%. Thus, in the case of CS/n-Hap, the adsorption of Cd(II) was essentially governed by ion-exchange mechanism. Since the Cd(II) ionic radius (0.095 nm) is slightly smaller than Ca(II) radius (0.099 nm) [61], the latter can be easily replaced by Cd(II) in the n-Hap crystal lattice:

$$Ca_{10}(PO_4)_6(OH)_2 + xCd^{2+} \rightarrow Ca_{10-x}Cd_x(PO_4)_6(OH)_2 + xCa^{2+} \tag{8}$$

Furthermore, as stated in the literature, neutral hydroxylated and negatively charged species of Hap ($\equiv$POH, $\equiv$CaOH, $\equiv$PO$^-$) could as well be involved in the Cd(II) adsorption mechanism according to the surface complexation reactions [62] that are next depicted:

$$\equiv POH + Cd^{2+} \rightarrow \equiv POCd^+ + H^+ \tag{9}$$

$$\equiv PO^- + Cd^{2+} \rightarrow \equiv POCd^+ \tag{10}$$

$$\equiv CaOH + Cd^{2+} \rightarrow \equiv CaOCd^+ + H^+ \tag{11}$$

Hence, we can conclude that the Cd(II) adsorption on CS/n-Hap involves electrostatic interaction, chelation, ion-exchange, and surface complexation mechanisms.

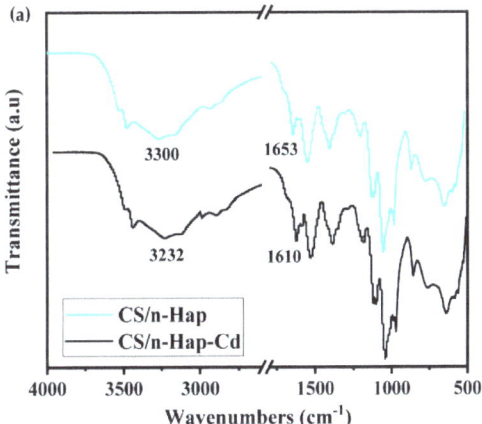

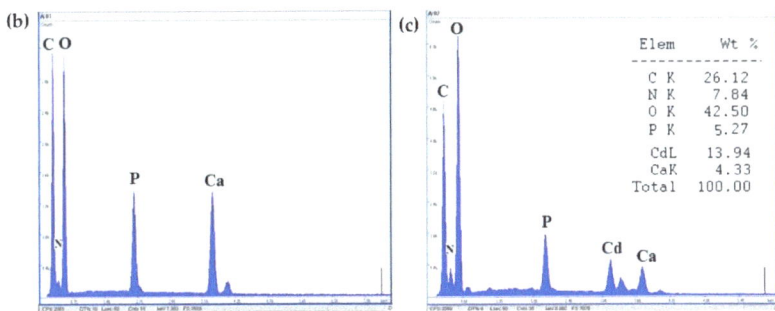

Figure 6. (a) FT-IR spectra, EDX of CS/n-Hap, (b) before adsorption, and (c) after Cd(II) adsorption.

3.3. Regeneration Study

The possibility of an adsorbent being easily regenerated and stable through regeneration allows for its reutilization, which is highly valued from an economic point of view. In order to find if this was the case of the CS/n-Hap composite, it was subjected to successive adsorption/desorption cycles. As illustrated in Figure 7, the removal percentage of Cd(II) by CS/n-Hap was initially 97.65%, which still maintained after five regeneration cycles, testifying the great stability and excellent regeneration ability of the here synthesized CS/n-

Hap composite. After the fifth regeneration cycle, the adsorption performance of CS/n-Hap progressively decreased, reaching 56.5% after ten successive regeneration cycles due to the decline in active adsorption sites after repeated regeneration cycles [63]. Nevertheless, even after five regeneration cycles, CS/n-Hap was able to adsorb Cd(II). Hence, it was evidenced that this bio-composite can serve as an economical adsorbent for an efficient removal of Cd(II) from contaminated water.

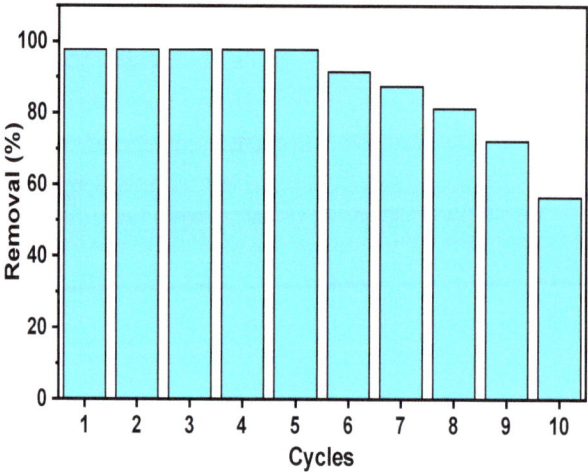

Figure 7. Regeneration-reutilization plot of CS/n-Hap.

4. Conclusions

A CS/n-Hap bio-composite was efficiently synthesized with morphological and physicochemical analysis showing excellent interaction between CS and n-Hap. The incorporation of n-Hap improved not only the SSA and thermal stability of the synthesized composite but also the Cd(II) uptake by the bio-composite, which provided an experimental maximum uptake (128.65 mg/g) that was about twice that of CS (67.5 mg/g). Additionally, the pH study revealed that the functional groups of the composite could be highly affected by the acidity of the medium, in which the higher the pH, the higher the degree of deprotonation of these functional groups will be, and consequently, the more significant the number of negative binding sites directed towards interaction with Cd(II) ions. This finding, along with the infra-red and elemental analyses results after Cd(II) adsorption, actively demonstrated that the adsorption mechanism onto CS/n-Hap was essentially an electrostatic interaction combined with chelation, ion-exchange, and surface complexation. The synthesized material also demonstrated excellent reusability by maintaining high removal percentages (>95%) after five consecutive regeneration cycles. The present study demonstrated that the CS/n-Hap composite could be used as an efficient, green, and cost-effective adsorbent for the removal of Cd(II) from contaminated water.

Supplementary Materials: The following supporting information can be downloaded at: https://www.mdpi.com/article/10.3390/polym15061524/s1, Figure S1: Schematic presentation of Cs/n-Hap bio-composite synthesis; Figure S2: SEM micrographs and EDX analysis of (a,c) n-Hap, and (b,d) Cs/n-Hap bio-composite; Figure S3: N2 adsorption/desorption isotherms of Cs, n-Hap, and Cs/n-Hap; Figure S4: Effect of the adsorbents dosage on Cd(II) removal; Figure S5: (a) Effect of temperature on Cd(II) adsorption onto Cs and Cs/n-Hap, (b) Van't Hoff plot for Cd(II) adsorption onto Cs and Cs/n-Hap; Table S1: Kinetic models used; Table S2: Used isotherm models; Table S3: Comparison of maximum monolayer adsorption capacities of Cd(II) on various adsorbents. References [59–63] are cited in the Supplementary Materials.

Author Contributions: Conceptualization, R.E.K.B.; methodology, R.E.K.B.; software, I.A.; validation, Y.A., M.A.K., M.O. and B.-H.J.; formal analysis, Z.K.; investigation, R.E.K.B. and Z.K.; resources, M.A. and A.S.; data curation, I.A. and Y.A.; writing—original draft preparation, R.E.K.B., I.A. and Z.K.; writing—review and editing, I.A., Y.A., M.A.K., M.O. and B.-H.J.; visualization, Y.A.; supervision, Y.A. and M.A.K.; project administration, M.A. and A.S.; funding acquisition, M.A.K., M.A., A.S. and M.O. All authors have read and agreed to the published version of the manuscript.

Funding: This research was funded through Researchers Supporting Project number (RSP2023R345), King Saud University, Riyadh, Saudi Arabia.

Institutional Review Board Statement: Not applicable.

Data Availability Statement: Data is available from the corresponding author on reasonable request.

Acknowledgments: Moonis Ali Khan acknowledges the financial support through Researchers Supporting Project number (RSP2023R345), King Saud University, Riyadh, Saudi Arabia.

Conflicts of Interest: Ikrame Ayouch is employed by MASCIR Foundation. The remaining Authors declare that the research was conducted in the absence of any commercial or financial relationships that could be construed as a potential conflict of interest.

References

1. Abdellaoui, Y.; El Ibrahimi, B.; Abou Oualid, H.; Kassab, Z.; Quintal-Franco, C.; Giácoman-Vallejos, G.; Gemero-Melo, P. Iron-Zirconium Microwave-Assisted Modification of Small-Pore Zeolite W and Its Alginate Composites for Enhanced Aqueous Removal of As (V) Ions: Experimental and Theoretical Studies. *Chem. Eng. J.* **2021**, *421*, 129909. [CrossRef]
2. Matović, V.; Buha, A.; Bulat, Z.; Đukić-ćosić, D. Cadmium Toxicity Revisited: Focus on Oxidative Stress Induction and Interactions with Zinc Znd Magnesium. *Arh. Za Hig. Rada I Toksikol.* **2011**, *62*, 65–76. [CrossRef]
3. Salah, T.A.; Mohammad, A.M.; Hassan, M.A.; El-Anadouli, B.E. Development of Nano-Hydroxyapatite/Chitosan Composite for Cadmium Ions Removal in Wastewater Treatment. *J. Taiwan Inst. Chem. Eng.* **2014**, *45*, 1571–1577. [CrossRef]
4. Hayat, M.T.; Nauman, M.; Nazir, N.; Ali, S.; Bangash, N. *Environmental Hazards of Cadmium: Past, Present, and Future*; Elsevier Inc.: Amsterdam, The Netherlands, 2018; ISBN 9780128148655.
5. Joseph, P. Mechanisms of Cadmium Carcinogenesis. *Toxicol. Appl. Pharmacol.* **2009**, *238*, 272–279. [CrossRef]
6. Azimi, A.; Azari, A.; Rezakazemi, M.; Ansarpour, M. Removal of Heavy Metals from Industrial Wastewaters: A Review. *ChemBioEng Rev.* **2017**, *4*, 37–59. [CrossRef]
7. Zamri, M.F.M.A.; Kamaruddin, M.A.; Yusoff, M.S.; Aziz, H.A.; Foo, K.Y. Semi-Aerobic Stabilized Landfill Leachate Treatment by Ion Exchange Resin: Isotherm and Kinetic Study. *Appl. Water Sci.* **2017**, *7*, 581–590. [CrossRef]
8. Abdellaoui, Y.; Gamero-Melo, P.; Díaz-Jiménez, L.; Ponce-Caballero, C.; Giácoman-Vallejos, G. Synthesis and Surface Modification of Small Pore Size Zeolite W for Improving Removal Efficiency of Anionic Contaminants from Water. *Bull. Environ. Contam. Toxicol.* **2020**, *105*, 934–940. [CrossRef]
9. Zaidi, N.A.H.M.; Lim, L.B.L.; Usman, A.; Kooh, M.R.R. Efficient Adsorption of Malachite Green Dye Using Artocarpus Odoratissimus Leaves with Artificial Neural Network Modelling. *Desalination Water Treat.* **2018**, *101*, 313–324. [CrossRef]
10. Crini, G.; Lichtfouse, E.; Wilson, L.; Morin-crini, N.; Crini, G.; Lichtfouse, E.; Wilson, L.; Conventional, N.M. Conventional and Non-Conventional Adsorbents for Wastewater Treatment. *Environ. Chem. Lett.* **2019**, *17*, 195–213. [CrossRef]
11. Panić, V.V.; Šešlija, S.I.; Nešić, A.R.; Veličković, S.J. Adsorption of Azo Dyes on Polymer Materials. *Hem. Ind.* **2013**, *67*, 881–900. [CrossRef]
12. Zhang, L.; Sellaoui, L.; Franco, D.; Dotto, G.L.; Bajahzar, A.; Belmabrouk, H.; Bonilla-Petriciolet, A.; Oliveira, M.L.S.; Li, Z. Adsorption of Dyes Brilliant Blue, Sunset Yellow and Tartrazine from Aqueous Solution on Chitosan: Analytical Interpretation via Multilayer Statistical Physics Model. *Chem. Eng. J.* **2020**, *382*, 122952. [CrossRef]
13. El Achaby, M.; Ruesgas-Ramón, M.; Fayoud, N.E.H.; Figueroa-Espinoza, M.C.; Trabadelo, V.; Draoui, K.; Ben Youcef, H. Bio-Sourced Porous Cellulose Microfibrils from Coffee Pulp for Wastewater Treatment. *Cellulose* **2019**, *26*, 3873–3889. [CrossRef]
14. Tang, X.; Gan, L.; Duan, Y.; Sun, Y.; Zhang, Y.; Zhang, Z. A Novel Cd^{2+}-Imprinted Chitosan-Based Composite Membrane for Cd^{2+} Removal from Aqueous Solution. *Mater. Lett.* **2017**, *198*, 121–123. [CrossRef]
15. Ayouch, I.; Barrak, I.; Kassab, Z.; El Achaby, M.; Barhoun, A.; Draoui, K. Impact of the Drying Process on the Efficiency of Alginate Beads for Cadmium Removal from Water: Kinetic, Isotherm and Thermodynamic Study. *Environ. Technol. Innov.* **2020**, *20*, 539–547. [CrossRef]
16. Libio, I.C.; Demori, R.; Ferrão, M.F.; Lionzo, M.I.Z.; da Silveira, N.P. Films Based on Neutralized Chitosan Citrate as Innovative Composition for Cosmetic Application. *Mater. Sci. Eng. C* **2016**, *67*, 115–124. [CrossRef] [PubMed]
17. Shahid-ul-Islam; Butola, B.S. Recent Advances in Chitosan Polysaccharide and Its Derivatives in Antimicrobial Modification of Textile Materials. *Int. J. Biol. Macromol.* **2019**, *121*, 905–912. [CrossRef] [PubMed]
18. Mujtaba, M.; Morsi, R.E.; Kerch, G.; Elsabee, M.Z.; Kaya, M.; Labidi, J.; Khawar, K.M. Current Advancements in Chitosan-Based Film Production for Food Technology; A Review. *Int. J. Biol. Macromol.* **2019**, *121*, 889–904. [CrossRef]

19. Liu, J.; Liu, S.; Zhang, X.; Kan, J.; Jin, C. Effect of Gallic Acid Grafted Chitosan Film Packaging on the Postharvest Quality of White Button Mushroom (*Agaricus bisporus*). *Postharvest Biol. Technol.* **2019**, *147*, 39–47. [CrossRef]
20. Smets, G.; Rüdelsheim, P. Biotechnologically Produced Chitosan for Nanoscale Products. A Legal Analysis. *New Biotechnol.* **2018**, *42*, 42–47. [CrossRef]
21. Luo, Q.; Huang, X.; Luo, Y.; Yuan, H.; Ren, T.; Li, X.; Xu, D.; Guo, X.; Wu, Y. Fluorescent Chitosan-Based Hydrogel Incorporating Titanate and Cellulose Nanofibers Modified with Carbon Dots for Adsorption and Detection of Cr (VI). *Chem. Eng. J.* **2021**, *407*, 127050. [CrossRef]
22. Hu, C.; Zhu, P.; Cai, M.; Hu, H.; Fu, Q. Comparative Adsorption of Pb (II), Cu (II) and Cd (II) on Chitosan Saturated Montmorillonite: Kinetic, Thermodynamic and Equilibrium Studies. *Appl. Clay Sci.* **2017**, *143*, 320–326. [CrossRef]
23. Sutirman, Z.A.; Sanagi, M.M.; Abd Karim, K.J.; Wan Ibrahim, W.A.; Jume, B.H. Equilibrium, Kinetic and Mechanism Studies of Cu (II) and Cd (II) Ions Adsorption by Modified Chitosan Beads. *Int. J. Biol. Macromol.* **2018**, *116*, 255–263. [CrossRef] [PubMed]
24. Wu, D.; Wang, Y.; Li, Y.; Wei, Q.; Hu, L.; Yan, T.; Feng, R.; Yan, L.; Du, B. Phosphorylated Chitosan/CoFe$_2$O$_4$ Composite for the Efficient Removal of Pb (II) and Cd (II) from Aqueous Solution: Adsorption Performance and Mechanism Studies. *J. Mol. Liq.* **2019**, *277*, 181–188. [CrossRef]
25. Ebisike, K.; Okoronkwo, A.E.; Alaneme, K.K. Adsorption of Cd (II) on Chitosan–Silica Hybrid Aerogel from Aqueous Solution. *Environ. Technol. Innov.* **2019**, *14*, 100337. [CrossRef]
26. Rathinam, K.; Singh, S.P.; Arnusch, C.J.; Kasher, R. An Environmentally-Friendly Chitosan-Lysozyme Biocomposite for the Effective Removal of Dyes and Heavy Metals from Aqueous Solutions. *Carbohydr. Polym.* **2018**, *199*, 506–515. [CrossRef]
27. Chen, Y.; Li, M.; Li, Y.; Liu, Y.; Chen, Y.; Li, H.; Li, L.; Xu, F.; Jiang, H.; Chen, L. Hydroxyapatite Modified Sludge-Based Biochar for the Adsorption of Cu^{2+} and Cd^{2+}: Adsorption Behavior and Mechanisms. *Bioresour. Technol.* **2021**, *321*, 124413. [CrossRef]
28. Long, Y.; Jiang, J.; Hu, J.; Hu, X.; Yang, Q.; Zhou, S. Removal of Pb (II) from Aqueous Solution by Hydroxyapatite/Carbon Composite: Preparation and Adsorption Behavior. *Colloids Surf. A Physicochem. Eng. Asp.* **2019**, *577*, 471–479. [CrossRef]
29. Gardea-Torresdey, J.L.; De La Rosa, G.; Peralta-Videa, J.R. Use of Phytofiltration Technologies in the Removal of Heavy Metals: A Review. *Pure Appl. Chem.* **2004**, *76*, 801–813. [CrossRef]
30. Wang, Y.Y.; Liu, Y.X.; Lu, H.H.; Yang, R.Q.; Yang, S.M. Competitive Adsorption of Pb (II), Cu (II), and Zn (II) Ions onto Hydroxyapatite-Biochar Nanocomposite in Aqueous Solutions. *J. Solid State Chem.* **2018**, *261*, 53–61. [CrossRef]
31. El, R.; Billah, K.; Islam, M.A.; Lgaz, H.; Lima, E.C.; Abdellaoui, Y.; Rakhila, Y.; Goudali, O.; Majdoubi, H.; Alrashdi, A.A.; et al. Shellfish Waste-Derived Mesoporous Chitosan for Impressive Removal of Arsenic (V) from Aqueous Solutions: A Combined Experimental and Computational Approach. *Arab. J. Chem.* **2022**, *15*, 104123. [CrossRef]
32. Abidi, S.S.A.; Murtaza, Q. Synthesis and Characterization of Nano-Hydroxyapatite Powder Using Wet Chemical Precipitation Reaction. *J. Mater. Sci. Technol.* **2014**, *30*, 307–310. [CrossRef]
33. Zaini, M.A.A.; Okayama, R.; Machida, M. Adsorption of Aqueous Metal Ions on Cattle-Manure-Compost Based Activated Carbons. *J. Hazard. Mater.* **2009**, *170*, 1119–1124. [CrossRef]
34. Ragab, A.; Ahmed, I.; Bader, D. The Removal of Brilliant Green Dye from Aqueous Solution Using Nano Hydroxyapatite/Chitosan Composite as a Sorbent. *Molecules* **2019**, *24*, 847. [CrossRef]
35. Tripathi, A.; Saravanan, S.; Pattnaik, S.; Moorthi, A.; Partridge, N.C.; Selvamurugan, N. Bio-Composite Scaffolds Containing Chitosan/Nano-Hydroxyapatite/Nano-Copper-Zinc for Bone Tissue Engineering. *Int. J. Biol. Macromol.* **2012**, *50*, 294–299. [CrossRef]
36. Rajiv Gandhi, M.; Kousalya, G.N.; Meenakshi, S. Removal of Copper(II) Using Chitin/Chitosan Nano-Hydroxyapatite Composite. *Int. J. Biol. Macromol.* **2011**, *48*, 119–124. [CrossRef] [PubMed]
37. Sairam Sundaram, C.; Viswanathan, N.; Meenakshi, S. Uptake of Fluoride by Nano-Hydroxyapatite/Chitosan, a Bioinorganic Composite. *Bioresour. Technol.* **2008**, *99*, 8226–8230. [CrossRef]
38. Kousalya, G.N.; Rajiv Gandhi, M.; Sairam Sundaram, C.; Meenakshi, S. Synthesis of Nano-Hydroxyapatite Chitin/Chitosan Hybrid Biocomposites for the Removal of Fe (III). *Carbohydr. Polym.* **2010**, *82*, 594–599. [CrossRef]
39. Said, H.A.; Noukrati, H.; Ben Youcef, H.; Bayoussef, A.; Oudadesse, H.; Barroug, A. Mechanical Behavior of Hydroxyapatite-Chitosan Composite: Effect of Processing Parameters. *Minerals* **2021**, *11*, 213. [CrossRef]
40. Nikpour, M.R.; Rabiee, S.M.; Jahanshahi, M. Synthesis and Characterization of Hydroxyapatite/Chitosan Nanocomposite Materials for Medical Engineering Applications. *Compos. B Eng.* **2012**, *43*, 1881–1886. [CrossRef]
41. Attar Nosrati, S.; Alizadeh, R.; Ahmadi, S.J.; Erfani, M. Optimized Precipitation Process for Efficient and Size-Controlled Synthesis of Hydroxyapatite–Chitosan Nanocomposite. *J. Korean Ceram. Soc.* **2020**, *57*, 632–644. [CrossRef]
42. Ran, J.; Jiang, P.; Sun, G.; Ma, Z.; Hu, J.; Shen, X.; Tong, H. Comparisons among Mg, Zn, Sr, and Si Doped Nano-Hydroxyapatite/Chitosan Composites for Load-Bearing Bone Tissue Engineering Applications. *Mater. Chem. Front.* **2017**, *1*, 900–910. [CrossRef]
43. Nazeer, M.A.; Yilgör, E.; Yilgör, I. Intercalated Chitosan/Hydroxyapatite Nanocomposites: Promising Materials for Bone Tissue Engineering Applications. *Carbohydr. Polym.* **2017**, *175*, 38–46. [CrossRef] [PubMed]
44. Li, C.; Lu, J.; Li, S.; Tong, Y.; Ye, B. Synthesis of Magnetic Microspheres with Sodium Alginate and Activated Carbon for Removal of Methylene Blue. *Materials* **2017**, *10*, 84. [CrossRef] [PubMed]

45. Thommes, M.; Kaneko, K.; Neimark, A.V.; Olivier, J.P.; Rodriguez-Reinoso, F.; Rouquerol, J.; Sing, K.S.W. Physisorption of Gases, with Special Reference to the Evaluation of Surface Area and Pore Size Distribution (IUPAC Technical Report). *Pure Appl. Chem.* **2015**, *87*, 1051–1069. [CrossRef]
46. Ba Mohammed, B.; Hsini, A.; Abdellaoui, Y.; Abou Oualid, H.; Laabd, M.; El Ouardi, M.; Ait Addi, A.; Yamni, K.; Tijani, N. Fe-ZSM-5 Zeolite for Efficient Removal of Basic Fuchsin Dye from Aqueous Solutions: Synthesis, Characterization and Adsorption Process Optimization Using BBD-RSM Modeling. *J. Environ. Chem. Eng.* **2020**, *8*, 104419. [CrossRef]
47. Mahmoud, M.E.; Yakout, A.A.; Abdel-Aal, H.; Osman, M.M. Enhanced Biosorptive Removal of Cadmium from Aqueous Solutions by Silicon Dioxide Nano-Powder, Heat Inactivated and Immobilized Aspergillus Ustus. *Desalination* **2011**, *279*, 291–297. [CrossRef]
48. Abdellaoui, Y.; Abou Oualid, H.; Hsini, A.; El Ibrahimi, B.; Laabd, M.; El Ouardi, M.; Giácoman-Vallejos, G.; Gamero-Melo, P. Synthesis of Zirconium-Modified Merlinoite from Fly Ash for Enhanced Removal of Phosphate in Aqueous Medium: Experimental Studies Supported by Monte Carlo/SA Simulations. *Chem. Eng. J.* **2020**, *404*, 126600. [CrossRef]
49. Mittal, H.; Maity, A.; Ray, S.S. Gum Karaya Based Hydrogel Nanocomposites for the Effective Removal of Cationic Dyes from Aqueous Solutions. *Appl. Surf. Sci.* **2016**, *364*, 917–930. [CrossRef]
50. Xia, C.; Jing, Y.; Jia, Y.; Yue, D.; Ma, J.; Yin, X. Adsorption Properties of Congo Red from Aqueous Solution on Modified Hectorite: Kinetic and Thermodynamic Studies. *Desalination* **2011**, *265*, 81–87. [CrossRef]
51. Mobasherpour, I.; Salahi, E.; Pazouki, M. Removal of Divalent Cadmium Cations by Means of Synthetic Nano Crystallite Hydroxyapatite. *Desalination* **2011**, *266*, 142–148. [CrossRef]
52. Aksu, Z.; Tezer, S. Biosorption of Reactive Dyes on the Green Alga Chlorella Vulgaris. *Process Biochem.* **2005**, *40*, 1347–1361. [CrossRef]
53. Ayouch, I.; Barrak, I.; Kassab, Z.; El Achaby, M.; Barhoun, A.; Draoui, K. Improved Recovery of Cadmium from Aqueous Medium by Alginate Composite Beads Filled by Bentonite and Phosphate Washing Sludge. *Colloids Surf. A Physicochem. Eng. Asp.* **2020**, *604*, 125305. [CrossRef]
54. Pelalak, R.; Heidari, Z.; Khatami, S.M.; Kurniawan, T.A.; Marjani, A.; Shirazian, S. Oak wood ash/GO/Fe3O4 adsorption efficiencies for cadmium and lead removal from aqueous solution: Kinetics, equilibrium and thermodynamic evaluation. *Arab. J. Chem.* **2021**, *14*, 102991. [CrossRef]
55. Fu, C.; Zhu, X.; Dong, X.; Zhao, P.; Wang, Z. Study of adsorption property and mechanism of lead(II) and cadmium(II) onto sulfhydryl modified attapulgite. *Arab. J. Chem.* **2021**, *14*, 102960. [CrossRef]
56. Fosso-Kankeu, E.; Mittal, H.; Waanders, F.; Sinha, S. Thermodynamic properties and adsorption behaviour of hydrogel nanocomposites for cadmium removal from mine effluents. *J. Ind. Eng. Chem.* **2017**, *48*, 151–161. [CrossRef]
57. Bagheri, S.; Amini, M.M.; Behbahani, M.; Rabiee, G. Low cost thiol-functionalized mesoporous silica, KIT-6-SH, as a useful adsorbent for cadmium ions removal: A study on the adsorption isotherms and kinetics of KIT-6-SH. *Microchem. J.* **2019**, *145*, 460–469. [CrossRef]
58. Soltani, R.; Pishnamazi, M.; Pelalak, R.; Rezakazemi, M.; Marjani, A.; Dinari, M.; Sarkar, S.M.; Shirazian, S. Preparation of COOH-KCC-1/polyamide 6 composite by in situ ring-opening polymerization: Synthesis, characterization, and Cd(II) adsorption study. *J. Environ. Chem. Eng.* **2021**, *9*, 104683. [CrossRef]
59. Alkan, M.; Çelikçapa, S.; Demirbaş, Ö.; Dogan, M. Removal of Reactive Blue 221 and Acid Blue 62 Anionic Dyes from Aqueous Solutions by Sepiolite. *Dye. Pigment.* **2005**, *65*, 251–259. [CrossRef]
60. Vilela, P.B.; Matias, C.A.; Dalalibera, A.; Becegato, V.A.; Paulino, A.T. Polyacrylic Acid-Based and Chitosan-Based Hydrogels for Adsorption of Cadmium: Equilibrium Isotherm, Kinetic and Thermodynamic Studies. *J. Environ. Chem. Eng.* **2019**, *7*, 103327. [CrossRef]
61. Shen, X.; Gao, X.; Wei, W.; Zhang, Y.; Zhang, Y.; Ma, L.; Liu, H.; Han, R.; Lin, J. Combined Performance of Hydroxyapatite Adsorption and Magnetic Separation Processes for Cd(II) Removal from Aqueous Solution. *J. Dispers. Sci. Technol.* **2021**, *42*, 664–676. [CrossRef]
62. Xu, Y.; Schwartz, F.W.; Traina, S.J. Sorption of Zn^{2+} and Cd^{2+} on Hydroxyapatite Surfaces. *Environ. Sci. Technol.* **1994**, *28*, 1472–1480. [CrossRef] [PubMed]
63. Basu, S.; Ghosh, G.; Saha, S. Adsorption Characteristics of Phosphoric Acid Induced Activation of Bio-Carbon: Equilibrium, Kinetics, Thermodynamics and Batch Adsorber Design. *Process Saf. Environ. Prot.* **2018**, *117*, 125–142. [CrossRef]

Disclaimer/Publisher's Note: The statements, opinions and data contained in all publications are solely those of the individual author(s) and contributor(s) and not of MDPI and/or the editor(s). MDPI and/or the editor(s) disclaim responsibility for any injury to people or property resulting from any ideas, methods, instructions or products referred to in the content.

Article

Development of Poly(acrylamide)-Based Hydrogel Composites with Powdered Activated Carbon for Controlled Sorption of PFOA and PFOS in Aqueous Systems

Maria Victoria X. Klaus [1,2], Angela M. Gutierrez [1,2] and J. Zach Hilt [1,2,*]

[1] Department of Chemical and Materials Engineering, University of Kentucky, Lexington, KY 40506, USA; mxklaus@uky.edu (M.V.X.K.); amgu232@g.uky.edu (A.M.G.)
[2] Superfund Research Center, University of Kentucky, Lexington, KY 40506, USA
* Correspondence: zach.hilt@uky.edu

Abstract: Per- and polyfluoroalkyl substances (PFAS) are anthropogenic compounds developed for various applications; some are connected to adverse health impacts including immunosuppression and higher susceptibility to some cancers. Current PFAS remediation treatments from aqueous sources include granular activated carbon (GAC) adsorption, membrane separation, and anion-exchange resin (AER) removal. Each has specific disadvantages, hence the need for a new and efficient technology. Herein, acrylamide-based hydrogel composites were synthesized with powdered activated carbon (PAC) and characterized to determine their affinity for PFAS. Physicochemical characterization included Fourier-Transform infrared spectroscopy (FTIR) to identify chemical composition, thermogravimetric analysis (TGA) to confirm PAC loading percentage, and aqueous swelling studies to measure the effect of crosslinking density. FTIR showed successful conversion of carbonyl and amine groups, and TGA analysis confirmed the presence of PAC within the network. Surface characterization also confirmed carbon-rich areas within composite networks, and the swelling ratio decreased with increasing crosslinking density. Finally, sorption of PFAS was detected via liquid chromatography with tandem mass spectrometry (LC-MS/MS), with removal efficiencies of up to 98% for perfluorooctanoic sulfonic acid (PFOS) and 96% for perfluorooctanoic acid (PFOA). The developed hydrogel composites exhibited great potential as advanced materials with tunable levers that can increase affinity towards specific compounds in water.

Keywords: hydrogels; PFAS; PAC

Citation: Klaus, M.V.X.; Gutierrez, A.M.; Hilt, J.Z. Development of Poly(acrylamide)-Based Hydrogel Composites with Powdered Activated Carbon for Controlled Sorption of PFOA and PFOS in Aqueous Systems. *Polymers* **2023**, *15*, 4384. https://doi.org/10.3390/polym15224384

Academic Editors: Marta Otero and Ricardo N. Coimbra

Received: 16 October 2023
Revised: 4 November 2023
Accepted: 7 November 2023
Published: 11 November 2023

Copyright: © 2023 by the authors. Licensee MDPI, Basel, Switzerland. This article is an open access article distributed under the terms and conditions of the Creative Commons Attribution (CC BY) license (https://creativecommons.org/licenses/by/4.0/).

1. Introduction

For decades, per- and polyfluoroalkyl substances (PFAS) have been manufactured for a plenitude of products due to their repellent nature and have, since, persisted in nature in large concentrations causing worldwide concern [1]. The industrialization of PFAS has plagued the environment and our drinking water; therefore, exposure to humans and animals is inevitable [2]. Perfluorooctanoic acid (PFOA) and perfluorooctanesulfonic acid (PFOS) are two of the most prominent "legacy" PFAS compounds. With both, there is sufficient evidence of association to several detrimental effects on human health such as dyslipidemia, decreased response to antibodies, and increased risk of kidney cancer [3,4]. In 2022, the United States Environmental Protection Agency (EPA) set a new 0.004- and 0.02-parts-per-trillion (ppt) interim health advisory level (HAL) for PFOA and PFOS in drinking water, respectively (compared to the previous HAL of 70 ppt for the pair), but levels beyond parts-per-billion (ppb) have been reported in various aqueous sources as well as the human body [5,6]. Current treatments for PFAS remediation in aqueous systems include granular activated carbon (GAC) adsorption, reverse osmosis and nanofiltration membrane separation, and anion-exchange resin (AER) removal [7–9]. Adsorption by GAC relies on hydrophobic interactions facilitated by a large surface area that can capture upwards of 90% of PFAS in certain cases, but has shown unselective behavior when competing species are present [10]. Membrane separation is based on its semi-permeable nature through size exclusion and on the manipulation of chemical properties through functionalization to increase affinity towards PFAS; however, materials that do not pass through the membrane can accumulate if they are unable to be removed from the surface, which leads to fouling and loss of efficiency [11]. The electrostatic interaction between certain negatively charged PFAS molecules and the cationic AERs results in ionic bonds, and the polymeric resin produces Van der Waals forces with the hydrophobic tail of the contaminants [7,11]. The performance of AERs in removing PFAS is dependent on several factors including polymer matrix and functional groups; they have also proven costly [7]. In conclusion, the drawbacks of these methods are significant; hence, the need for a new cost-effective and efficient technology for PFAS removal is of utmost importance [7–10].

Novel materials based on natural and synthetic polymer sorbents have recently gained attention for remediation of PFAS in aqueous systems due to their high water-retention capacity and low production cost [10,12]. In particular, hydrogels, i.e., crosslinked hydrophilic polymers that swell in aqueous systems whilst maintaining their physical and chemical integrity, have been investigated [13]. These materials can be easily functionalized with co-monomers and/or hydrophobic crosslinkers that enhance their sorptive properties [14]. Furthermore, hydrogel composites are polymer networks containing embedded particulates that can also introduce hydrophobic interactions and are of high interest for environmental applications [15,16]. Among natural compounds, cellulose is the most abundant polymer in nature and can be extracted into its microcrystalline form and functionalized with poly(ethylenimine) (PEI) for increased sorption of the negatively charged PFAS head. Ateia et al. demonstrated that the adsorption capacity of cellulose for PFOA and PFOS at the initial capacity of 1 µg/L was within 70–80% removal in the first 100 s [17]. Moreover, researchers have explored the incorporation of ionic fluorogels, through thermally initiated radical copolymerization of perfluoropolyethers (PFPE), with methacrylate chain-end functionality (Fluorolink MD 700) and an amine-containing monomer (2-dimethylaminoethyl methacrylate, DMAEMA) for further affinity for PFAS. A variety of both short- and long-chain compounds, including PFOA and PFOS, were removed from wastewater treatment plant-collected sample water with efficiencies of >95% [18]. Furthermore, chitosan, an amine-containing polysaccharide, has also exhibited adsorptive properties towards these contaminants due to the ionic interactions between the carboxylic/sulfonic groups of PFAS and the cationic groups of the polymer [19]. Affinity for specific contaminants can be enhanced by crosslinking with epoxy epichlorohydrin (ECH) in the presence of the contaminant, forming a molecularly imprinted hydrogel that still adsorbs over 50% of PFAS in the presence of competing pollutants [20]. Ateia et al. have reported on the development of poly(N-[3-(dimethylamino)propyl] acrylamide, methyl chloride quaternary) (DMAPAA-Q) hydrogels for PFAS sequestration [12]. Again, both short- and long-chain compounds were analyzed in a range of pH solutions with removal efficiencies beyond 80%, and full removal of select sulfonic contaminants was also achieved. Although novel adsorbent materials are promising platforms for PFAS removal from water, uncertainty regarding potential toxic effects from the bare adsorbents and the possible limitations of scalability have hindered research in this area; therefore, they need to be further explored [21].

Even though GAC is well-known for its sorptive properties and the binding capacity of PFAS, powdered activated carbon (PAC) exhibits higher binding capacity due to its greater surface area, shorter internal diffusion, better site accessibility, and faster adsorption kinetics than GAC. Son et al. calculated the removal efficiency of PAC from various sources and activation methods by removing nine PFAS compounds, which exhibited removal exceeding 80% of most carbons for five analytes, including PFOA and PFOS, and at least 10% for short-chain compounds [22]. Nonetheless, the smaller diameter size (normally below 0.1 mm) of the powdered activated carbon limited these materials from being implemented in large-scale columns as separation and regeneration have proven inefficient [7]. Researchers have developed a stable aqueous suspension of PAC within a poly(diallyldimethylammonium chloride) (DADMAC) network to remove PFAS from water systems, with adsorption capacities of 379.8 mg/g and 495.2 mg/g for PFOA and PFOS, respectively [23]. The combination of a hydrophobic material within a hydrogel network allows for improved contact between sorbent and contaminants in aqueous systems [16]. Herein, we report on the robust synthesis of a poly(acrylamide) (PAm)-powdered activated carbon hydrogel composite with an affinity for PFAS. A series of crosslinking densities were explored to understand the impacts on polymer properties; increasing PAC loading percentages was also explored to measure the effects of hydrophobic interactions, as shown in Figure 1. Physicochemical characterization of the materials was performed using infrared spectroscopy and thermogravimetric analysis to determine conversion of the chemical bonds of monomer to polymer and PAC loading percentage, respectively. Surface characterization was also analyzed for most systems using high imaging microscopy and energy-dispersive X-ray spectroscopy. In addition, the swelling capacity of the composites and the removal efficiency of two PFAS contaminants (PFOS and PFOA) were also studied.

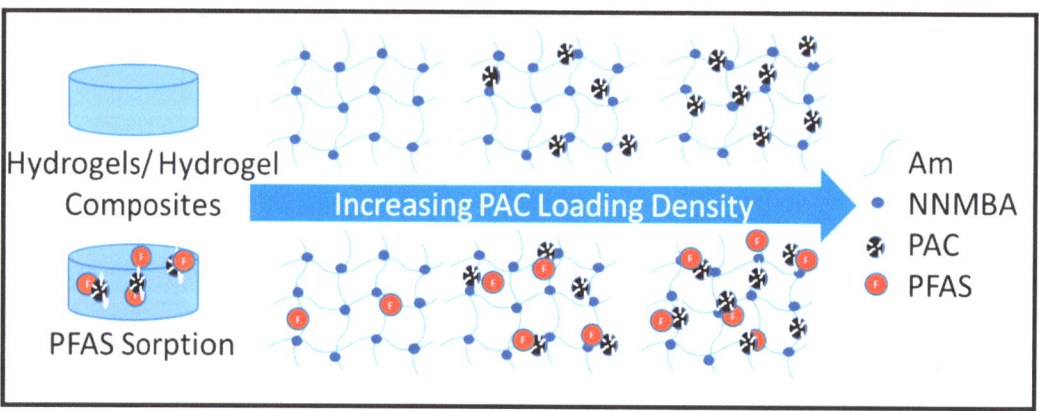

Figure 1. Visual innovation of materials which show an increase in PFAS removal efficiency with increasing PAC loading density.

2. Materials and Methods

2.1. Materials

Acrylamide (Am) (CAS No. 79-06-1), N,N-methylenebisacrylamide (NNMBA) (CAS No. 110-26-9), sodium hydroxide (NaOH) (CAS No. 1310-73-2), and ammonium persulfate (APS) (CAS No. 7727-54-0) were purchased from Sigma-Aldrich (Milwaukee, WI, USA). Tetramethylethylenediamine (TEMED) (CAS No. 110-18-9) was received from TCI Chemicals (Tokyo, Japan). Carbon powder, activated, Norit GSX, steam-activated, acid-washed (CAS No. 7440-44-0) was purchased from Alfa Aesar, with a surface area of 694.7 ± 93.9 nm (determined using dynamic light scattering). Finally, PFAS such as PFOA (95% purity) (CAS No. 335-67-1) and PFOS (97% purity) (CAS No. 1763-23-1) were received from Alfa Aesar and Strem Chemicals (Milwaukee, WI, USA), respectively. All the chemical components were used as received. To construct a synthesis reaction cell, a pair of microscope glass plates, a polytetrafluoroethylene (PTFE) spacer, and binder clips were used.

2.2. Preparation and Physicochemical Characterization of Hydrogels and Hydrogel Composites

To synthesize the hydrogel, a solution was prepared by mixing the acrylamide monomer, the chemical crosslinker, NNMBA and, when applicable, the PAC particulate, together in a 20 mL borosilicate scintillation vial using a 10 mol basis of feed in a 1:2 ratio of deionized (DI) water. By varying the feed ratio of the monomer to the crosslinker and PAC loading percentage, a series of hydrogels and hydrogel composites were synthesized to investigate the size-restriction capability of the network and PAC effects. The feeding compositions of the prepared gel samples are summarized in Table 1. The resulting solutions were thoroughly mixed using a vortex, and the hydrogel composites were additionally subjected to probe sonication using a Model 500 Sonic Dismembrator (Fisher Scientific, Hampton, VA, USA) for increased dispersion of PAC for 1 min with 5 s on pulses and 10 s off in between. To initiate the free-radical polymerization reaction, a redox system consisting of APS and TEMED was applied. A 10 wt% APS solution in deionized (DI) water at concentrations of 0.2, 0.8, and 1 wt% of the total feed mass for the neat (0% PAC), 1% PAC, and 5% PAC systems, respectively, was added simultaneously with 1/5 of the APS volume of TEMED. The polymerization was carried out at 25 °C. This exothermic reaction can be altered by controlling the amount of the initiator/catalyst composition, which can increase the kinetics for the composites as the reaction needs to occur faster to keep the particles from settling. For the neat acrylamide systems, reaction occurred within 30 min while, for the composites, the reaction occurred within 2 min. The solution was then immediately transferred into the reaction cell, where a PTFE frame of 1.6 mm thickness was sandwiched between a pair of parallel microscope glass slides with a disposable pipette.

Table 1. Feed compositions of hydrogels and hydrogel composites with mass yield percentage ranging from 69% to 99% and varying PAC final loading percentages.

Hydrogel/ Hydrogel Composite Samples	Crosslinking Density Am:NNMBA (mol%)	PAC Loading Percentage (mass %)	Mass Yield %	PAC Final Loading %
1NNT	99:1	-	91	-
5NNT	95:5	-	98	-
10NNT	90:10	-	99	-
1N1P	99:1	1	90	2.44
5N1P	95:5	1	95	1.49
10N1P	90:10	1	96	1.00
1N5P	99:1	5	78	7.50
5N5P	95:5	5	69	3.87
10N5P	90:10	5	79	5.89

Following polymerization, the hydrogel films were then removed from the reaction cell mold and subsequently immersed in DI water to remove any unreacted chemical components and for spontaneous hydration. After 24 h, the materials were then punched into circular disks using a 6 mm diameter metal cutter, and the water was changed. The water was replaced one more time, after 48 h, to obtain the satisfactory removal of any unreacted monomers and initiators from the solution. Finally, the swollen polymers were dried at 80 °C for 48 h and stored in a scintillation vial until further use as shown in Figure 2.

Fourier transform infrared (FTIR) spectra were obtained on a 7000e FTIR spectrometer, with attenuated total reflectance sampling technology (Varian Inc., Palo Alto, CA, USA). For analyzing polymers, a fine powder is desired for direct contact with the crystal; this was achieved here using a pestle and mortar. The scanning range was 700–4000 cm^{-1} and the resolution was 8 cm^{-1}. Thermogravimetric analysis (TGA) was performed on a Q600 SDT dynamic scattering calorimetry (DSC)-TGA (TA Instruments, New Castle, DE, USA), under nitrogen gas from 20 °C to 110 °C at 20 °C/min; The temperature remained isothermal for 15 min to evaporate any free water or physically bound molecules. After the water was removed, the temperature was then ramped at 10 °C/min to reach 800 °C and held isothermal for 15 min before the completion of the experiment. TGA was used to determine the incorporation of PAC loading percentage of hydrogels and hydrogel composites. The amount of PAC that was determined to be incorporated into the polymeric network can be calculated by the following equation, Equation (1):

$$PAC\,composite\,final\,wt\% - neat\,hydrogel\,final\,wt\% = PAC\,loading\% \tag{1}$$

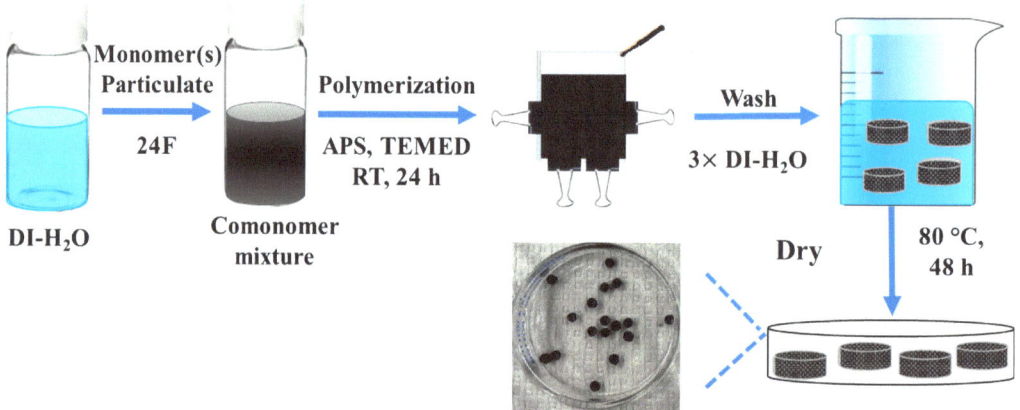

Figure 2. Schematic of hydrogel and hydrogel composite synthesis.

2.3. Mass Yield of Monomer and Crosslinker to Polymer Conversion

Mass yield was investigated to determine how much of the monomer/crosslinker converted into a final polymeric material by comparing the feed mass of Am and NNMBA with the final mass of dry polymers. The mass yield was calculated by using Equation (2):

$$\left(\frac{\text{final dry mass}}{\text{initial dry mass}}\right) \times 100\% = \%\text{Mass Yield} \tag{2}$$

Initial dry mass was calculated by summing the feed components with the exception of the solvent (DI water).

2.4. Surface Characterization of Sorbent Materials

Surface characterization of hydrogels and hydrogel composites was conducted using a Quanta FEG-250 scanning electron microscope (SEM) (FEI, Lausanne, Switzerland), for high-resolution surface imaging and by analyzing the composition by energy-dispersive X-ray microanalysis (EDS). SEM and EDS were conducted for 1NNT, 5NNT, 10NNT, 10N1P, and 10N5P materials. Samples were prepared by placing materials on conductive carbon tape and sputter coating with 5 nm of platinum utilizing an EM ACE600 sputter coater (Leica, Wetzlar, Germany).

2.5. Aqueous Swelling Experiments of Hydrogels and Hydrogel Composites

Swelling behavior indicates the expansibility of the materials in water; this was studied under environmental conditions following a common gravimetric method. The dry hydrogel circular disks were immersed into a 1M sodium hydroxide-titrated deionized water solution (pH = 7.0 ± 0.2) at 25 °C. During the swelling process, the samples ($n = 3$) were taken out of the solution at scheduled time points, blotted with Kimwipe tissues to remove excess surface water, weighed, and re-immersed into the water. These steps were repeated at the desired time point(s) to obtain the complete kinetic swelling profile. For kinetic studies, samples were massed at 1, 4, 8, 12, and 24 h; equilibrium studies were performed at 4 h, which were determined by the kinetic experiments. The mass swelling ratio (q) was calculated by using Equation (3). To calculate the equilibrium mass swelling ratio of composites, the following equation was used:

$$q = \frac{W_s}{W_d} \tag{3}$$

where W_s and W_d represent the weights of the swollen gel and the dry gel, respectively.

2.6. PFAS Sorption Studies

The sorption capability of the materials to bind PFAS was determined by an individual batch experiment. A mixture of PFOA and PFOS at initial concentrations of 100 µg/L each, with an adsorbent concentration of 0.5 mg/mL, was exposed to 30 mL of 1 M NaOH-titrated deionized water solution (pH = 7.0 ± 0.2) in 50 mL centrifuge tubes at 25 °C. All hydrogels and hydrogel composites were assessed, as well as controls comprised of solvent only, neat hydrogels and solvent, and PAC only. The mixtures were shaken at room temperature (25 °C) on a temperature-controlled orbit shaker at 200 rpm. After 24 h of interaction, samples were collected and filtered using a Whatman 4 filter with 90mm diameter; PFOA and PFOS were analyzed via a Vanquish Core high-performance liquid chromatography (HPLC) system coupled with a TSQ Altis Plus Triple Quadrupole Mass Spectrometer (MS) (Thermo Fisher Scientific, Waltham, MA, USA). The mobile phase consisted of (A) LCMS-grade water and (B) LCMS-grade methanol: water (85:15), both with 20 mM ammonium acetate (Sigma Aldrich). The column was heated and held constant at 40 °C. The LCMS was operated in negative polarity mode with electrospray ionization. The limit of detection (LOD) for target analytes was 1 ng/L; seven calibration points with linear dynamic range (LDR) were within 1–150 µg/L and had R^2 values of 0.9989 and 0.9981 for PFOA and PFOS, respectively. Calibration standards and instrument blanks were run at the beginning and end of the analytical run. Batch experiments were performed in triplicates (n = 3) and the results have been reported as an average ± standard deviation. The removal efficiency was calculated using Equation (4) as follows:

$$RE = \frac{(Co - Ct)}{Co} * 100\% \qquad (4)$$

where Co and Ct are the initial and residual concentrations of PFAS analytes in (µg/L), respectively.

3. Results and Discussion

Acrylamide-based hydrogel and hydrogel composites were successfully synthesized using free-radical polymerization to obtain a series of materials with varying crosslinking densities and PAC loading percentages to determine impacts on physicochemical properties. The sorbent systems were also characterized and analyzed for their binding affinity to PFAS.

3.1. Synthesis Confirmation and Physicochemical Characterizations

Mass conversion was calculated for all systems, with >90% yield for both neat and 1% PAC hydrogels and hydrogel composites, whereas the 5% PAC loaded systems drastically decreased in yield, as shown in Table 1. This indicates a disruption of the polymerization reaction with increasing particulate concentration, which could be attributed to the adsorption of the initiator/catalyst (APS and TEMED) by the PAC during the reaction. Additionally, since the PAC is a solid particulate that is physically entrapped within the hydrogel network, it can impact the network formation during polymerization and consequently impact yield. As shown in Figures 3 and S1, the IR spectra of all PAm systems confirmed successful polymerization of the monomer and crosslinker into polyacrylamide through the absorption of expected peaks. Characteristic carbonyl and amide functionalities were indicated by the absorption peaks between 1640 and 1660 cm^{-1} corresponding to the dashed line in between those peaks which revealed the formation of the primary bending amide (NH$_2$) band and the stretching carbonyl (C=O) group. The doublet peaks between 3100 and 3400 cm^{-1} further account for amide functionality through the stretching of both primary (NH2) and secondary (NH) groups as highlighted by the two dashed lines within the corresponding peaks [24–26]. Additionally, the absence of the alkene peak (C=C) around 1610 cm^{-1} indicates the high conversion of the acrylamide monomer to polymer [25]. Figure 3 elucidates the impacts of PAC loading density by comparing 5% crosslinked hydrogel composites, with 1 and 5% PAC, with the 5NNT system. All characteristic peaks, i.e., carbonyl peaks between 1640 and 1660 cm^{-1} and amide peaks between

3100 and 3400 cm^{-1}, reduced in intensity as the amount of PAC was increased, which can again be attributed to the adsorption of APS and TEMED by PAC and the subsequent inhibition of polymerization. In contrast, FTIR spectra were also analyzed to compare the influence of varying crosslinking densities between the three neat systems (1, 5, and 10%), as displayed in Figure S1. No significant trends were observed between the three differing crosslinking concentrations within diagnostic peaks.

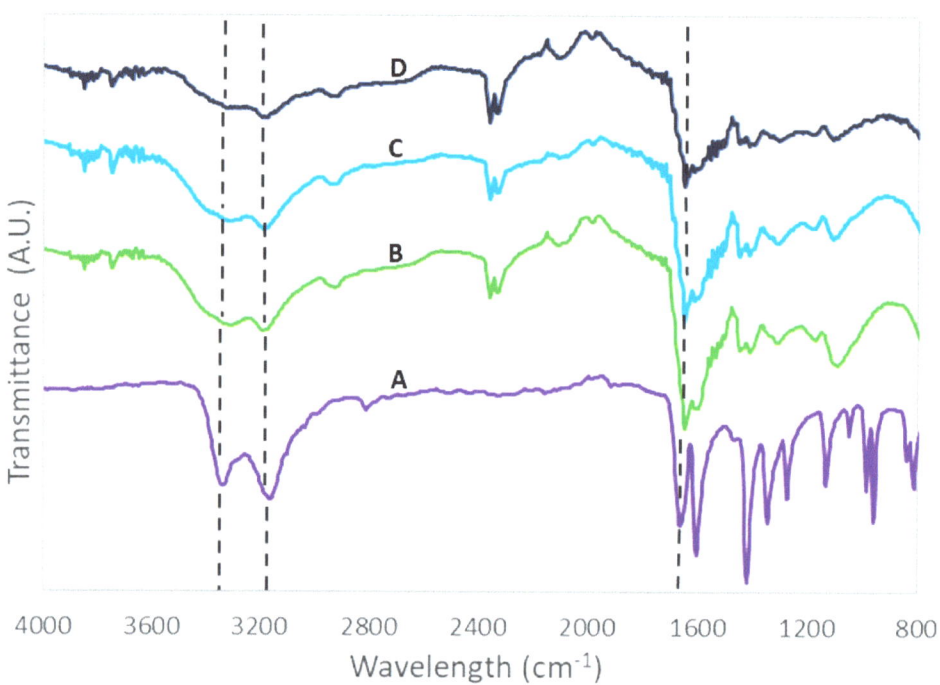

Figure 3. FTIR spectra performed with ATR technologies of (A) plain acrylamide, and hydrogels and hydrogel composites with increasing PAC loading density, (B) 5NNT hydrogels, (C) 5N1P composites, and (D) 5N5P hydrogels, respectively.

The TGA curves for the hydrogels and hydrogel composites were normalized to weight at 150 °C to account for water evaporation in order to obtain the hydrogels' and composites' degradation as shown in Figure 4. The normalized graphs shown in Figure 4 display a two-stage decomposition, where the first mass loss, from 250 to 375 °C, is assigned to the loss of ammonia as a result of the imidization reaction of the amide groups, which accounted for roughly a 20% weight decrease; the second sharp mass loss zone, from 375 to 450 °C, indicates the breakdown of imide groups formed in the first stage and polymer backbone by the liberation of H_2, CO, and NH_3, resulting in the final carbonaceous residues [12,25,27]. The residual amount of carbon stabilized out at roughly 10% of the initial weight, and this amount was used as the baseline to determine actual PAC loading percentage. After normalizing the neat weight (%) that remained following degradation, the percentage of PAC was determined for the 1% and 5% loaded systems seen in Table 1. Both the 5% and 10% crosslinked systems had PAC loading percentages within 1.13% max deviation from the expected values of 1 and 5 initial PAC loading percentages, whereas the 1% crosslinked systems had a higher deviation comparatively (upwards of 2.5% for the system with 5% PAC), which is consistent with the assumption that the particulate PAC is disrupting the polymerization reaction. Since the 1% systems have fewer crosslinked networks and less conversion of polymeric material, the loading percentage appears higher given that there is less polymeric feed.

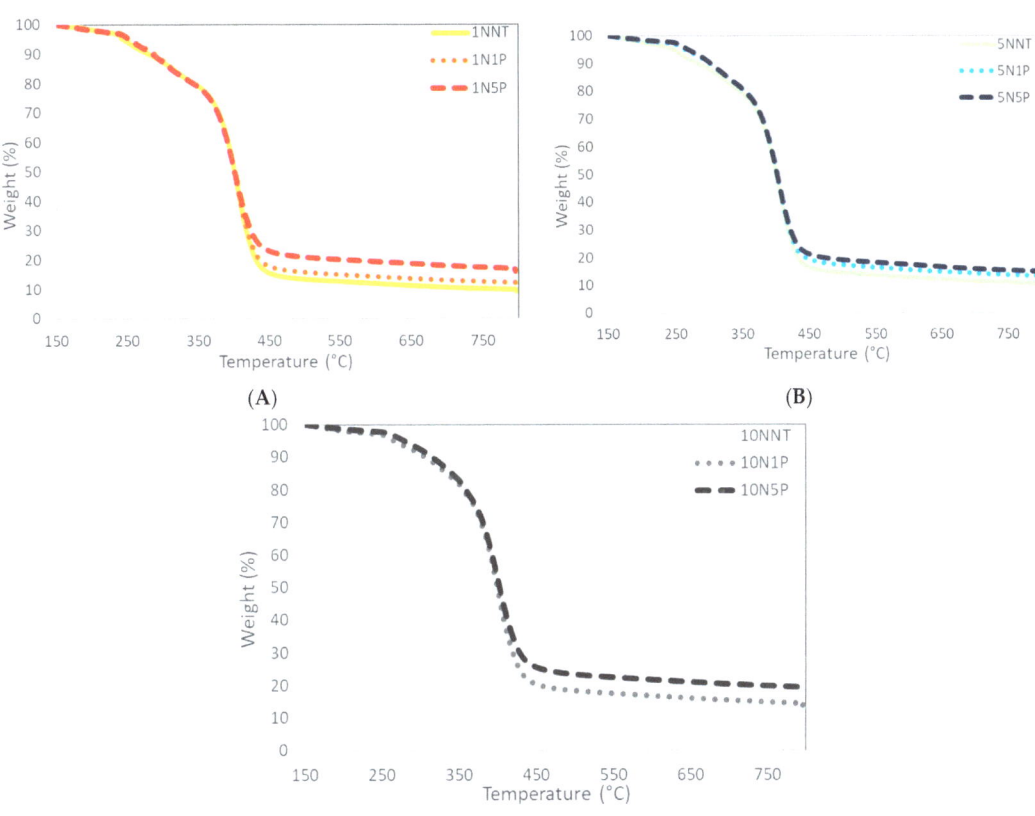

Figure 4. TGA analyses of hydrogels and hydrogel composites in Nitrogen gas from 150 °C to 800 °C to determine PAC loading percentages in (**A**) 1% crosslinked systems, (**B**) 5% crosslinked hydrogels, and (**C**) 10% crosslinked polymers, all with varying PAC loading percentages.

3.2. Surface Characterization

SEM and EDS were analyzed for 1NNT, 5NNT, 10NNT, 10N1P, and 10N5P materials to determine any impacts on morphology due to varying crosslinking densities (1, 5, and 10%) and PAC loading percentages (0, 1, and 5%). High-resolution images for all systems are shown in Figure 5. Figure 5A–C compare the morphology of the three neat materials with different crosslinking densities, i.e., 1NNT, 5NNT, and 10NNT, at two magnifications; no significant differences in morphology were observed. Additionally, Figure 5C–E compare how the varying PAC loading concentrations impact morphology; again, there were no significant differences in morphology noted; however, EDS microanalysis of 10% crosslinked systems with and without PAC indicates the inclusion of carbon-rich spots with increasing PAC loading percentages, represented by the concentrated blue marks on the EDS maps of composites in Figure S2.

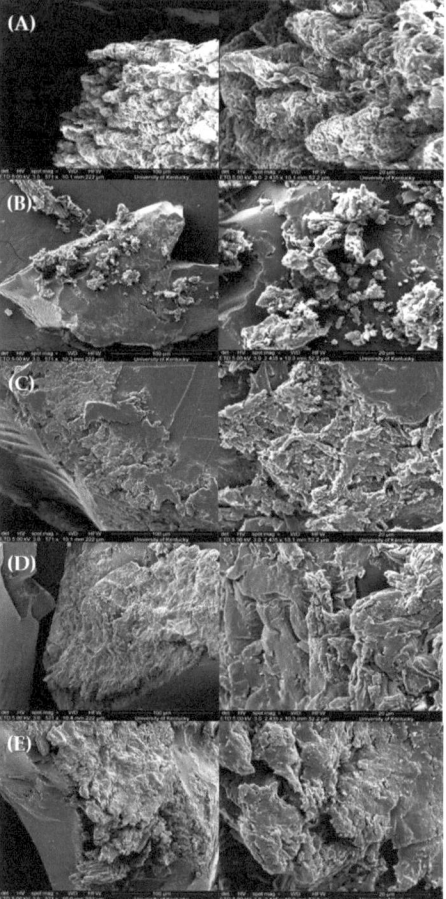

Figure 5. Scanning electron microscopy images of neat hydrogels with (**A**) 1%, (**B**) 5%, and (**C**) 10% NNMBA to measure impacts of increasing crosslinking density, and (**D**) 10N1P and (**E**) 5% PAC composites to determine the effect of PAC inclusion within the network with scale bars of 100 μm on the left, at lower magnification, and 20 μm on the panels on the right.

3.3. Aqueous Swelling Studies

3.3.1. Kinetic Swelling Studies

The swelling kinetics of the hydrogels and hydrogel composites containing various NNMBA and PAC contents at 1, 4, 8, 12, and 24 h are displayed in Figure 6. For all cases, the swelling profiles exhibited rapid swelling with equilibrium swelling reached by approximately 4 h. The initial swelling was dominated by the capillary adsorption interaction from the porous structure of the hydrogels and hydrogel composites. The second phase was attributed to the saturation of the polymeric network by free diffusion of the water molecules until the swelling equilibrium was reached [28].

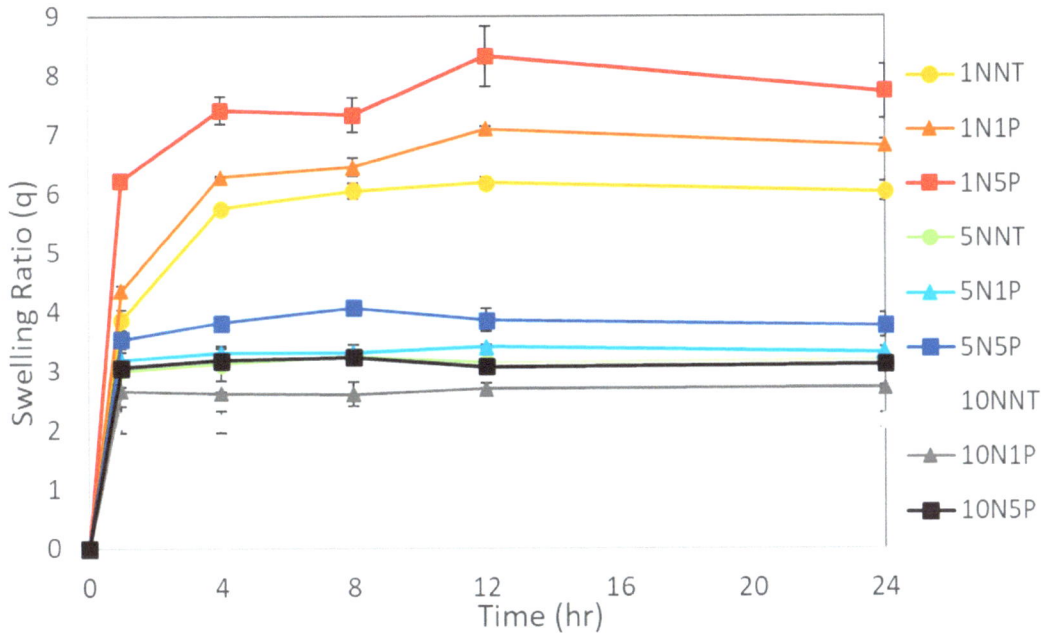

Figure 6. Kinetic swelling behavior of hydrogels and hydrogel composites in DI water after 1, 4, 8, 12, and 24 h.

3.3.2. Equilibrium Swelling Studies

Swelling behavior measures the expansibility of materials and determines water-retention properties. A high swelling ratio indicates a large amount of water that the network can absorb and a more open network, which can impact the ability of molecules of different sizes to absorb into the system. As expected, hydrogels with increased NNMBA contents exhibited a decrease in water uptake due to a higher crosslinking density and a more restricted network, as illustrated in Figure 7. Moreover, there was an increase in equilibrium swelling with increasing PAC loading percentage, which can be attributed to a decrease in polymer conversion due to the physical inhibition of the polymerization reaction by the particulate PAC addition. The hydrogels experienced swelling ratios in the range of 2–8 where the least adsorption was displayed as 2.16 ± 0.12 for the neat hydrogel with the highest crosslinking density (10N), and the highest swelling ratio was displayed by the least crosslinked system (1N) with the largest PAC loading (5P), with 7.73 ± 0.46.

3.4. Equilibrium Sorption Studies

The removal efficiency of the hydrogels and hydrogel composites for PFAS was determined for all systems using a mixture of PFOA and PFOS at an initial concentration of 100 ppb of each, as shown in Figure 8. No sorption was observed for the systems with no sorbent, demonstrating that the filters are not absorbing any contaminant and, therefore, are not interfering with the reported results. Neat hydrogels revealed limited sorption ability with a maximum binding capacity of PFAS analytes with a maximum of 57% removal for PFOS for the 5% crosslinked system and 6.93% PFOA from the 1NNT system, with no significant trends between varying crosslinking densities, as seen in Figure 8A. The sorption mechanism of PFAS binding to the neat hydrogels is likely attributable to the physisorption of the PFAS analytes through Van der Waals electrostatic interactions [29,30]. The addition of PAC within the hydrogel network showed significantly higher binding percentages than those seen for the neat hydrogels visible in Figure 8B,C. PFAS removal by composites with 1% PAC increased by >10% for all three crosslinking densities, with the 5% crosslinking density exhibiting a slightly higher removal efficiency between all of the three polymers for PFOS. Hydrogel composites with the highest PAC loading percentages (5% PAC) exhibited removal efficiencies upwards of 72% for PFOA and 82% for PFOS, with a maximum PFOA and PFOS removal of 96% and 98%, respectively, by the 5N5P system. Although there was not a significant trend between the 1 and 5% crosslinked systems, the 10N5P sorbent showed the potential restriction of these longer-chain PFAS due to the increase in crosslinking density (i.e., higher NNMBA concentration). The PAC-only systems showed virtually 100% removal efficiency for both PFOS and PFOA. The improved binding of the PAC systems is explained by the expected hydrophobic interactions of the PFAS tail and activated carbon surface. All composite systems exhibited higher removal of PFOS compared to PFOA, which is consistent with the literature, which has found that the removal of long-chain and sulfonic PFAS is dominated by hydrophobic interactions [12]. Moreover, the sorption studies were performed in NaOH-titrated DI-water solution, which introduced Na^+ ions following dissolution and increased the ionic charge, which enhanced PFOS binding and had little effect on PFOA sorption [31].

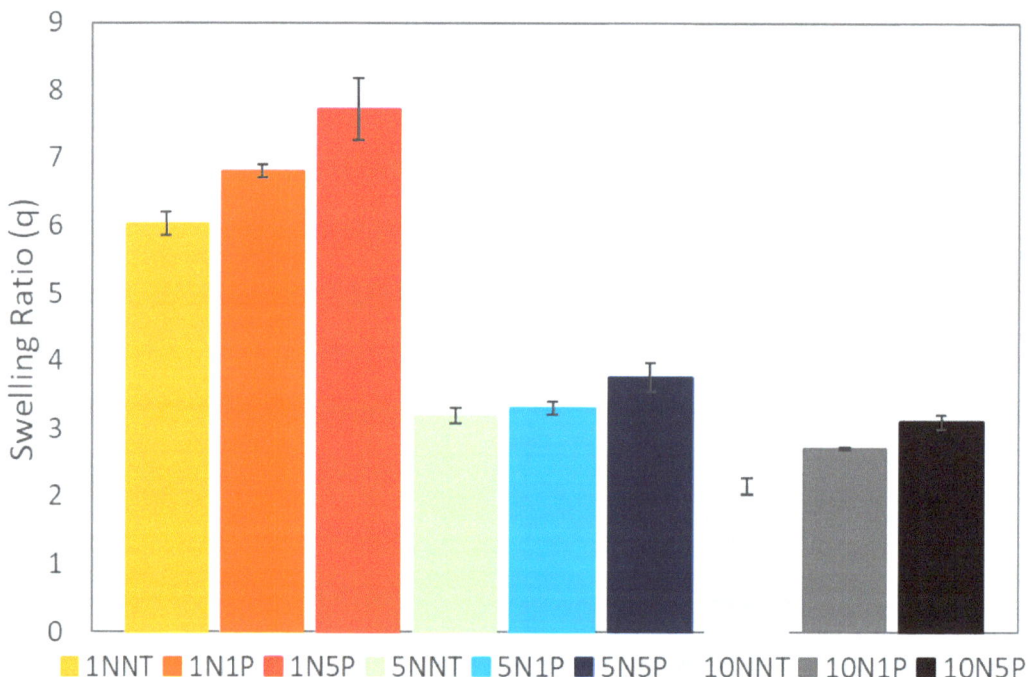

Figure 7. Aqueous equilibrium swelling behavior of all hydrogels and hydrogel composites in DI water after 24 h.

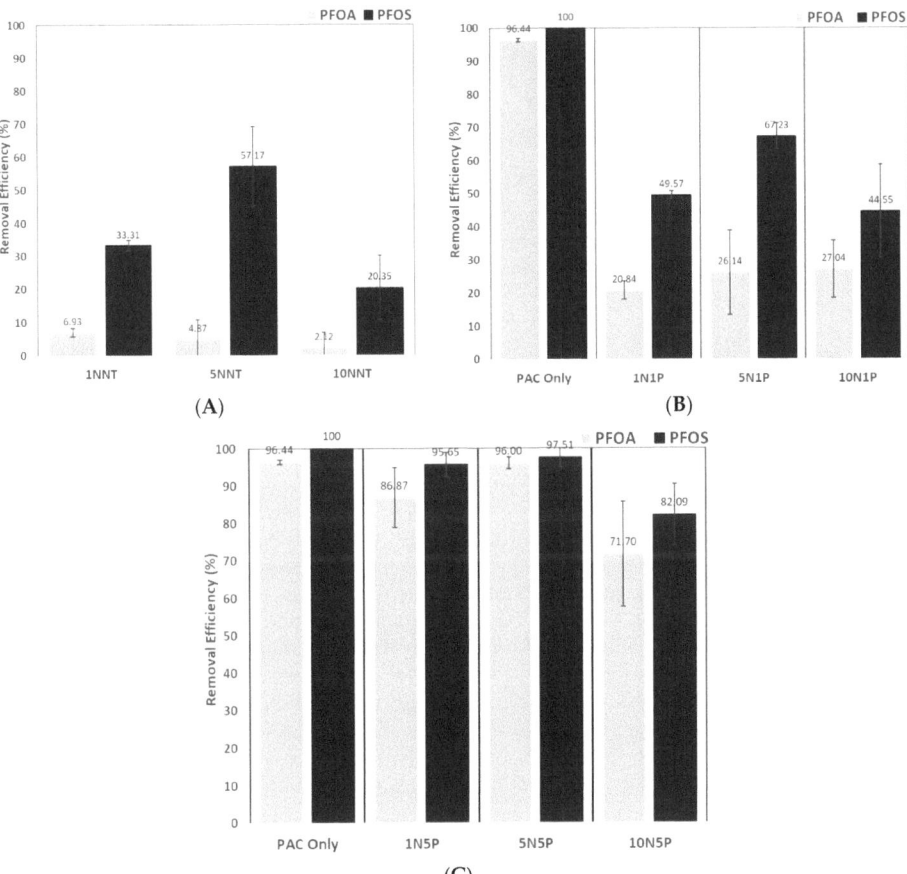

Figure 8. PFOA and PFOS removal efficiency of sorbent materials of PFAS at a starting concentration of 100 µg/mL in a 12 mL DI-water solution titrated to obtain a pH 7 with NaOH and 0.5 mg/mL. A PAC-only control was also tested and is compared against (**A**) the neat hydrogels, (**B**) PAC composites with 1% PAC loading, and (**C**) 5% PAC systems.

Although PAC outperformed all sorbents in removing PFOA and PFOS, it has proved to be highly unselective towards short-chain compounds when competing with legacy analytes as its major pores are saturated with legacy compounds and therefore block the binding of short-chain contaminants. It was hypothesized that modifying the NNMBA concentration of the material and increasing its crosslinking density would limit the removal percentage of larger PFAS molecules due to a more restrictive network and, therefore, allow for more sorption of short-chain compounds. The binding data suggest that this hypothesis could be accurate, given that the systems with the highest crosslinking densities appear to limit some of the PFOA and PFOS sorption. This hypothesis will be tested further with follow-up studies that explore additional crosslinking densities and binding studies, including competing short-chain PFAS. The findings from this study indicate the high sorption potential of PFAS contaminants and introduce an exciting platform that can be further modified for increased selectivity of contaminants, such as a composite with added cationic functionality.

4. Conclusions

Overall, acrylamide-based hydrogel and hydrogel composite sorbents were successfully synthesized using free-radical polymerization and characterized using several methods including mass yield, FTIR, and TGA. Mass yield was the primary determining factor for polymer conversion by making sure the feed materials were transferring to the final product. Congruently, FTIR confirmed the presence of the carbonyl and amide peaks within the polymeric systems that were also present in the analogous monomer. TGA analysis further indicates polymeric degradation between 300 and 450 °C, which is consistent with hydrogel breakdown, and the residual weight observed at 800 °C confirmed the incorporation of PAC within the hydrogel composites at each respective loading capacity, which increased with increased loading percentage. SEM analysis concluded that there were no significant changes between the images with varying crosslinking densities but confirms the inclusion of PAC within hydrogel composites by carbon-rich spots in EDS maps. Kinetic swelling studies were performed to determine the time required for the systems to reach equilibrium, which was shown to be reached within 4 h for all systems. This swelling time was then used to calculate equilibrium swelling ratios for all systems, where the highest swelling ratio was attributed to that with the least crosslinked network, as expected, with a decrease as the crosslinking density was increased. The addition of PAC particulate also enhanced the swelling capacity of the hydrogels, with the 5% PAC loaded systems having a higher swelling ratio than the 1% PAC, both being higher than the neat systems.

Finally, neat hydrogels showed limited sorption of the PFAS analytes with only 50% capacity for the 5NNT systems, whereas the composites showed greater sorption of both PFOA and PFOS. The higher sorption of PFOS above PFOA was attributed to the electronegativity of the sulfonic group of PFOS being higher than the carboxylic group in PFOA. The findings from this study indicate the high sorption potential of PFAS contaminants in these hydrogel composites, and these materials can be further modified for increased selectivity and investigated for a wider range of analytes.

Supplementary Materials: The following supporting information can be downloaded at: https://www.mdpi.com/article/10.3390/polym15224384/s1, Figure S1: FTIR spectra of neat hydrogels with increasing crosslinking density as follows, (a) 1NNT, (b) 5NNT, and (c) 10NNT. Figure S2: Scanning electron microscopy energy-dispersive X-ray microanalysis maps of 10% crosslinked systems with (a) no activated carbon, (b) 1% PAC, and (c) 5% PAC.

Author Contributions: Conceptualization, M.V.X.K. and J.Z.H.; methodology, A.M.G. and M.V.X.K.; writing—original draft preparation, M.V.X.K.; writing—review and editing, A.M.G. and J.Z.H.; supervision, J.Z.H.; project administration, J.Z.H.; proof-reading, A.M.G. and J.Z.H. All authors have read and agreed to the published version of the manuscript.

Funding: This research is funded by the National Institute of Environmental Health Sciences (NIEHS) of the National Institutes of Health (NIH), grant number P42ES007380. The content is solely the responsibility of the authors and does not necessarily represent the official views of the National Institutes of Health.

Institutional Review Board Statement: Not applicable.

Informed Consent Statement: Not applicable.

Data Availability Statement: The data presented in this study are available on request from the corresponding authors.

Acknowledgments: The authors graciously acknowledge that the research reported in this publication was supported by the National Institute of Environmental Health Sciences of the National Institutes of Health under Award Number P42ES007380. The content is solely the responsibility of the authors and does not necessarily represent the official views of the National Institutes of Health.

Conflicts of Interest: The authors declare no conflict of interest.

References

1. Dickman, R.A.; Aga, D.S. A Review of Recent Studies on Toxicity, Sequestration, and Degradation of per- and Polyfluoroalkyl Substances (PFAS). *J. Hazard. Mater.* **2022**, *436*, 129120. [CrossRef]
2. Evich, M.G.; Davis, M.J.B.; McCord, J.P.; Acrey, B.; Awkerman, J.A.; Knappe, D.R.U.; Lindstrom, A.B.; Speth, T.F.; Tebes-Stevens, C.; Strynar, M.J.; et al. Per- and Polyfluoroalkyl Substances in the Environment. *Science* **2022**, *375*, eabg9065. [CrossRef] [PubMed]
3. Sunderland, E.M.; Hu, X.C.; Dassuncao, C.; Tokranov, A.K.; Wagner, C.C.; Allen, J.G. A Review of the Pathways of Human Exposure to Poly- and Perfluoroalkyl Substances (PFASs) and Present Understanding of Health Effects. *J. Expo. Sci. Environ. Epidemiol.* **2019**, *29*, 131–147. [CrossRef]
4. Committee on the Guidance on PFAS Testing and Health Outcomes; Board on Environmental Studies and Toxicology; Board on Population Health and Public Health Practice; Division on Earth and Life Studies; Health and Medicine Division; National Academies of Sciences, Engineering, and Medicine. *Guidance on PFAS Exposure, Testing, and Clinical Follow-Up*; National Academies Press: Washington, DC, USA, 2022; p. 26156. [CrossRef]
5. Abunada, Z.; Alazaiza, M.Y.D.; Bashir, M.J.K. An Overview of Per- and Polyfluoroalkyl Substances (PFAS) in the Environment: Source, Fate, Risk and Regulations. *Water* **2020**, *12*, 3590. [CrossRef]
6. Pérez, F.; Nadal, M.; Navarro-Ortega, A.; Fàbrega, F.; Domingo, J.L.; Barceló, D.; Farré, M. Accumulation of Perfluoroalkyl Substances in Human Tissues. *Environ. Int.* **2013**, *59*, 354–362. [CrossRef] [PubMed]
7. Crone, B.C.; Speth, T.F.; Wahman, D.G.; Smith, S.J.; Abulikemu, G.; Kleiner, E.J.; Pressman, J.G. Occurrence of Per- and Polyfluoroalkyl Substances (PFAS) in Source Water and Their Treatment in Drinking Water. *Crit. Rev. Environ. Sci. Technol.* **2019**, *49*, 2359–2396. [CrossRef] [PubMed]
8. Shahsavari, E.; Rouch, D.; Khudur, L.S.; Thomas, D.; Aburto-Medina, A.; Ball, A.S. Challenges and Current Status of the Biological Treatment of PFAS-Contaminated Soils. *Front. Bioeng. Biotechnol.* **2021**, *8*, 602040. [CrossRef]
9. Gagliano, E.; Sgroi, M.; Falciglia, P.P.; Vagliasindi, F.G.A.; Roccaro, P. Removal of Poly- and Perfluoroalkyl Substances (PFAS) from Water by Adsorption: Role of PFAS Chain Length, Effect of Organic Matter and Challenges in Adsorbent Regeneration. *Water Res.* **2020**, *171*, 115381. [CrossRef]
10. Kucharzyk, K.; Darlington, R.; Benotti, M.; Deeb, R.; Hawley, E. Novel Treatment Technologies for PFAS Compounds: A Critical Review. *J. Environ. Manag.* **2017**, *204*, 757–764. [CrossRef]
11. Interstate Technology and Regulatory Council. Treatment Technologies and Methods for Per- and Polyfluoroalkyl Substances (PFAS). Available online: https://pfas-1.itrcweb.org/wp-content/uploads/2020/10/treatment_tech_508_Aug-2020-Final.pdf (accessed on 11 November 2020).
12. Ateia, M.; Arifuzzaman, M.; Pellizzeri, S.; Attia, M.F.; Tharayil, N.; Anker, J.N.; Karanfil, T. Cationic Polymer for Selective Removal of GenX and Short-Chain PFAS from Surface Waters and Wastewaters at Ng/L Levels. *Water Res.* **2019**, *163*, 114874. [CrossRef]
13. Frazar, E.M.; Shah, R.A.; Dziubla, T.D.; Hilt, J.Z. Multifunctional Temperature-responsive Polymers as Advanced Biomaterials and Beyond. *J. Appl. Polym. Sci.* **2020**, *137*, 48770. [CrossRef] [PubMed]
14. Tang, S.; Bhandari, R.; Delaney, S.P.; Munson, E.J.; Dziubla, T.D.; Hilt, J.Z. Synthesis and Characterization of Thermally Responsive N-Isopropylacrylamide Hydrogels Copolymerized with Novel Hydrophobic Polyphenolic Crosslinkers. *Mater. Today Commun.* **2017**, *10*, 46–53. [CrossRef] [PubMed]
15. Gutierrez, A.M.; Frazar, E.M.; X. Klaus, M.V.; Paul, P.; Hilt, J.Z. Hydrogels and Hydrogel Nanocomposites: Enhancing Healthcare through Human and Environmental Treatment. *Adv. Healthc. Mater.* **2022**, *11*, 2101820. [CrossRef]
16. Uspenskaya, M.V.; Sitnikova, V.E.; Dovbeta, M.A.; Olekhnovich, R.O.; Denisyuk, I.Y. Sorption Properties of Clay and Pectin-Containing Hydrogels. In *Recent Research in Polymerization*; Cankaya, N., Ed.; InTech: St. Petersburg, Russia, 2018. [CrossRef]
17. Ateia, M.; Alsbaiee, A.; Karanfil, T.; Dichtel, W. Efficient PFAS Removal by Amine-Functionalized Sorbents: Critical Review of the Current Literature. *Environ. Sci. Technol. Lett.* **2019**, *6*, 688–695. [CrossRef]
18. Kumarasamy, E.; Manning, I.M.; Collins, L.B.; Coronell, O.; Leibfarth, F.A. Ionic Fluorogels for Remediation of Per- and Polyfluorinated Alkyl Substances from Water. *ACS Cent. Sci.* **2020**, *6*, 487–492. [CrossRef]
19. Zhang, Q.; Deng, S.; Yu, G.; Huang, J. Removal of Perfluorooctane Sulfonate from Aqueous Solution by Crosslinked Chitosan Beads: Sorption Kinetics and Uptake Mechanism. *Bioresour. Technol.* **2011**, *102*, 2265–2271. [CrossRef]
20. Yu, Q.; Deng, S.; Yu, G. Selective Removal of Perfluorooctane Sulfonate from Aqueous Solution Using Chitosan-Based Molecularly Imprinted Polymer Adsorbents. *Water Res.* **2008**, *42*, 3089–3097. [CrossRef] [PubMed]
21. Garg, S.; Wang, J.; Kumar, P.; Mishra, V.; Arafat, H.; Sharma, R.S.; Dumée, L.F. Remediation of Water from Per-/Poly-Fluoroalkyl Substances (PFAS)—Challenges and Perspectives. *J. Environ. Chem. Eng.* **2021**, *9*, 105784. [CrossRef]
22. Son, H.; Kim, T.; Yoom, H.-S.; Zhao, D.; An, B. The Adsorption Selectivity of Short and Long Per- and Polyfluoroalkyl Substances (PFASs) from Surface Water Using Powder-Activated Carbon. *Water* **2020**, *12*, 3287. [CrossRef]
23. Liu, C.; Hatton, J.; Arnold, W.A.; Simcik, M.F.; Pennell, K.D. In Situ Sequestration of Perfluoroalkyl Substances Using Polymer-Stabilized Powdered Activated Carbon. *Environ. Sci. Technol.* **2020**, *54*, 6929–6936. [CrossRef]
24. Asgharzadehahmadi, S.A.; Muhamad, I.I.; Zaidel, D.N.A.; Supriyanto, E. Synthesis and Characterization of Polyacrylamide Based Hydrogel Containing Magnesium Oxide Nanoparticles for Antibacterial Applications. 6. Available online: https://www.semanticscholar.org/paper/Synthesis-and-characterization-of-polyacrylamide-Asgharzadehahmadi/c84a283b93362aae45e3d65795627009bf16e0fe (accessed on 1 January 2021).

25. Shatat, R.S.; Niazi, S.K.; Ariffin, A. Synthesis and Characterization of Different Molecular Weights Polyacrylamide. *IOSR J. Appl. Chem.* **2017**, *10*, 67–73. [CrossRef]
26. Abdelkader, R.; Mohammed, B. Green Synthesis of Cationic Polyacrylamide Composite Catalyzed by An Ecologically Catalyst Clay Called Maghnite-H+ (Algerian MMT) Under Microwave Irradiation. *Bull. Chem. React. Eng. Catal.* **2016**, *11*, 170. [CrossRef]
27. Leung, W.M.; Axelson, D.E.; Van Dyke, J.D. Thermal Degradation of Polyacrylamide and Poly(Acrylamide-Co-Acrylate). *J. Polym. Sci. Part Polym. Chem.* **1987**, *25*, 1825–1846. [CrossRef]
28. Wang, X.; Jing, S.; Qiu, X.; Zhao, S.; Liu, Y.; Tan, Y. Novel Bile Acid Sequestrant: A Biodegradable Hydrogel Based on Amphiphilic Allylamine Copolymer. *Chem. Eng. J.* **2016**, *304*, 493–502. [CrossRef]
29. Lebkiri, I.; Abbou, B.; Kadiri, L.; Ouass, A.; Elamri, A.; Ouaddari, H.; Elkhattabi, O.; Lebkiri, A.; Rifi, E.H. Swelling Properties and Basic Dye Adsorption Studies of Polyacrylamide Hydrogel. *Desalination Water Treat.* **2021**, *233*, 361–376. [CrossRef]
30. Sato, N.; Aoyama, Y.; Yamanaka, J.; Toyotama, A.; Okuzono, T. Particle Adsorption on Hydrogel Surfaces in Aqueous Media Due to van Der Waals Attraction. *Sci. Rep.* **2017**, *7*, 6099. [CrossRef]
31. Zhao, L.; Bian, J.; Zhang, Y.; Zhu, L.; Liu, Z. Comparison of the Sorption Behaviors and Mechanisms of Perfluorosulfonates and Perfluorocarboxylic Acids on Three Kinds of Clay Minerals. *Chemosphere* **2014**, *114*, 51–58. [CrossRef]

Disclaimer/Publisher's Note: The statements, opinions and data contained in all publications are solely those of the individual author(s) and contributor(s) and not of MDPI and/or the editor(s). MDPI and/or the editor(s) disclaim responsibility for any injury to people or property resulting from any ideas, methods, instructions or products referred to in the content.

Article

Modeling of Hexavalent Chromium Removal with Hydrophobically Modified Cellulose Nanofibers

Francisco de Borja Ojembarrena [1,*], Jose Luis Sánchez-Salvador [1], Sergio Mateo [1], Ana Balea [1], Angeles Blanco [1], Noemí Merayo [2] and Carlos Negro [1,*]

[1] Department of Chemical Engineering and Materials, Complutense University of Madrid, Avda. Complutense s/n, 28040 Madrid, Spain
[2] Department of Mechanical, Chemical and Industrial Design Engineering, ETSIDI, Universidad Politécnica de Madrid, Ronda de Valencia 3, 28012 Madrid, Spain
* Correspondence: frojemba@ucm.es (F.d.B.O.); cnegro@ucm.es (C.N.)

Citation: Ojembarrena, F.d.B.; Sánchez-Salvador, J.L.; Mateo, S.; Balea, A.; Blanco, A.; Merayo, N.; Negro, C. Modeling of Hexavalent Chromium Removal with Hydrophobically Modified Cellulose Nanofibers. *Polymers* **2022**, *14*, 3425. https://doi.org/10.3390/polym14163425

Academic Editor: Marta Otero

Received: 29 July 2022
Accepted: 16 August 2022
Published: 22 August 2022

Publisher's Note: MDPI stays neutral with regard to jurisdictional claims in published maps and institutional affiliations.

Copyright: © 2022 by the authors. Licensee MDPI, Basel, Switzerland. This article is an open access article distributed under the terms and conditions of the Creative Commons Attribution (CC BY) license (https://creativecommons.org/licenses/by/4.0/).

Abstract: Cellulose nanofibers (CNF) are sustainable nanomaterials, obtained by the mechanical disintegration of cellulose, whose properties make them an interesting adsorbent material due to their high specific area and active groups. CNF are easily functionalized to optimize the performance for different uses. The hypothesis of this work is that hydrophobization can be used to improve their ability as adsorbents. Therefore, hydrophobic CNF was applied to adsorb hexavalent chromium from wastewater. CNF was synthetized by TEMPO-mediated oxidation, followed by mechanical disintegration. Hydrophobization was performed using methyl trimetoxysilane (MTMS) as a hydrophobic coating agent. The adsorption treatment of hexavalent chromium with hydrophobic CNF was optimized by studying the influence of contact time, MTMS dosage (0–3 mmol·g^{-1} CNF), initial pH of the wastewater (3–9), initial chromium concentration (0.10–50 mg·L^{-1}), and adsorbent dosage (250–1000 mg CNF·L^{-1}). Furthermore, the corresponding adsorption mechanism was identified. Complete adsorption of hexavalent chromium was achieved with CNF hydrophobized with 1.5 mmol MTMS·g^{-1} CNF with the faster adsorption kinetic, which proved the initial hypothesis that hydrophobic CNF improves the adsorption capacity of hydrophilic CNF. The optimal adsorption conditions were pH 3 and the adsorbent dosage was over 500 mg·L^{-1}. The maximum removal was found for the initial concentrations of hexavalent chromium below 1 mg·L^{-1} and a maximum adsorption capacity of 70.38 mg·g^{-1} was achieved. The kinetic study revealed that pseudo-second order kinetics was the best fitting model at a low concentration while the intraparticle diffusion model fit better for higher concentrations, describing a multi-step mechanism of hexavalent chromium onto the adsorbent surface. The Freundlich isotherm was the best adjustment model.

Keywords: wastewater treatment; adsorption; nanocellulose; cellulose nanofibers; hexavalent chromium; hydrophobization process

1. Introduction

Chromium is a harmful water pollutant. Several environmental effects have been associated with the presence of trivalent and hexavalent chromium in natural water bodies. Trivalent chromium shows lower toxicity due to its low cell permeability, whereas hexavalent chromium presents acute toxicity to many species, causing carcinogenicity and mutagenicity [1–3], aside from being neurotoxic [4]. Furthermore, Cr(VI) is highly toxic and one of the most common environmental contaminants and is non-easily biodegradable in nature, thus staying in the environment for a long time, polluting the soil and water, with the subsequent health risks to humans and wildlife [5]. Several industries generate hexavalent chromium in their effluents including tanneries, electroplating, hardware, textile, cement, or mining industries [3,6]. The limit of discharge of this contaminant varies depending on the country, but most of European countries have low limits in water, between 0.05 and 0.1 mg·L^{-1} [7].

The most relevant hexavalent chromium removal treatments include adsorption, ion exchange, membrane filtration, electrochemical treatments, coagulation–flocculation, chemical precipitation, biological treatments, and photocatalysis [6,8–10]. Among them, adsorption is the most widely used technology due to the low operational costs, ease of operation, low environmental impact associated with the process, and its effectiveness in removing trace levels of chromium [11]. Some of these adsorption treatments are based on microbial bioadsorption [12], since some microbes are able to reduce hexavalent chromium to trivalent chromium; as the latter is much less harmful, their excretion by microbes is not an issue [5]. Common adsorbents such as activated carbons or zeolites have been successfully applied, but there is a need for new materials from renewable sources, with better biodegradability and whose raw materials have less environmental impact. Adsorbents coming from renewable sources include those obtained from waste phytomass, which could be valorized to adsorb hexavalent chromium, achieving 87.2% removal, with the possibility of regeneration for at least four cycle runs [13]. Agro-based biomass has also been used as a bioadsorbent, which contributes to significant sustainable waste management, with an adsorption capacity of more than 10 mg/g [12]. Other waste materials used to produce adsorbents are fishbone waste to produce hydroxyapatite, which successfully removed Ni^{2+}, Cu^{2+}, and Zn^{2+} (more than 95% removal at the optimum conditions) [14]. Furthermore, nanomaterials can be used as adsorbents such as the polyacrylonitrile nanofiber membrane modified with bovine serum albumin used to remove Ca^{2+} from process streams, achieving removal efficiencies of about 62% [15]. Other membrane structure materials used as adsorbents are, for example, cellulose acetate-based membranes with glass nanoparticles for CO_2 separation [16]. The surfaces of the nanomaterial adsorbents are usually functionalized or chemically modified to improve Cr(VI) adsorption such as the attachment of active binding sites [17].

As cellulose is the most abundant biopolymer on Earth, and it is an inexpensive and renewable material, it is an excellent option as a raw material to produce new adsorbents [18,19] as well as its derivates such as sulfate cellulose [20].

Nanocelluloses are cellulosic materials with at least one dimension in the nanometer scale. There are different types of nanocelluloses including bacterial cellulose (BC), cellulose nanocrystals (CNC), and cellulose nanofibers (CNF). These materials show good adsorptive properties due to their high specific area and the presence of active groups such as carboxyl groups over their surface [21]. These nanomaterials have been proven as excellent adsorbents of different heavy metals such as cadmium, nickel, copper, lead, arsenic, iron, silver, cobalt, or mercury, thanks to their ease of functionalization and application, which enhances the adaptation to each heavy metal [22,23]. Celluloses are commonly anionic charged, caused by the pulping and bleaching processes [24]. Most of the heavy metals are cationic species, facilitating the direct application of nanocelluloses without any surface modification to obtain both a high removal yield and high adsorption capacity as well as fast adsorption rates. Khoo et al. [25] confirmed that there was a large affinity of CNC for Cu^{2+}, Cd^{2+}, Ni^{2+}, Pb^{2+}, Zn^{2+}, Fe^{3+}, and Ag^+ adsorption, reached with none or low modifications. Compared to these cationic heavy metals, the adsorption capacity of CNC for anionic As(V) was reduced by an order of magnitude. The same trend was also observed by Liu et al. [26], who applied both untreated CNF and 2,2,6,6-tetramethylpiperidin-1-yl-oxyl (TEMPO)-mediated oxidized CNF to a list of 15 heavy metals, with hexavalent chromium the only anionic species. The authors found a maximum adsorption capacity of 87.5 mg·g^{-1} using TEMPO-oxidized CNF-PAN membranes to adsorb Cr(VI), which was lower than the adsorption capacity reached when cationic Pb(II) was adsorbed.

The adsorption of anionic hexavalent chromium species with cellulosic materials usually requires other strategies and one of the most common options is surface modification. These surface modifications often include a chemical reaction, which forces a change in the surface groups present in the cellulose. Cationization, oxidation, esterification, alkaline treatment, or halogenation are some of the most typical treatments [25]. These modifications have demonstrated their applicability, and some of them such as dialdehyde oxidation or the cationization of celluloses have reached the large yields and adsorption capacities of hexavalent chromium [27,28]. Nevertheless, they usually imply a purification process that needs washing steps and the loss of reagents after treatment.

Another option could be the coating of the nanocelluloses. In this case, the nanocelluloses undergo a reaction where the coating reagent is completely mixed with the nanocellulose in suspension and the nanocellulose surface becomes covered by the coating agent. As this chemical is directly added to the final nanocellulose suspension, there is no need for separation after the reaction, avoiding the loss of material. Among the possible hydrophobization reactions, silanization becomes a great option because of the maintenance of large specific area of the fibrils and the enhanced hydrophobicity of cellulose fibers [29]. This fact allows for subsequent advanced treatments such as lyophilization to obtain aerogels [30]. Other silanized materials such as graphene oxide have been successfully applied to hexavalent chromium adsorption [31,32], but little information has been developed on the application of silanized cellulose-based materials to this contaminant. The closest approach is by Jamroz et al. [33], who indicated that amino-silanized celluloses could become an option for the ultra-trace determination of hexavalent chromium in water, which is not a direct wastewater treatment.

The state-of-the-art shows that there is a lack of studies about the use of nanocellulose to remove Cr(VI) from water streams, and particularly, the efficiency of hydrophobic nanocelluloses is not known. Furthermore, the kinetic and adsorptive behavior of these modified materials have not been well-studied yet. On the other hand, in this study, the applied hydrophobization method was based on a coating process to reduce product losses.

Therefore, in this study, the main objective was the synthesis and characterization of CNF hydrogels modified with methyl trimetoxysilane (MTMS) as a hydrophobic coating agent and their application to Cr(VI) adsorption from wastewater. The novelty of this approach is related to the application of hydrophobic CNF, hydrophobized by a coating method, to improve the adsorption performance of cellulosic nanofibers. Furthermore, the proposed adsorbent material application was optimized for the removal of Cr(VI) by performing a set of experiments on a batch system to evaluate the effect of system variables, MTMS dosage, pH value, initial chromium concentration, adsorbent dosage, and contact time. The kinetic and isotherm data were modeled to increase the knowledge related to the process, making easier its future application. Furthermore, different parameters have been obtained from a wide variety of models, which are the key factors in understanding the way hexavalent chromium interacts with the surface of the tested nanocelluloses.

2. Materials and Methods

2.1. Materials

Bleached pine kraft pulp was supplied by Arauco (Chile) and used as a raw material to synthetize CNF hydrogels. All of the experiments were performed using Milli-Q quality water. The chemicals used for the production and characterization of nanocelluloses were hydroxylamine hydrochloride, silver nitrate, methyl trimetoxysilane (MTMS), and 2,2,6,6-tetramethylpiperidin-1-yl-oxyl (TEMPO) supplied by Sigma Aldrich; sodium bromide, potassium chromate, sodium chloride, sodium hydroxide pellets, and sodium hypochlorite solution (10 w/v%) supplied by Panreac; and hydrochloric acid (37% v/v) and sulfuric acid (98% v/v) were supplied by Labkem. All of the chemicals used were of analytical grade.

To analyze hexavalent chromium, standard solutions were used. Hexavalent chromium standard solution (50 mg·L^{-1} as Cr(VI)) was used as a calibration standard for the spectrophotometric method. Analytical reagents for hexavalent chromium determination were purchased from Macherey Nagel (Dueren, Germany) following Standard Method 3500 Cr B. Poly(diallyldimethylammonium chloride) (PDADMAC) and polyethylenesulfonate (PesNA) solutions with a concentration of 0.00025 N were used as standard titration reagents during cationic and anionic demand determination.

2.2. CNF Hydrogel Synthesis

Briefly, bleached pine kraft pulp was disintegrated at 30,000 revolutions and diluted at 1% of consistency. Then, TEMPO-mediated oxidation was carried out through the addition of 5 mmol of NaClO per gram of cellulose using TEMPO and NaBr at 0.1 mmol and 1 mmol per gram of cellulose, respectively [34–37]. The oxidized cellulose suspension, after a cleaning process, was then treated under three-step mechanical homogenization at 600 bar pressure. Finally, the homogenized suspension underwent surface modification with MTMS to perform the hydrophobization, as indicated by Zhang et al. [30].

2.3. CNF Characterization

The characterization of the CNF hydrogels differs from the typical physical-chemical characterization of adsorbents as it is a hydrogel and must be characterized as a nanocellulose suspension instead of the common solid characterization. This characterization included the determination of the consistency of the suspension as well as the anionic demand and transmittance of the CNF suspension at $\lambda = 800$ nm, as described by Balea et al. [38], and the zeta potential of cellulose nanofibers and the amount of carboxyl groups in the oxidized pulp suspension determined by conductimetric titration as described by Sanchez-Salvador et al. [39].

2.4. Experimental Procedure: Batch Adsorption of Hexavalent Chromium Solution

The experimental installation consists of a batch, stirred laboratory beakers of 250 mL filled with 100 mL of sample of synthetic wastewater placed on multiple position hot plates under temperature control. Different operating conditions were studied: contact time, pH, chromium concentration, and adsorbent dosage. In the case of the CNF hydrogel, the dosage of hydrophobic reagent MTMS in the synthesis route was also optimized regarding adsorption efficiency.

During the kinetic experiments, the adsorption contact times were considered, taking samples at 1, 5, 15, 30, 60 min, 2, 3, 4, 24, and 48 h. The samples were filtered by cellulose acetate syringe filters (pore size 0.45 µm) to separate the insoluble fraction of the adsorbent from the soluble fraction as this material did not interact or adsorb chromium while passing through. Each experiment was considered finished (equilibrium time) when the pollutant level remained constant in 2–3 samples in a row. The hexavalent chromium concentration was determined by using a spectrophotometer calibrated with diluted solutions from a standard of chromium of 50 mg·L^{-1} between 0.02 and 0.50 mg·L^{-1} and the measurement was performed at the peak wavelength $\lambda = 540$ nm, as indicated in Standard Method 3500 Cr B [40]. Synthetic hexavalent chromium solution samples were diluted when the concentration exceeded the top of the calibration curve.

The pH effect was evaluated from acidic (pH 3) to neutral and alkaline (pH 9) conditions through the addition of HCl and NaOH, both at 0.1 mol·L^{-1}. Hexavalent chromium concentration in the kinetic and isotherm studies varied from 0.1 to 50 mg·L^{-1}. Hydrophobized CNF hydrogels were tested in a dosage range 250–1000 mg CNF·L^{-1}. The dosage of the hydrophobic reagent MTMS during CNF hydrogel synthesis was varied from 0 to 5 mmol MTMS·g^{-1} CNF hydrogel, according to the indications from Zhang et al. [30]. Each test was repeated three times.

2.5. Isotherm and Kinetic Studies

The conversion of the hexavalent chromium concentration into adsorption capacity values was determined. Efficiency of the Cr(VI) removal and adsorption capacity was calculated according to Equations (1) and (2) [41].

$$\% \text{ Cr(VI) removal} = \frac{(C_0 - C_t)}{C_0} \cdot 100 \quad (1)$$

$$q_{Cr(VI)} \left[\frac{\text{mg Cr(VI)adsorbed}}{\text{g CNF}} \right] = \frac{(C_0 - C_t)}{m_{ads}} \cdot V_{water} \quad (2)$$

where C_0 and C_t represent the initial and equilibrium concentrations of pollutants in solution, respectively (mg·L^{-1}); V_{water} is the volume (L) of the solution; m_{ads} is the adsorbent mass (g CNF). The adsorbent mass (m_{ads}) can be calculated by the following Equation (3), which relates to the volume dose of the adsorbent and the consistency measured as indicated before.

$$m_{ads} \text{ [g CNF]} = V_{ads} \cdot \text{Consistency} \quad (3)$$

where V_{ads} is the volume of adsorbent and the consistency is the dried mass of adsorbent at 60 °C [42]. These compiled data were subsequently analyzed through different kinetic models.

The experimental kinetic and isotherm data were fitted through different kinetic and isotherm models to identify the adsorption mechanisms of the hexavalent chromium onto the surface of the CNF. The selected kinetic and isotherm equations and their linearized forms are summarized in Tables 1 and 2.

Table 1. The nonlinear and linearized equations of the analyzed kinetic models.

Model	Nonlinearized Equations		Linearization		Ref.
Pseudo-first order [1]	$q = q_e \cdot \left(1 - e^{-k_1 \cdot t}\right)$	(4)	$\ln(q_e - q) - \ln(q_e) = -k_1 \cdot t$	(8)	[43]
Pseudo-second order	$q = \dfrac{q_e^2 \cdot k_2 \cdot t}{1 + q_e \cdot k_2 \cdot t}$	(5)	$\dfrac{t}{q} = \left(\dfrac{1}{q_e^2 \cdot k_2}\right) + \left(\dfrac{1}{q_e}\right) \cdot t$	(9)	[43]
			$\dfrac{1}{q} = \left(\dfrac{1}{q_e^2 \cdot k_2}\right) \cdot \dfrac{1}{t} + \left(\dfrac{1}{q_e}\right)$	(10)	[44]
Elovich	$q = \dfrac{1}{\beta} \cdot \ln(t) + \dfrac{1}{\beta} \ln(\alpha \cdot \beta)$	(6)	-		[45]
Weber and Morris (Intraparticle)	$q = k_i \cdot t^{0.5} + C$	(7)	-		[46]

[1] The value of q_e must be previously obtained by estimation or experimentally.

Table 2. The nonlinear and linearized equations of the analyzed isotherm models.

Model	Nonlinearized Equations		Linearization		Ref.
Langmuir	$q_e = \dfrac{K_L \cdot q_{max} \cdot C_e}{1 + K_L \cdot C_e}$	(11)	Type I: $\dfrac{C_e}{q_e} = \dfrac{1}{K_L \cdot q_{max}} + \left(\dfrac{1}{q_{max}}\right) \cdot C_e$	(20)	
			Type II: $\dfrac{1}{q_e} = \dfrac{1}{q_{max}} + \left(\dfrac{1}{K_L \cdot q_{max}}\right) \cdot \left(\dfrac{1}{C_e}\right)$	(21)	
			Type III: $q_e = \left(-\dfrac{1}{K_L}\right) \cdot \left(\dfrac{q_e}{C_e}\right) + q_{max}$	(22)	
			Type IV: $\dfrac{q_e}{C_e} = (-K_L) \cdot q_e + K_L \cdot q_{max}$	(23)	
	$R_L = \dfrac{1}{1 + K_L \cdot C_0}$	(12)	Type V: $\dfrac{1}{C_e} = (K_L \cdot q_{max}) \cdot \dfrac{1}{q_e} - K_L$	(24)	
Freundlich	$q_e = K_F \cdot C_e^{1/n_F}$	(13)	$\ln(q_e) = \dfrac{1}{n_F} \cdot \ln(C_e) + \ln(K_F)$	(25)	[47]
Temkin	$q_e = B_T \cdot \ln(A_T \cdot C_e)$	(14)	$q_e = B_T \cdot \ln(A_T) + B_T \cdot \ln(C_e)$	(26)	
	$B_T = \dfrac{R \cdot T}{b_T}$	(15)			
Dubinin–Raduskevich	$q_e = q_{max} \cdot \exp(-B_{DR} \cdot \varepsilon^2)$	(16)	$\ln(q_e) = \ln(q_{max}) - B_{DR} \cdot \varepsilon^2$	(27)	
	$\varepsilon = R \cdot T \cdot \ln\left(1 + \dfrac{1}{C_e}\right)$	(17)			
	$E = \dfrac{1}{\sqrt{2 \cdot B_{DR}}}$	(18)			
Sips [1]	$q_e = \dfrac{K_S \cdot q_{max} \cdot C_e^{1/n_S}}{1 + K_S \cdot C_e^{1/n_S}}$	(19)	$\ln\left(\dfrac{q_e}{q_{max} - q_e}\right) = \dfrac{1}{n_s} \cdot \ln(C_e) + \ln(K_S)$	(28)	

[1] The value of q_{max} can be first estimated from the q_{max} obtained from the Langmuir model as the first input for optimization using a calculation software.

3. Results and Discussion

3.1. CNF Characterization

The cellulose-based physical-chemical characterization begins with the determination of the carboxyl group content on the oxidized cellulose before homogenization. Carboxylic acids created from primary hydroxyl groups on the cellulose leads to an increase in the repulsion between individual fibrils that form the cellulose fiber, which subsequently favors the homogenization process. Therefore, the number of carboxylic groups formed is indicative of the technical feasibility of the homogenization process, particularly to avoid clogging the device. The oxidized cellulose achieved 1.099 mmol COOH·g^{-1}. This content was higher than the values reported by Patiño-Masó et al. [48] (0.75 mmol COOH·g^{-1}), Lu et al. [49] (0.73 mmol COOH·g^{-1}) and Balea et al. [38] (0.50 and 0.25 mmol COOH·g^{-1}) using bleached kraft eucalyptus pulp, bleached bagasse, recycled newspaper, and corrugated container as raw materials under equivalent NaClO dosages, respectively. These differences are related to the quantity of impurities and lignin present in the source of cellulose, which has a strong influence on the amount of carboxylic groups formed for a specific amount of NaClO [38]. Therefore, the high number of carboxylic groups that formed was due to the reduced amount of impurities and lignin in the bleached pine kraft pulp used in this case, which agrees with the good performance of the oxidized cellulose during the homogenization process to obtain the CNF, and explains the low number of homogenization cycles (three).

Once the CNF hydrogel was obtained by homogenization, it was characterized by measuring the zeta potential of the CNF suspension. Zeta potential is indicative of the stability and separation of individual cellulose nanofibers, negatively charged; therefore, as its value becomes more negative, the suspension has more stability due to the repulsion between nanofibers that avoid aggregation between them. The value of the zeta potential for this suspension was −41.33 mV, which revealed the negative charge on the nanocellulose surface and the stability of the CNF suspension. In the literature, zeta potential values below −30 mV are indicators of bilateral repulsion and colloidal stability [50,51]. Other authors have obtained similar results for nanocellulose from acid hydrolysis (−38.2 mV) and TEMPO-oxidized nanocellulose (−46.5 mV) caused by the sulfonate and carboxyl groups, respectively [52].

Aside from the zeta potential, the cationic demand of the CNF hydrogel was measured, resulting in 838.5 μeq·g^{-1}. This value was higher than the one obtained by Balea et al. [38] for recycled fibers (200–600 μeq·g^{-1}) and lower than those obtained by Patiño-Masó et al. [48] (1000 μeq·g^{-1}) for bleached kraft pulp from eucalyptus under the same testing conditions. Cationic demand is indicative of the degree of defibrillation achieved by homogenization. As the cellulose surface is negatively charged, a higher amount of negatively charged groups in the suspension means that the specific surface of the cellulose material/nanomaterial was higher. Therefore, the cationic demand will be higher as the specific surface increases, which implies that a higher number of individual nanofibers was achieved during the homogenization process. The degree of nanofibrillation achieved was directly related to both the cellulose source and the cycles applied during the homogenization, which explains the higher value achieved regarding recycled fibers such as the one obtained with virgin fibers, as the amount of cellulose in virgin sources is higher than the cellulose in recycled ones.

To evaluate the amount of nanofibers in the CNF suspension, the degree of nanofibrillation in the CNF, the transmittance at 800 nm was determined [53]. The value of transmittance achieved was 95.6%, indicating that the synthetized CNF were highly nanofibrillated. As a comparison, Patiño-Masó et al. [48] obtained a 99.13% of nanofibrillation yield with a suspension of 88.0% of transmittance. This fact shows that most of the CNF can be considered disaggregated and separated in small individual nanofibers through this treatment.

3.2. Kinetics of Cr(VI) Adsorption with Hydrophobic CNF

3.2.1. Effect of MTMS Dosage

The effect of MTMS dosage, added to coat the surface of the CNF, on the hexavalent chromium adsorption was evaluated. Three MTMS dosages were tested (0, 1.5, and 3 mmol MTMS·g^{-1} cellulose). These adsorption batch experiments were evaluated at pH 3, 1000 mg·L^{-1} of adsorbent dosage, and 0.1 mg·L^{-1} of initial chromium concentration. Hexavalent chromium concentration in the water was measured along time, using the three different hydrophobic CNF used as adsorbents (Figure 1). The trends of chromium decreased for each dose of MTMS, indicating a variation in both the adsorption rate and the maximum adsorption capacity. Whereas the CNF hydrogel without the hydrophobic coating reached a maximum chromium removal of about 80%, after 75 h of contact time, both the MTMS doped CNF hydrogels showed a higher adsorption capacity as they adsorbed hexavalent chromium up to complete abatement, after 75 h of contact time. However, the lowest MTMS dosage applied to the CNF enhanced the adsorption rate (more than 90% of chromium was removed before 6 h of treatment) in comparison with the CNF without the MTMS coating and the CNF with the highest MTMS dosage. Other modified CNF (i.e., cationized [54], carboxylated [55,56], acid treated [57], and diethylenetriamine [58] modifications) achieved similar removal yields, over 90% of hexavalent chromium removal after 120 min of contact time using doses of the adsorbent between 0.3 and 3 g adsorbent·L^{-1} at acidic pH values from 1 to 5.5.

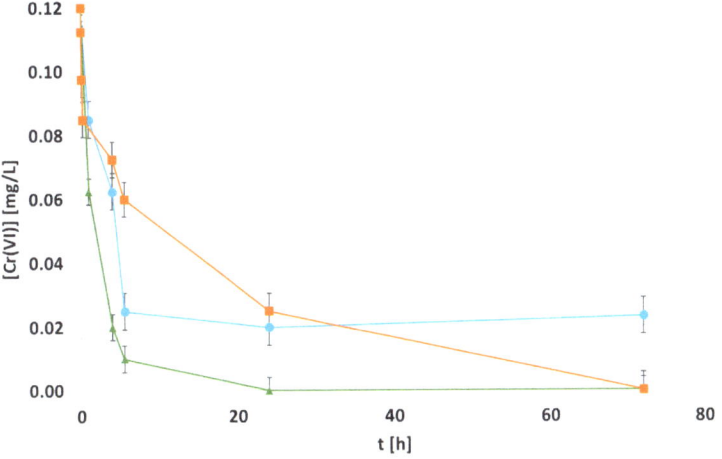

Figure 1. The evolution of the hexavalent chromium concentration [mg·L^{-1}] at 0.1 mg·L^{-1} of the initial chromium concentration, pH 3, and 1000 mg CNF·L^{-1} of dosage during adsorption with different doses of MTMS in the CNF hydrogels.

Following the calculation methodology described in Section 2.5, the adsorption capacity was determined and plotted in each experiment. Then, these data were fitted according to the different kinetic equations proposed in Table 1 (Figure 2).

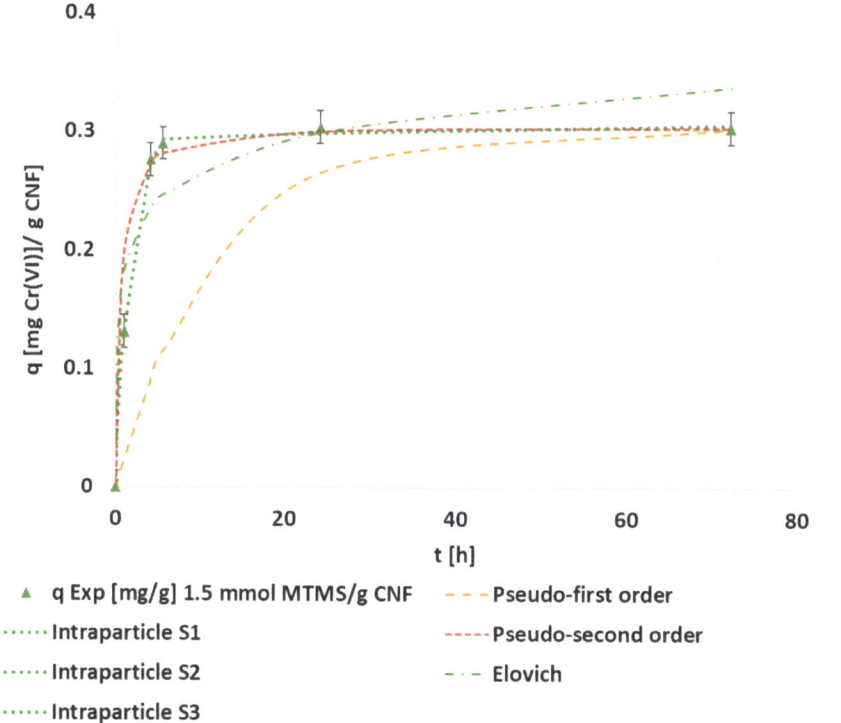

Figure 2. The evolution of the kinetic adsorption experiment at 0.1 mg·L^{-1} of the initial chromium concentration, pH 3, and 1000 mg CNF·L^{-1} hydrophobized with 1.5 mmol MTMS·g^{-1} cellulose and kinetic fitting of the pseudo-first, pseudo-second, Elovich, and intraparticle models.

The best fitting model in this case was the pseudo-second order kinetic equation, which corresponded to a saturation mechanism. The fitted kinetic parameters achieved through the evaluated models are supplied in the Appendix A (Table A1). The adjustment of the intraparticle diffusion model showed a multi-step adsorption process. The first linear adjustment of all experiments showed a high correlation coefficient and the intercept was close to the origin, meaning that internal diffusion is a rate-limiting step due to both the linearity and the number of steps, and the low boundary layer effect was considered according to the intercept [59]. Pseudo-second order kinetics and multi-step kinetic models were also indicated by Xu et al. [60], who used black wattle tannin-modified dialdehyde nanocellulose to adsorb hexavalent chromium. These authors associated this pseudo-second order fitting to the possibility of diffusion as a rate-limiting step.

Comparing the nonlinear adjustment of the pseudo-second order for each MTMS dosage applied to CNF, the maximum adsorption capacity corresponded to 0.30 mg Cr(VI)·g^{-1} CNF when applying the 1.5 mmol MTMS·g^{-1} CNF hydrogel (Figure 3), confirming the experimental results plotted in Figure 2. Non-modified CNF and hydrophobized CNF with 1.5 mmol MTMS·g^{-1} worked similarly, both reaching high adsorption capacities and fast saturation, while hydrophobized CNF with 3 mmol MTMS·g^{-1} showed slower adsorption, and saturation of the adsorbent was not found at 24 h of operation. The results of the adsorption capacity in the equilibrium and contact time to equilibrium were close to those found by other silanized cellulose applied by Jamroz et al. [33] to adsorb hexavalent chromium of 0.30 mg·g^{-1} at 300 min, respectively. However, these authors applied hydrophobic cellulose to measure ultra-traces of Cr(VI) in water. They also found that the pseudo-second order kinetic was the best fitting model.

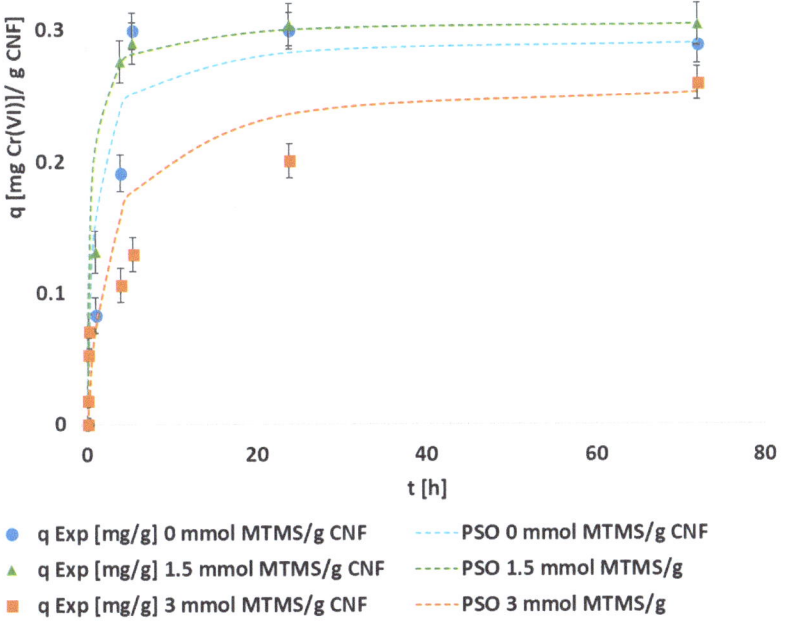

Figure 3. The evolution of the kinetic adsorption experiment at the previously indicated conditions under different MTMS dosages and kinetic fitting of the pseudo-second order model.

The slightly MTMS-coated CNF adsorbed chromium better than the naked CNF as the CNF surface groups were hidden by the coating layer, thus reducing the interaction of these groups with water. Furthermore, this technique prevented electrostatic repulsion between the negative CNF-surface charges and the hexavalent chromate ions. Nevertheless, excessive coating of MTMS over the surface of CNF supposed a mass transfer limitation, minimizing the adsorption rate. For these reasons, the selected MTMS dosage for the resting adsorption tests was 1.5 mmol MTMS·g^{-1}. Compared to other silanized materials applied for hexavalent chromium removal, this amount of silanization agent was considerably lower. Around 40 mmol·g^{-1} of silanization reagents ((3-aminopropyl)trimethoxysilane, [3-(2-aminoethylamino)propyl]trimethoxysilane, and 3-[2-(2-aminoethylamino)ethylamino]propyl-trimethoxysilane) were added to coat the graphene oxide to adsorb hexavalent chromium from water [32].

3.2.2. Effect of pH

The experimental evolution of the hexavalent chromium concentration along the batch adsorption tests by modifying the pH from pH 3 to pH 9 is shown in Figure 4. These adsorption tests were carried out by adding the best tested operating condition of MTMS dosage, 0.1 mg·L^{-1} of initial chromium concentration, and 1000 mg·L^{-1} of CNF dosage. The trend of the Cr(VI) removal rate was similar under neutral (pH 7) and alkali (pH 9) conditions, while the adsorption was faster under acidic (pH 3) conditions, with a total removal of 80% after 1 h of operation (Figure 4). The high adsorption capacity found at pH 3 was associated with the equilibrium changes of hexavalent chromium under acidic conditions. While divalent chromate is the predominant specie when pH > 6, the monovalent specie is mainly present between pH 2 and 4. Therefore, the amount of adsorbate, hexavalent chromium, is doubled at acid pH because only one active site is required per anion. A similar result of pH optimization can be found with independence of the kind of adsorbent, as indicated by Owlad et al. [9] and Saha and Orvig [61]. It can be concluded that the pH effect is related to the adsorbate ionic forms, being the effect on the adsorbent negligible. After these experiments, the pH 3 condition was selected for the rest of the optimization process.

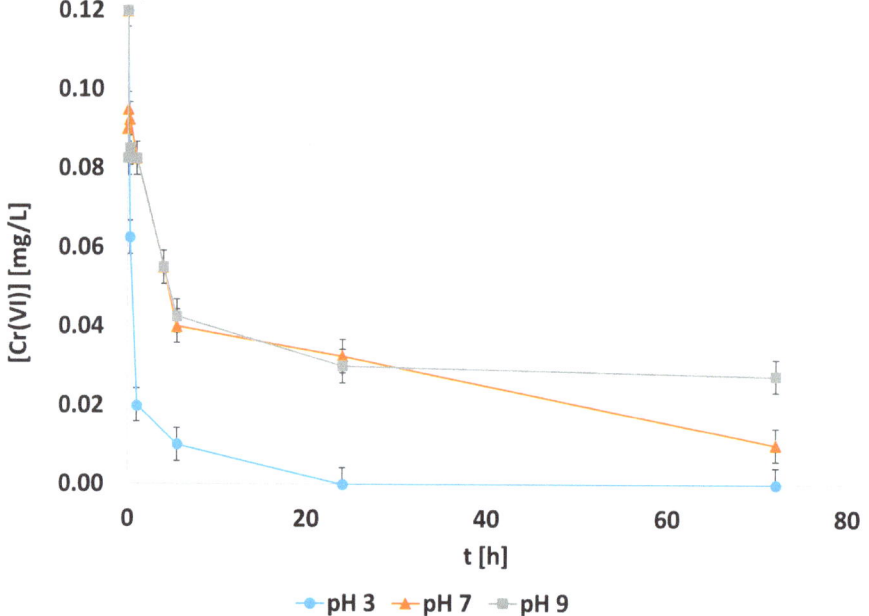

Figure 4. The evolution of the hexavalent chromium concentration [mg·L^{-1}] during adsorption with the CNF hydrogel at 0.1 mg·L^{-1} of the initial chromium concentration, 1.5 mmol MTMS·g^{-1} CNF applied during hydrophobization, and 1000 mg CNF·L^{-1} of dosage under different pH conditions.

The obtained adsorption capacity data were evaluated through different kinetic models. The result of plotting each fitted kinetic to the pH 3 experimental data can be observed in Figure 5.

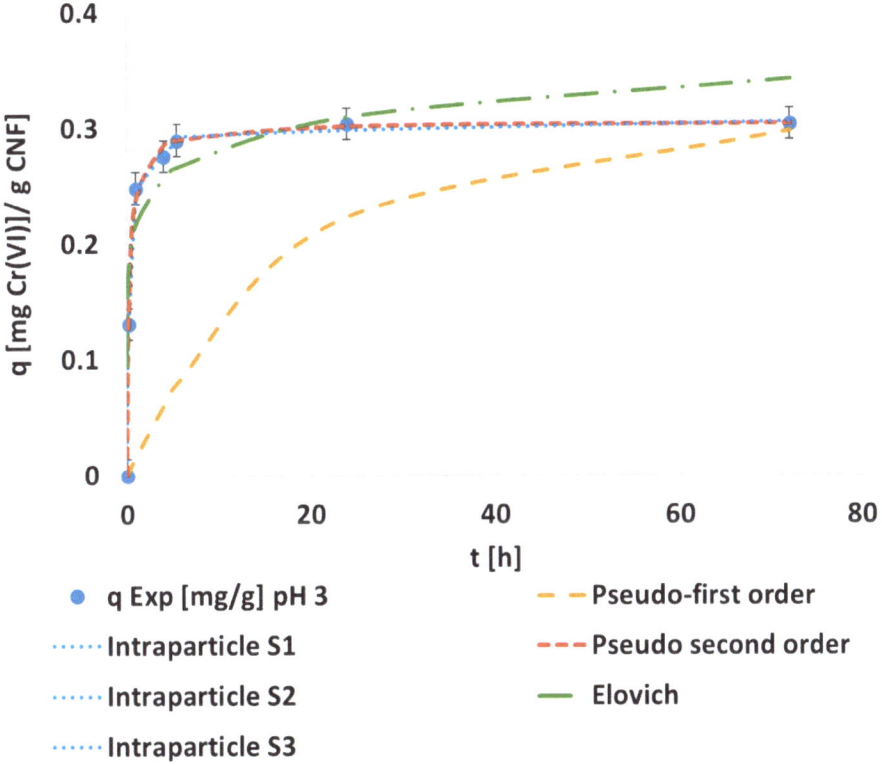

Figure 5. The evolution of the kinetic adsorption experiment at 0.1 mg·L^{-1} of the initial chromium concentration, 1.5 mmol MTMS·g^{-1} CNF applied during hydrophobization and 1000 mg CNF·L^{-1} of dosage under pH 3 conditions and the kinetic fitting of the pseudo-first, pseudo-second, Elovich, and intraparticle models.

The adjustments demonstrate that the most representative model was the pseudo-second kinetic model (Figure 5). The fitted kinetic parameters obtained in the different pH experiments are shown in Table A2. The intraparticle model analysis showed a multi-step adsorption mechanism with independence in the pH, as seen in the MTMS optimization. The representation of the pseudo-second order kinetic adjustment to the experimental data from each experiment is seen in Figure 6. The maximum adsorption capacity was achieved while operating under pH 3 conditions, when 0.30 mg·g^{-1} of the adsorption capacity and 80% of the maximum removal was reached in 1 h. The operation under neutral and alkaline media was similar in adsorption rate, but with a slightly lower adsorption capacity when operating at alkaline conditions.

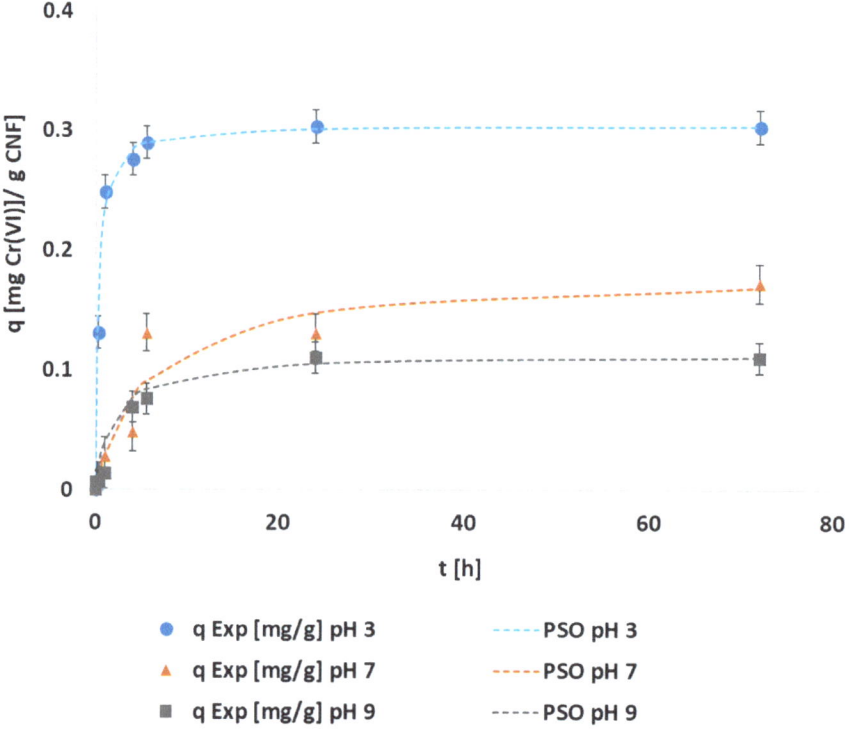

Figure 6. The evolution of the kinetic adsorption experiment at 0.1 mg·L^{-1} of the initial chromium concentration, 1.5 mmol MTMS·g^{-1} CNF applied during hydrophobization, and 1000 mg CNF·L^{-1} of dosage under different pH and kinetic fitting of the pseudo-second order model.

3.2.3. Effect of Adsorbent Dosage

Following the same experimental procedure of the previous optimization processes, the adsorbent dosage effect was also evaluated from 250 up to 1000 mg CNF·L^{-1} (Figure 7). The kinetic curves showed a clear tendency of increasing the adsorption rate as the dosage of CNF increased. Among the studied CNF dosages, the minimum one needed to obtain a 100% of adsorption of hexavalent chromium was 500 mg·L^{-1} whereas the lowest adsorbent dosage studied showed a fast saturation of the CNF and a total removal below 10% (Figure 7). The highest dosage studied, 1000 mg·L^{-1}, achieved the complete removal of hexavalent chromium after 25 h of contact time, the same as the dosage of 500 mg·L^{-1}. However, the higher the dosage, the faster the adsorption, as expected. In terms of the adsorption capacity, the optimum value was found while applying 500 mg·L^{-1}, reaching the largest efficiency in chromium removal per gram of CNF.

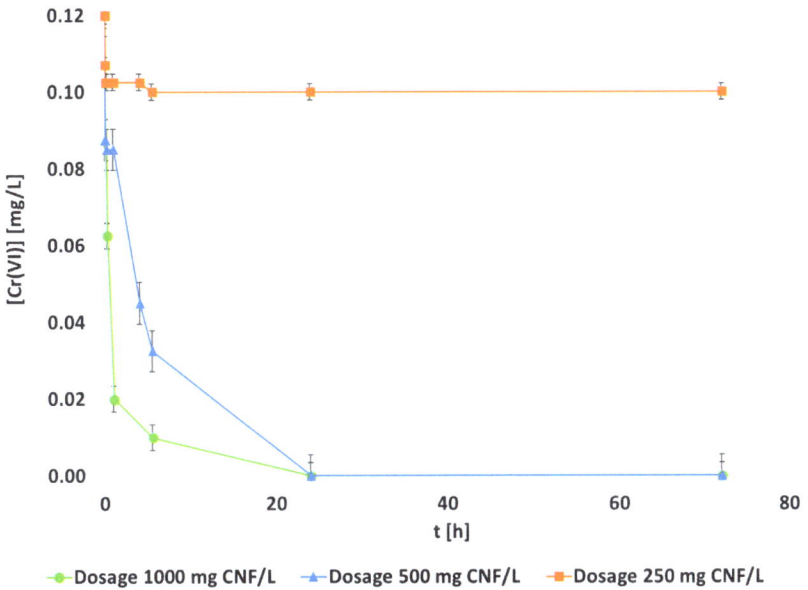

Figure 7. The evolution of the hexavalent chromium concentration [mg·L^{-1}] during adsorption with the CNF hydrogel at 0.1 mg·L^{-1} of the chromium initial concentration, 1.5 mmol MTMS·g^{-1} CNF, and pH 3 conditions under different adsorbent dosages.

The quantity of hydrophobic CNF needed to reach the complete removal of hexavalent chromium is in most cases lower, compared to other cellulosic materials. Xu et al. [60] reached the maximum adsorption capacity of hexavalent chromium using 500 mg·L^{-1} with black wattle tannin-modified dialdehyde nanocellulose. Qiu et al. [62] and Huang et al. [58] achieved the complete depletion of hexavalent chromium employing 3000 mg·L^{-1} of polyethylenimine facilitated ethyl cellulose and diethylenetriamine-modified hydroxypropyl methylcellulose, respectively. In the study developed by Singh et al. [63], a dosage of 5000 mg·L^{-1} of aminated cellulose nanocrystals adsorbed 98.33% of hexavalent chromium. These results depict that hydrophobic CNF reached equivalent yields compared to other cellulosic materials, reducing the dosage by an order of magnitude.

Compared to other kinds of adsorbents, the dosage of hydrophobic CNF can still be considered reduced. In the case of waste material adsorbents such as treated sawdust [64], pristine almond green hull [65], and *Hibiscus cannabinus* kenaf [66] were applied to treat hexavalent chromium under optimal conditions, 1600, 2000 and 3000 mg·L^{-1} were necessary, respectively. Other typical materials used to adsorb hexavalent chromium are both commercial and mango kernel-synthetized activated carbons [67,68], and chitosan microparticles and nanoparticles [69], but they require larger adsorbent dosages than the hydrophobic CNF. The optimal dosage needed for the activated carbons was 2000 mg·L^{-1}, while 800 mg·L^{-1} was the minimum required dose of the chitosan microparticles and nanoparticles to reach hexavalent chromium removal.

The results of adjusting the selected kinetic models to a dosage of 500 mg·L^{-1} are shown in Figure 8. The optimal kinetic model corresponded to a pseudo-second order equation. All of the kinetic parameters obtained through the kinetic fittings from the dosage optimization experiments are shown in detail in Table A3. The results of the intraparticle diffusion model adjustments were similar for each dosage, corresponding to multi-step adsorption.

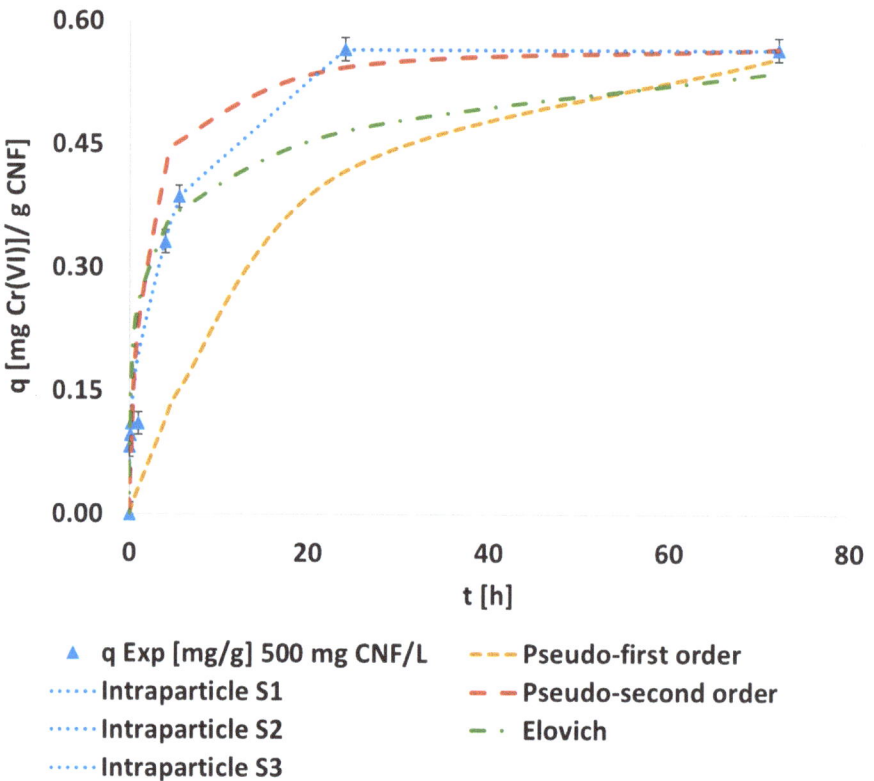

Figure 8. The evolution of the kinetic adsorption experiment at 0.1 mg·L^{-1} of the chromium initial concentration, 1.5 mmol MTMS·g^{-1} CNF, and pH 3 conditions under 500 mg CNF·L^{-1} of dosage and the kinetic fitting of the pseudo-first, pseudo-second, Elovich, and intraparticle models.

Figure 9 represents the final pseudo-second order kinetic curves fitted to the experimental data from adsorbent dosage evaluation. The evolution observed revealed a maximum adsorption capacity of 0.58 mg·g^{-1} when 500 mg CNF·L^{-1} was dosed compared to 0.30 mg·g^{-1} in the case of 1000 mg CNF·L^{-1}. However, the pseudo-second order kinetic constant was fourteen-times higher in the case of the highest dosage. The contrast between reaching fast-equilibrium with high adsorbent dosages and increasing the adsorption efficiency can be observed through the use of other nanomaterial adsorbents such as Fe_2O_3 nanoparticles [70]. This fact would suggest that the optimal dosage would depend on the objective between reaching the rapid total removal or maximizing the total capacity of the adsorbent.

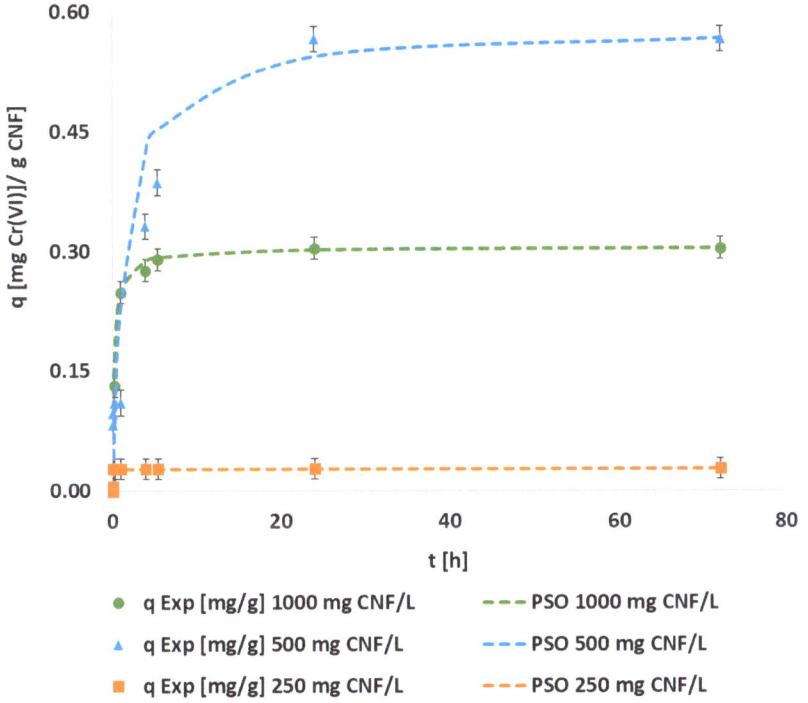

Figure 9. The evolution of the kinetic adsorption experiment at 0.1 mg·L^{-1} of the chromium initial concentration, 1.5 mmol MTMS·g^{-1} CNF, and pH 3 conditions under different adsorbent dosages and kinetic fitting of the pseudo-second order model.

3.2.4. Effect of Initial Chromium Concentration

Hexavalent chromium concentration was modified to study the adsorption capacity of CNF when the initial chromium concentration was between 0.1 and 5 mg·L^{-1} and when the initial chromium concentration drastically increased to the range from 10 to 50 mg·L^{-1} (Figure 10a,b, respectively). These kinetic experiments were developed by implementing the optimal tested MTMS dosage, the optimal tested pH, and the maximum adsorbent dosage (1000 mg CNF·L^{-1}), thus prioritizing the process kinetic over the adsorption capacity. Furthermore, this CNF hydrogel dosage was selected to prevent the CNF from extremely fast saturation.

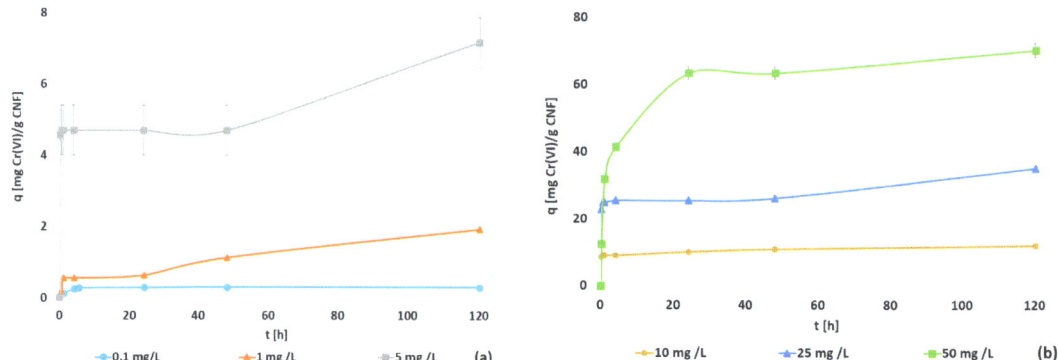

Figure 10. (a) The evolution of the hexavalent chromium concentration [mg·L^{-1}] during the adsorption with CNF hydrogel at 1000 mg·L^{-1} of dosage, 1.5 mmol MTMS·g^{-1} CNF, and pH 3 conditions under 0.1 to 5 mg·L^{-1}; (b) 10 to 50 mg·L^{-1} initial hexavalent chromium concentrations.

All of the studied initial chromium concentrations showed similar trends, with an initial fast adsorption of chromium, a second stationary step, and a final slow adsorption step (Figure 10). This situation corresponded to a multilayer adsorption. At the beginning, the free active sites were easy to find, so a first layer was generated rapidly. Once the first layer was formed, there was an accumulation of anions in the CNF surface, which led to the minimization of both the driving force between the liquid and solid phases and the electrostatic force interaction. Therefore, the stationary and intermediate step is caused by the slow reduction of anionic hexavalent chromium to less toxic cationic trivalent chromium. This reduction of chromium through the oxidation of C–OH surface groups to carboxylic groups was studied deeply by Wang and Lee [19]. Afterward, the attachment of the second layer of hexavalent chromium onto the CNF surface begins once the surface gradually becomes positively charged and the attraction force of anionic hexavalent chromate from the wastewater to the cationic surface of the CNF overpasses the mass transfer limitations. From the results shown in Figure 10, the resulting kinetic curves of hydrophobic CNF adsorption when modifying the initial chromium concentration showed an increase in the adsorption capacity (from 0.30 to 70.38) as the initial concentration of chromium increased, coupled with a reduction in the adsorption efficiency (from 100% to 51%) as the initial chromium concentration increased. This fact demonstrates that both parameters, adsorption capacity and adsorption efficiency, showed opposite trends, in agreement with Pourfadakari et al. [57], as for a selected dosage, increasing the initial concentration involves both a lesser percentage of chromium removal and a higher driving force between the liquid and the solid surface. This is due to the difference in the concentrations, which enhances the efficient usage per mass unit of material, reaching a higher amount of adsorbed chromium per active site. These authors [57] also showed the change in the shape of the curves while increasing the initial concentration, where the presence of different steps became clearer as the concentration rose.

The achieved experimental data were fitted through the kinetic models applied previously, with the initial hexavalent chromium concentration of 25 mg Cr(VI)·L^{-1} (Figure 11). At this initial concentration, as it happened at 1 mg·L^{-1}, the only model that allowed an adequate interpretation and simulation of the experimental results was the intraparticle diffusion model. Each kinetic parameter obtained by kinetic analysis for all the kinetic models can be checked in Table A4. The intraparticle diffusion model predicts a three-step adsorption mechanism including the first high adsorption rate step, a steady-state step, and the last slow adsorption step, confirming the trend seen in the experimental data. Evaluating the intraparticle adjustment of the first step in the experiments using an initial concentration of 5 mg·L^{-1} and above, the intercept is extremely high, in some cases, close to the maximum adsorption capacity. This fact indicates a strong effect of the boundary layer, which means that external diffusion limitation will play a major role in the overall adsorption rate [59].

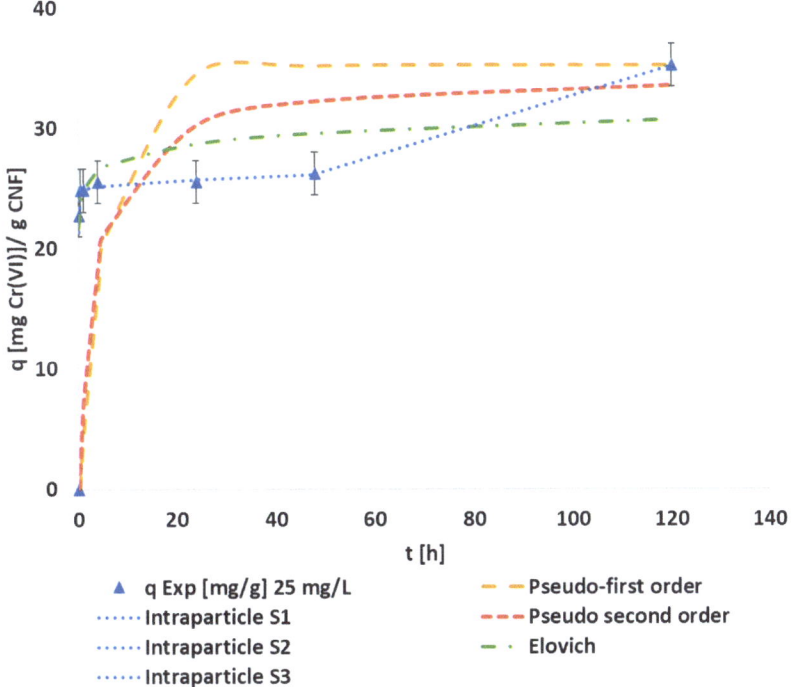

Figure 11. The evolution of the kinetic adsorption experiment of the CNF hydrogel at 1000 mg·L^{-1} of dosage, 1.5 mmol MTMS·g^{-1} CNF, and pH 3 conditions under 25 mg·L^{-1} of the initial hexavalent chromium concentration and kinetic fitting of the pseudo-first, pseudo-second, Elovich, and intraparticle models.

The resulting curves of each initial chromium concentration experiment adjusted to the pseudo-second order kinetics (Figure 12a,b) showed the change in the mechanism of adsorption. At the lower initial concentration of hexavalent chromium, the adsorption mechanism was mainly saturation, whereas as the initial concentration increased, the multistep adsorption mechanism was more noticed. The final values of both equilibrium concentrations of chromium and the adsorption capacities of CNF achieved in these tests were used to plot the isotherm curves. The low correlation coefficients for the kinetic models were due to the assumption that all of the selected kinetic equations made, as they suppose the gradual increase in the adsorption capacity or continuous adsorption up to saturation of the adsorbent, but none of these models assumed a two-layer adsorption. Nevertheless, as it was previously pointed out, the kinetic data showed a third step in the adsorption process, corresponding to the adsorption at long contact times, longer than 20 h of batch adsorption. These long contact times are not feasible at an industrial scale; therefore, the selection of experimental data with shorter contact times (first and second steps) would enhance the adjustment of these models.

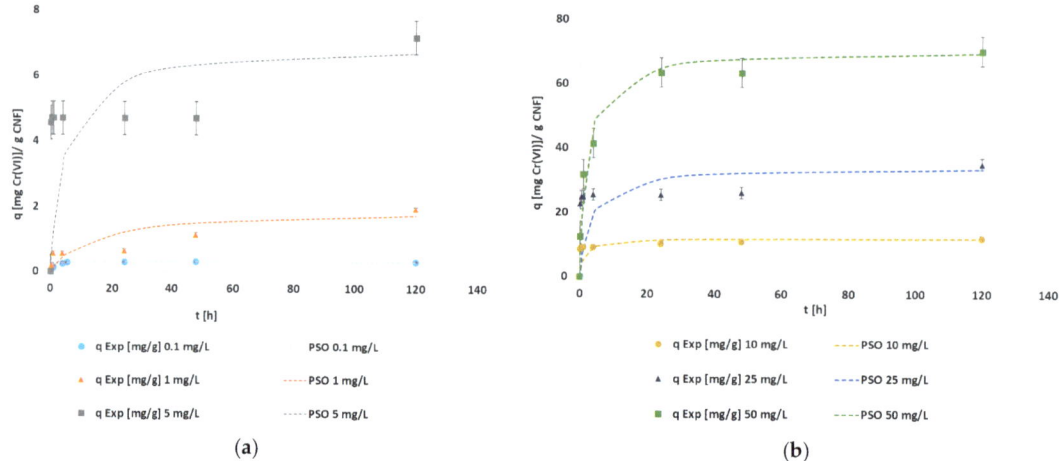

Figure 12. (a) The evolution of the kinetic adsorption experiment of the CNF hydrogel at 1000 mg·L^{-1} of dosage, 1.5 mmol MTMS·g^{-1} CNF, and pH 3 conditions under 0.1 to 5 mg·L^{-1}; (b) 10 to 50 mg·L^{-1} of the initial hexavalent chromium concentration and kinetic fitting of the pseudo-second order model.

There are a large variety of adsorbents to remove hexavalent chromium and many results related to how the contact between water and adsorbent is made, and to the different surface modifications, which affects the mechanism of adsorption. The hydrophobic CNF hydrogel showed the slowest equilibrium time regarding other adsorbent materials based on nanocellulose or on activated carbon (Table 3), but it reached the complete abatement of chromium and good adsorption capacity. Similar results of the maximum adsorption capacity in the equilibrium using polyacrylonitrile-modified CNF membranes were reported by Yang et al. [71] (87.5 mg·g^{-1}). The equilibrium time of the hydrophobic CNF was similar to the one of another silanized cellulosic material applied by Jamroz et al. [33] (300 min) to detect ultra-trace concentrations of hexavalent chromium in water, suggesting that silanization processes lead to larger contact times than other kinds of adsorbents (Table 4) and could be associated with relevant mass transfer limitations due to the silane reaction with celluloses. The applied dosage was in the order of magnitude of other cellulosic adsorbents such as polypyrrole-bacterial CNF and polyaniline-functionalized CNF and was lower than that of activated carbons [72–76]. This comparison suggests that the hydrophobized CNF hydrogel is an efficient material for hexavalent chromium adsorption from wastewater compared to other adsorbents including other nanocellulosic materials.

Table 3. A comparison of the hexavalent chromium adsorption through different NC and activated carbon adsorbents.

Adsorbent	Contact Time [min]	Adsorbent Dosage [mg·L^{-1}]	Initial Cr(VI) Concentration [mg·L^{-1}]	pH	q_{max} [mg·g^{-1}]	Maximum Removal Yield [%]	Ref.
CNF from rice husk	100	1500	30	6	3.76	92.99	[57]
Polypyrrole-bacterial CNF	180	250	300	2	555.6	97.5	[72]
Thiol-modified CNF composite	20		50	4	87.5	96	[71]
Citric acid-incorporated CNF	120	40	50	2	11	23	[77]
Amino-silanized cellulose membranes	300	5000	50	4	34.7		[33]
Polyaniline-functionalized CNC	40	500	30	2.5	48.92	97.84	[73]
Microwave-assisted H_3PO_4/Fe-modified activated carbon	200	1000	30	3	34.39	100	[74]
$ZnCl_2$-modified tamarind wood activated carbon	70	3000	10	3	28.02	99	[75]
Acid-base surface modified activated carbon	180	2000	50		13.89		[76]
Hydrophobized CNF Hydrogel (MTMS dosage = 1.5 mmol·g^{-1})	330	500	50	3	70.38	>97.14	This work

3.3. Isotherm Analysis

The isotherm curves were graphed with the equilibrium data of the chromium concentration. The equilibrium data of the maximum adsorption capacities and final concentration at the equilibrium stage of hexavalent chromium treated with the hydrophobic CNF hydrogel were analyzed and adjusted to different isotherm models. The isotherm data showed an exponential increase trend instead of a saturation trend (Figure 13). According to the classification established by McCabe, Smith, and Harriott [78], this concave upward curve involves a large effect of mass transfer limitations and indicates unfavorable adsorption. This mass transfer limitation can be associated with the coating of MTMS on the surface of the material, which implies more tortuosity to reach the active sites. The parameters achieved from the adjustment of the different isotherm models are summarized in Table 3. The optimal correlation parameters are found when the Freundlich isotherm is applied to experimental data, which means the distribution of the energy through an exponential equation, and considering the heterogeneous dispersion of the active sites, happens over the adsorbent surface [79]. This mechanism supposes a multilayer adsorption [80], which fits the trend seen during the adsorption kinetics in Section 3.2.

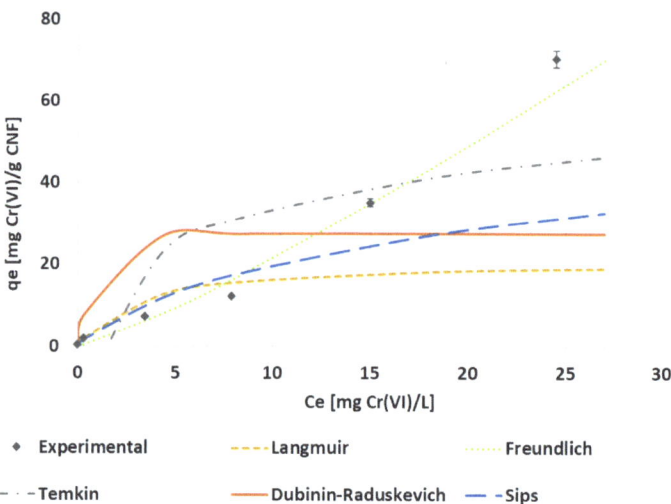

Figure 13. The isotherm experimental data and isotherm model adjustment of the Langmuir, Freundlich, Temkin, Dubinin–Raduskevich, and Sips equations.

Table 4. The results of the isotherm model adjustment to the adsorption equilibrium data of hexavalent chromium on a CNF hydrogel.

Model	Parameters	Values
Langmuir	Isotherm parameters	k_L [L·mg^{-1}] = 21.26 q_e [mg·g^{-1}] = 0.3417 R_L (C_0 = 0.1 mg·L^{-1}) [-] = 0.9670 R_L (C_0 = 50 mg·L^{-1}) [-] = 5.53·10^{-2}
	Correlation parameters	R^2 = 0.7420 RSS = 2949.55
Freundlich	Isotherm parameters	k_F [mg$^{(1-1/n)}$·L$^{(1/n)}$·g^{-1}] = 1.3914 n_F [-] = 0.8404
	Correlation parameters	R^2 = 0.9902 RSS = 108.01
Dubinin–Raduskevich	Isotherm parameters	B_{DR} [mol^2·J^{-2}] = 9.93·10^{-8} q_{max} [mg·g^{-1}] = 27.72
	Thermodynamic parameters	E_{DR} [J·mol^{-1}] = 2243.50
	Correlation parameters	R^2 = 0.5754 RSS = 2542.39
Temkin	Isotherm parameters	B_T [J·mol^{-1}] = 12.83 b_T [-J·mol^{-1}] = 188.08 A_T [L·g^{-1}] = 1.3759
	Correlation parameters	R^2 = 0.7481 RSS = 1415.93
Sips	Isotherm parameters	n_S [-] = 1.2442 k_S [L$^{(1/nS)}$·mol$^{-(1/nS)}$] = 6.16·10^{-2}
	Correlation parameters	R^2 = 0.9023 RSS = 1529.83

The value of the parameter n_F <1 indicates an unfavorable process as well as the small bond adsorbate–adsorbent compared to a favorable process [81]. A similar value of n_F was reported by Dawodu et al. [82] for hexavalent chromium adsorption onto the seed coat biomass, showing a cooperative adsorption between Cr(VI) and adsorbent surface.

As expected by the shape of the isotherm, the selected models, which supposed the saturation of the adsorbent such as Langmuir, Sips, or Dubinin–Raduskevich, showed low correlation coefficients and a high residual sum of squares. These could represent the initial part of the curve, but failed the clear concave shape adjustment in the end. For this reason, these models did not offer a good estimation of the maximum adsorption capacity reached by the hydrophobic CNF.

The Langmuir's separation factor values calculated for the minimum and maximum initial concentrations were in the interval of $0 < R_L < 1$, but close to the upper and lower limits of the interval, respectively. The variability of the separation factor indicates that while treating lower concentrations, the high value of R_L implies a reduced affinity adsorbate–adsorbent. On the other hand, the treatment of concentrated solutions showed a reduction in the R_L values close to 0, strengthening the chromium attachment onto the CNF surface [83].

The mean free energy of adsorption calculated through the Dubinin–Raduskevich model was 2.24 kJ·mol^{-1}, similar to the typical values indicated for the physisorption of chromium (<8 kJ·mol^{-1}) [84]. The value of the Temkin b_T parameter, which is related to the heat of sorption, was 0.19 kJ·mol^{-1}. Choudhary and Paul [85] indicated that values of b_T below 8 kJ·mol^{-1} revealed a weak interaction chromium-CNF surface. These low b_T values are related to physisorption processes, where the values of adsorption enthalpy are in the order of physical processes such as intermolecular forces.

4. Conclusions

The modified CNF hydrogels successfully removed more than 97% of Cr(VI), opening promising applications for these nanomaterials as heavy metal adsorbents. The nanofibers showed a large amount of active carboxylic and anionic groups. Thus, it was necessary to develop a surface modification through hydrophobization treatment. The inclusion of 1.5 mmol MTMS·g^{-1} as a hydrophobizing agent allowed for the increase in the Cr(VI) kinetic constant of adsorption k_2 by 84.97%. The pseudo-second order and intraparticle diffusion kinetic models were the best fitting models. These models revealed that both the sorption rate and hexavalent chromium diffusion played a major role as rate-limiting steps. The adsorption mechanism is ruled by the multi-step adsorption of hexavalent chromium on the CNF hydrogel dominated by internal diffusion at low concentrations and external diffusion with concentrations above 5 mg·L^{-1}. The optimized conditions were found to be pH 3 and a dosage over 500 mg·L^{-1}. More than 97% of hexavalent chromium removal was reached, treating concentrations below 1 mg·L^{-1} and the maximum adsorption capacity of 70.38 mg·g^{-1} was achieved at 50 mg·L^{-1}. The isotherm analysis showed that Freundlich was the best fitting model, meaning that multilayer adsorption and the heterogeneous dispersion of surface energy is the main adsorption mechanism of hexavalent chromium onto the surface of hydrophobized CNF. The Freundlich unfavorable isotherm predicts a multilayer adsorption and a weak interaction between hexavalent chromium and CNF, associated with a physical sorption mechanism. The relatively low values of mean free energy of adsorption (2.24 kJ·mol^{-1}) and heat of sorption (0.19 kJ·mol^{-1}) calculated through the Dubinin–Raduskevich and Temkin models are also indicators of a physical sorption process. In general terms, the hydrophobized CNF hydrogel reached relatively high adsorption capacities compared to previously developed cellulose nanomaterials and activated carbons, and the complete removal of chromium could be found, even at low adsorbent dosages.

Author Contributions: Conceptualization, C.N., N.M., and F.d.B.O.; Methodology, F.d.B.O., N.M., and C.N.; Software, F.d.B.O.; Validation, F.d.B.O., A.B. (Ana Balea) and J.L.S.-S.; Formal analysis, F.d.B.O., A.B. (Angeles Blanco), A.B. (Ana Balea), C.N., and N.M.; Investigation, F.d.B.O. and S.M.; Resources, C.N. and A.B. (Angeles Blanco); Data curation, F.d.B.O., A.B. (Ana Balea), and J.L.S.-S.; Writing—original draft preparation, F.d.B.O., A.B. (Ana Balea), and N.M.; Writing—review and editing, F.d.B.O., A.B. (Ana Balea), and C.N.; Supervision, C.N., N.M., and A.B. (Angeles Blanco);

Project administration, C.N. and A.B. (Angeles Blanco); Funding acquisition, C.N. and A.B. (Angeles Blanco) All authors have read and agreed to the published version of the manuscript.

Funding: This research was funded by the Community of Madrid, project S2018/EMT-4459 (RETO-PROSOST-CM) and the Economy and Competitiveness Ministry of Spain, project CTQ2017-85654-C2-2-R and predoctoral fellowship PRE2018-085034.

Institutional Review Board Statement: Not applicable.

Informed Consent Statement: Not applicable.

Data Availability Statement: Not applicable.

Acknowledgments: The authors wish to thank the Community of Madrid and the Economy and Competitiveness Ministry of Spain for their support to projects S2018/EMT-4459 (RETO-PROSOST-CM) and CTQ2017-85654-C2-2-R as well as the support of the Economy and Competitiveness Ministry of Spain for the grant provided to F.B. Ojembarrena (PRE2018-085034).

Conflicts of Interest: The authors declare no conflict of interest.

Appendix A

Table A1. The results of the kinetic equation adjustment to the adsorption experimental data of 0–3 mmol MTMS·g^{-1} CNF.

Kinetic Model		0 mmol MTMS·g^{-1} CNF	1.5 mmol MTMS·g^{-1} CNF	3 mmol MTMS·g^{-1} CNF
Pseudo-first order	Kinetic parameters	k_1 [h^{-1}] = 0.2378	k_1 [h^{-1}] = 8.63·10^{-2}	k_1 [h^{-1}] = 9.16·10^{-2}
	Correlation parameters	R^2 = 0.9653 RSS = 7.11·10^{-3}	R^2 = 0.7725 RSS = 7.84·10^{-2}	R^2 = 0.9674 RSS = 9.38·10^{-3}
Pseudo-second order	Kinetic parameters	k_2 [mg·g^{-1}·h^{-1}] = 3.7399 q_e [mg·g^{-1}] = 0.2933	k_2 [mg·g^{-1}·h^{-1}] = 6.9175 q_e [mg·g^{-1}] = 0.3058	k_2 [mg·g^{-1}·h^{-1}] = 1.4827 q_e [mg·g^{-1}] = 0.2610
	Correlation parameters	R^2 = 0.9440 RSS = 1.23·10^{-2}	R^2 = 0.9661 RSS = 5.96·10^{-3}	R^2 = 0.9430 RSS = 1.08·10^{-2}
Elovich	Kinetic parameters	α [h·mg·g^{-1}] = 0.8350 β [g·mg^{-1}] = 21.5517	α [h·mg·g^{-1}] = 6.6752 β [g·mg^{-1}] = 28.1690	α [h·mg·g^{-1}] = 1.6142 β [g·mg^{-1}] = 37.7358
	Correlation parameters	R^2 = 0.8109 RSS = 1.23·10^{-2}	R^2 = 0.8005 RSS = 7.74·10^{-3}	R^2 = 0.9540 RSS = 3.90·10^{-3}
Intraparticle diffusion	Kinetic parameters: Step 1	$k_{i,1}$ [mg·g^{-1}·min$^{-0.5}$] = 0.1193 $C_{i,1}$ [mg·g^{-1}] = $-1.62\cdot10^{-2}$	$k_{i,1}$ [mg·g^{-1}·min$^{-0.5}$] = 0.1462 $C_{i,1}$ [mg·g^{-1}] = $-2.30\cdot10^{-3}$	$k_{i,1}$ [mg·g^{-1}·min$^{-0.5}$] = 0.1462 $C_{i,1}$ [mg·g^{-1}] = 7.00·10^{-4}
	Correlation parameters	R^2 = 0.9709 RSS = 1.23·10^{-4}	R^2 = 0.9996 RSS = 3.17·10^{-5}	R^2 = 0.9918 RSS = 5.06·10^{-5}
	Kinetic parameters: Step 2	$k_{i,2}$ [mg·g^{-1}·min$^{-0.5}$] = 1·10^{-17} $C_{i,1}$ [mg·g^{-1}] = 0.299	$k_{i,2}$ [mg·g^{-1}·min$^{-0.5}$] = 4.00·10^{-2} $C_{i,1}$ [mg·g^{-1}] = 0.196	$k_{i,2}$ [mg·g^{-1}·min$^{-0.5}$] = 0.0239 $C_{i,1}$ [mg·g^{-1}] = 0.0658
	Correlation parameters	R^2 = 0.5477 RSS = 1.23·10^{-4}	R^2 = 0.9996 RSS = 4.25·10^{-8}	R^2 = 0.9839 RSS = 3.67·10^{-4}
	Kinetic parameters: Step 3		$k_{i,3}$ [mg·g^{-1}·min$^{-0.5}$] = 2.10·10^{-3} $C_{i,3}$ [mg·g^{-1}] = 0.2878	
	Correlation parameters		R^2 = 0.8137 RSS = 4.29·10^{-5}	

Appendix B

Table A2. The results of the kinetic equation adjustment to the adsorption experimental data of pH 3–9.

Kinetic Model			pH 3	pH 7	pH 9
Pseudo-first order		Kinetic parameters	$k_1\ [h^{-1}] = 5.55 \cdot 10^{-2}$	$k_1\ [h^{-1}] = 6.09 \cdot 10^{-2}$	$k_1\ [h^{-1}] = 0.2179$
		Correlation parameters	$R^2 = 0.6366$ $RSS = 0.167$	$R^2 = 0.9219$ $RSS = 7.25 \cdot 10^{-3}$	$R^2 = 0.9960$ $RSS = 1.49 \cdot 10^{-4}$
Pseudo-second order		Kinetic parameters	$k_2\ [mg \cdot g^{-1} \cdot h^{-1}] = 11.4405$ $q_e\ [mg \cdot g^{-1}] = 0.3050$	$k_2\ [mg \cdot g^{-1} \cdot h^{-1}] = 0.9889$ $q_e\ [mg \cdot g^{-1}] = 0.1821$	$k_2\ [mg \cdot g^{-1} \cdot h^{-1}] = 4.4596$ $q_e\ [mg \cdot g^{-1}] = 0.1139$
		Correlation parameters	$R^2 = 0.9980$ $RSS = 3.29 \cdot 10^{-4}$	$R^2 = 0.9612$ $RSS = 2.77 \cdot 10^{-3}$	$R^2 = 0.9801$ $RSS = 8.14 \cdot 10^{-4}$
Elovich		Kinetic parameters	$\alpha\ [h \cdot mg \cdot g^{-1}] = 43.5512$ $\beta\ [g \cdot mg^{-1}] = 33.7838$	$\alpha\ [h \cdot mg \cdot g^{-1}] = 0.3496$ $\beta\ [g \cdot mg^{-1}] = 46.9484$	$\alpha\ [h \cdot mg \cdot g^{-1}] = 0.3013$ $\beta\ [g \cdot mg^{-1}] = 66.2252$
		Correlation parameters	$R^2 = 0.8683$ $RSS = 5.44 \cdot 10^{-3}$	$R^2 = 0.8989$ $RSS = 6.16 \cdot 10^{-3}$	$R^2 = 0.9255$ $RSS = 2.18 \cdot 10^{-3}$
Intraparticle diffusion		Kinetic parameters: Step 1	$k_{i,1}\ [mg \cdot g^{-1} \cdot min^{-0.5}] = 0.2346$ $C_{i,1}\ [mg \cdot g^{-1}] = 1.38 \cdot 10^{-2}$	$k_{i,1}\ [mg \cdot g^{-1} \cdot min^{-0.5}] = 2.41 \cdot 10^{-2}$ $C_{i,1}\ [mg \cdot g^{-1}] = 4.75 \cdot 10^{-4}$	$k_{i,1}\ [mg \cdot g^{-1} \cdot min^{-0.5}] = 8.34 \cdot 10^{-3}$ $C_{i,1}\ [mg \cdot g^{-1}] = 4.63 \cdot 10^{-3}$
		Correlation parameters	$R^2 = 0.9995$ $RSS = 2.13 \cdot 10^{-10}$	$R^2 = 0.9841$ $RSS = 5.36 \cdot 10^{-5}$	$R^2 = 0.9055$ $RSS = 2.73 \cdot 10^{-5}$
		Kinetic parameters: Step 2	$k_{i,2}\ [mg \cdot g^{-1} \cdot min^{-0.5}] = 3.01 \cdot 10^{-2}$ $C_{i,2}\ [mg \cdot g^{-1}] = 0.2178$	$k_{i,2}\ [mg \cdot g^{-1} \cdot min^{-0.5}] = 0.2399$ $C_{i,2}\ [mg \cdot g^{-1}] = -0.4314$	$k_{i,2}\ [mg \cdot g^{-1} \cdot min^{-0.5}] = 4.82 \cdot 10^{-2}$ $C_{i,2}\ [mg \cdot g^{-1}] = -3.30 \cdot 10^{-2}$
		Correlation parameters	$R^2 = 0.9965$ $RSS = 6.23 \cdot 10^{-6}$	$R^2 = 0.9999$ $RSS = 2.13 \cdot 10^{-8}$	$R^2 = 0.9891$ $RSS = 5.05 \cdot 10^{-5}$
		Kinetic parameters: Step 3	$k_{i,3}\ [mg \cdot g^{-1} \cdot min^{-0.5}] = 2.10 \cdot 10^{-3}$ $C_{i,3}\ [mg \cdot g^{-1}] = 0.288$	$k_{i,3}\ [mg \cdot g^{-1} \cdot min^{-0.5}] = 7.05 \cdot 10^{-3}$ $C_{i,3}\ [mg \cdot g^{-1}] = 0.1079$	$k_{i,3}\ [mg \cdot g^{-1} \cdot min^{-0.5}] = 1.35 \cdot 10^{-2}$ $C_{i,3}\ [mg \cdot g^{-1}] = 4.42 \cdot 10^{-2}$
		Correlation parameters	$R^2 = 0.8137$ $RSS = 4.29 \cdot 10^{-5}$	$R^2 = 0.9103$ $RSS = 1.96 \cdot 10^{-4}$	$R^2 = 0.9999$ $RSS = 6.74 \cdot 10^{-9}$
		Kinetic parameters: Step 4			$k_{i,4}\ [mg \cdot g^{-1} \cdot min^{-0.5}] = 0$ $C_{i,4}\ [mg \cdot g^{-1}] = 0.1114$
		Correlation parameters			$RSS = 1.98 \cdot 10^{-6}$

Appendix C

Table A3. The results of the kinetic equation adjustment to the adsorption experimental data of 250 to 1000 mg CNF·L^{-1}.

Kinetic Model		250 mg·L^{-1}	500 mg·L^{-1}	1000 mg·L^{-1}
Pseudo-first order	Kinetic parameters	$k_1\ [h^{-1}] = 4.49 \cdot 10^{-2}$	$k_1\ [h^{-1}] = 5.56 \cdot 10^{-2}$	$k_1\ [h^{-1}] = 6.69 \cdot 10^{-2}$
	Correlation parameters	$R^2 = 0.3713$ $RSS = 3.31 \cdot 10^{-3}$	$R^2 = 0.9267$ $RSS = 0.159$	$R^2 = 0.6605$ $RSS = 0.152$
Pseudo-second order	Kinetic parameters	$k_2\ [mg \cdot g^{-1} \cdot h^{-1}] = 3822.25$ $q_e\ [mg \cdot g^{-1}] = 2.76 \cdot 10^{-2}$	$k_2\ [mg \cdot g^{-1} \cdot h^{-1}] = 1.1320$ $q_e\ [mg \cdot g^{-1}] = 0.5787$	$k_2\ [mg \cdot g^{-1} \cdot h^{-1}] = 14.3041$ $q_e\ [mg \cdot g^{-1}] = 0.3047$
	Correlation parameters	$R^2 = 0.9256$ $RSS = 1.55 \cdot 10^{-4}$	$R^2 = 0.9596$ $RSS = 3.78 \cdot 10^{-2}$	$R^2 = 0.9957$ $RSS = 9.31 \cdot 10^{-4}$
Elovich	Kinetic parameters	$\alpha\ [h \cdot mg \cdot g^{-1}] = 229.62$ $\beta\ [g \cdot mg^{-1}] = 462.96$	$\alpha\ [h \cdot mg \cdot g^{-1}] = 3.2448$ $\beta\ [g \cdot mg^{-1}] = 15.1976$	$\alpha\ [h \cdot mg \cdot g^{-1}] = 66.5435$ $\beta\ [g \cdot mg^{-1}] = 35.9712$
	Correlation parameters	$R^2 = 0.6241$ $RSS = 2.78 \cdot 10^{-4}$	$R^2 = 0.9246$ $RSS = 4.49 \cdot 10^{-2}$	$R^2 = 0.8683$ $RSS = 5.34 \cdot 10^{-3}$

Table A3. Cont.

Kinetic Model			250 mg·L^{-1}	500 mg·L^{-1}	1000 mg·L^{-1}
Intraparticle diffusion		Kinetic parameters: Step 1	$k_{i,1}$ [mg·g^{-1}·min$^{-0.5}$] = 0.1384 $C_{i,1}$ [mg·g^{-1}] = $-1.23·10^{-2}$	$k_{i,1}$ [mg·g^{-1}·min$^{-0.5}$] = 0.1397 $C_{i,1}$ [mg·g^{-1}] = $5.45·10^{-2}$	$k_{i,1}$ [mg·g^{-1}·min$^{-0.5}$] = 0.2346 $C_{i,1}$ [mg·g^{-1}] = $1.38·10^{-2}$
		Correlation parameters	R^2 = 0.9631 RSS = $1.51·10^{-4}$	R^2 = 0.9295 RSS = $1.03·10^{-2}$	R^2 = 0.9995 RSS = $2.13·10^{-10}$
		Kinetic parameters: Step 2	$k_{i,2}$ [mg·g^{-1}·min$^{-0.5}$] = $1·10^{-17}$ $C_{i,2}$ [mg·g^{-1}] = $2.76·10^{-2}$	$k_{i,2}$ [mg·g^{-1}·min$^{-0.5}$] = $7.03·10^{-2}$ $C_{i,2}$ [mg·g^{-1}] = 0.2210	$k_{i,2}$ [mg·g^{-1}·min$^{-0.5}$] = $3.01·10^{-2}$ $C_{i,2}$ [mg·g^{-1}] = 0.2178
		Correlation parameters	R^2 = 0.5117 RSS = $1.44·10^{-11}$	R^2 = 1.0000 RSS = $4.19·10^{-7}$	R^2 = 0.9965 RSS = $6.23·10^{-6}$
		Kinetic parameters: Step 3		$k_{i,3}$ [mg·g^{-1}·min$^{-0.5}$] = 0 $C_{i,3}$ [mg·g^{-1}] = 0.5658	$k_{i,3}$ [mg·g^{-1}·min$^{-0.5}$] = $2.10·10^{-3}$ $C_{i,3}$ [mg·g^{-1}] = 0.2880
		Correlation parameters		RSS = $1.73·10^{-9}$	R^2 = 0.8137 RSS = $4.29·10^{-5}$

Appendix D

Table A4. The results of the kinetic equation adjustment to the adsorption experimental data of the initial hexavalent chromium concentrations from 0.1 to 50 mg·L^{-1}.

Kinetic Model		0.1 mg·L^{-1}	1 mg·L^{-1}	5 mg·L^{-1}
Pseudo-first order	Kinetic parameters	k_1 [h^{-1}] = $8.00·10^{-2}$	k_1 [h^{-1}] = $4.76·10^{-2}$	k_1 [h^{-1}] = 1.5675
	Correlation parameters	R^2 = 0.8105 RSS = $6.81·10^{-2}$	R^2 = 0.8880 RSS = 1.1163	R^2 = 0.7040 RSS = 35.56
Pseudo-second order	Kinetic parameters	k_2 [mg·g^{-1}·h^{-1}] = 4.9284 q_e [mg·g^{-1}] = 0.3059	k_2 [mg·g^{-1}·h^{-1}] = $3.53·10^{-2}$ q_e [mg·g^{-1}] = 1.9361	k_2 [mg·g^{-1}·h^{-1}] = $2.99·10^{-2}$ q_e [mg·g^{-1}] = 6.9541
	Correlation parameters	R^2 = 0.9853 RSS = $3.05·10^{-3}$	R^2 = 0.8975 RSS = 0.7128	R^2 = 0.5884 RSS = 54.66
Elovich	Kinetic parameters	α [h·mg·g^{-1}] = 8.6542 β [g·mg^{-1}] = 31.257	α [h·mg·g^{-1}] = 1.6140 β [g·mg^{-1}] = 4.7996	α [h·mg·g^{-1}] = $1.20·10^9$ β [g·mg^{-1}] = 4.7547
	Correlation parameters	R^2 = 0.8606 RSS = $5.62·10^{-3}$	R^2 = 0.8760 RSS = 0.5550	R^2 = 0.5875 RSS = 3.5383
Intraparticle diffusion	Kinetic parameters: Step 1	$k_{i,1}$ [mg·g^{-1}·min$^{-0.5}$] = 0.1099 $C_{i,1}$ [mg·g^{-1}] = $2.88·10^{-2}$	$k_{i,1}$ [mg·g^{-1}·min$^{-0.5}$] = 0.6325 $C_{i,1}$ [mg·g^{-1}] = -0.1448	$k_{i,1}$ [mg·g^{-1}·min$^{-0.5}$] = 0.3298 $C_{i,1}$ [mg·g^{-1}] = 4.4590
	Correlation parameters	R^2 = 0.9982 RSS = $1.68·10^{-4}$	R^2 = 0.9271 RSS = $3.91·10^{-2}$	R^2 = 0.8235 RSS = 19.88
	Kinetic parameters: Step 2	$k_{i,2}$ [mg·g^{-1}·min$^{-0.5}$] = $6.00·10^{-3}$ $C_{i,2}$ [mg·g^{-1}] = 0.2615	$k_{i,2}$ [mg·g^{-1}·min$^{-0.5}$] = $2.29·10^{-2}$ $C_{i,2}$ [mg·g^{-1}] = 0.5194	$k_{i,2}$ [mg·g^{-1}·min$^{-0.5}$] = 0 $C_{i,2}$ [mg·g^{-1}] = 4.6922
	Correlation parameters	R^2 = 0.9978 RSS = $1.66·10^{-6}$	R^2 = 0.9690 RSS = $2.79·10^{-4}$	RSS = $9.68·10^{-9}$
	Kinetic parameters: Step 3	$k_{i,3}$ [mg·g^{-1}·min$^{-0.5}$] = 0 $C_{i,3}$ [mg·g^{-1}] = 0.3036	$k_{i,3}$ [mg·g^{-1}·min$^{-0.5}$] = 0.2121 $C_{i,3}$ [mg·g^{-1}] = -0.3775	$k_{i,3}$ [mg·g^{-1}·min$^{-0.5}$] = 0.617 $C_{i,3}$ [mg·g^{-1}] = 0.4177
	Correlation parameters	RSS = $1.68·10^{-4}$	R^2 = 0.9986 RSS = $2.48·10^{-3}$	R^2 = 0.9999 RSS = $7.46·10^{-8}$

Table A4. Cont.

Kinetic Model		0.1 mg·L^{-1}	1 mg·L^{-1}	5 mg·L^{-1}
Pseudo-first order	Kinetic parameters	k_1 [h^{-1}] = 3.00·10^{-2}	k_1 [h^{-1}] = 0.1664	k_1 [h^{-1}] = 3.86·10^{-2}
	Correlation parameters	R^2 = 0.5969 RSS = 307.88	R^2 = 0.5860 RSS = 1592.74	R^2 = 0.8982 RSS = 2433.18
Pseudo-second order	Kinetic parameters	k_2 [mg·g^{-1}·h^{-1}] = 5.40·10^{-2} q_e [mg·g^{-1}] = 12.115	k_2 [mg·g^{-1}·h^{-1}] = 8.70·10^{-3} q_e [mg·g^{-1}] = 34.51	k_2 [mg·g^{-1}·h^{-1}] = 6.16·10^{-3} q_e [mg·g^{-1}] = 70.92
	Correlation parameters	R^2 = 0.7242 RSS = 121.03	R^2 = 0.6332 RSS = 1286.27	R^2 = 0.9851 RSS = 236.03
Elovich	Kinetic parameters	α [h·mg·g^{-1}] = 3.82·10^8 β [g·mg^{-1}] = 2.2222	α [h·mg·g^{-1}] = 7.29·10^8 β [g·mg^{-1}] = 0.8143	α [h·mg·g^{-1}] = 153.74 β [g·mg^{-1}] = 0.1041
	Correlation parameters	R^2 = 0.9184 RSS = 1.5841	R^2 = 0.7408 RSS = 44.33	R^2 = 0.9934 RSS = 166.49
Intraparticle diffusion	Kinetic parameters: Step 1	$k_{i,1}$ [mg·g^{-1}·min$^{-0.5}$] = 1.3193 $C_{i,1}$ [mg·g^{-1}] = 8.1756	$k_{i,1}$ [mg·g^{-1}·min$^{-0.5}$] = 4.9473 $C_{i,1}$ [mg·g^{-1}] = 21.343	$k_{i,1}$ [mg·g^{-1}·min$^{-0.5}$] = 27.162 $C_{i,1}$ [mg·g^{-1}] = 4.5797
	Correlation parameters	R^2 = 0.8392 RSS = 66.84	R^2 = 0.8504 RSS = 455.52	R^2 = 0.9935 RSS = 20.97
	Kinetic parameters: Step 2	$k_{i,2}$ [mg·g^{-1}·min$^{-0.5}$] = 0 $C_{i,2}$ [mg·g^{-1}] = 9.1085	$k_{i,2}$ [mg·g^{-1}·min$^{-0.5}$] = 0.1966 $C_{i,2}$ [mg·g^{-1}] = 24.783	$k_{i,2}$ [mg·g^{-1}·min$^{-0.5}$] = 8.0237 $C_{i,2}$ [mg·g^{-1}] = 24.416
	Correlation parameters	RSS = 2.08·10^{-9}	R^2 = 0.9204 RSS = 0.2037	R^2 = 0.9987 RSS = 1.4264
	Kinetic parameters: Step 3	$k_{i,3}$ [mg·g^{-1}·min$^{-0.5}$] = 0.3399 $C_{i,3}$ [mg·g^{-1}] = 8.5209	$k_{i,3}$ [mg·g^{-1}·min$^{-0.5}$] = 2.2280 $C_{i,3}$ [mg·g^{-1}] = 10.785	$k_{i,3}$ [mg·g^{-1}·min$^{-0.5}$] = 1.2205 $C_{i,3}$ [mg·g^{-1}] = 56.515
	Correlation parameters	R^2 = 0.9953 RSS = 4.63·10^{-2}	R^2 = 0.9999 RSS = 2.05·10^{-7}	R^2 = 0.9443 RSS = 3.4399

References

1. Outridge, P.M.; Scheuhammer, A.M. Bioaccumulation and toxicology of chromium: Implications for wildlife. *Rev. Environ. Contam. Toxicol.* **1993**, *130*, 31–77.
2. Kumar, V.; Dwivedi, S.K. A review on accessible techniques for removal of hexavalent Chromium and divalent Nickel from industrial wastewater: Recent research and future outlook. *J. Clean. Prod.* **2021**, *295*, 126229. [CrossRef]
3. Mishra, S.; Bharagava, R.N. Toxic and genotoxic effects of hexavalent chromium in environment and its bioremediation strategies. *J. Environ. Sci. Health C Environ. Carcinog. Ecotoxicol. Rev.* **2016**, *34*, 1–32. [CrossRef] [PubMed]
4. Wise, J.P.; Young, J.L.; Cai, J.; Cai, L. Current understanding of hexavalent chromium [Cr(VI)] neurotoxicity and new perspectives. *Environ. Int.* **2022**, *158*, 106877. [CrossRef] [PubMed]
5. Sharma, P.; Singh, S.P.; Parakh, S.K.; Tong, Y.W. Health hazards of hexavalent chromium (Cr (VI)) and its microbial reduction. *Bioengineered* **2022**, *13*, 4923–4938. [CrossRef] [PubMed]
6. Chai, W.S.; Cheun, J.Y.; Kumar, P.S.; Mubashir, M.; Majeed, Z.; Banat, F.; Ho, S.-H.; Show, P.L. A review on conventional and novel materials towards heavy metal adsorption in wastewater treatment application. *J. Clean. Prod.* **2021**, *296*, 126589. [CrossRef]
7. Vaiopoulou, E.; Gikas, P. Regulations for chromium emissions to the aquatic environment in Europe and elsewhere. *Chemosphere* **2020**, *254*, 126876. [CrossRef]
8. Farooq, U.; Kozinski, J.A.; Khan, M.A.; Athar, M. Biosorption of heavy metal ions using wheat based biosorbents—A review of the recent literature. *Bioresour. Technol.* **2010**, *101*, 5043–5053. [CrossRef] [PubMed]
9. Owlad, M.; Aroua, M.K.; Daud, W.A.W.; Baroutian, S. Removal of Hexavalent Chromium-Contaminated Water and Wastewater: A Review. *Water Air Soil Pollut.* **2009**, *200*, 59–77. [CrossRef]
10. GracePavithra, K.; Jaikumar, V.; Kumar, P.S.; SundarRajan, P. A review on cleaner strategies for chromium industrial wastewater: Present research and future perspective. *J. Clean. Prod.* **2019**, *228*, 580–593. [CrossRef]
11. Islam, M.A.; Angove, M.J.; Morton, D.W. Recent innovative research on chromium (VI) adsorption mechanism. *Environ. Nanotechnol. Monit. Manag.* **2019**, *12*, 100267. [CrossRef]
12. Singh, P.; Itankar, N.; Patil, Y. Biomanagement of hexavalent chromium: Current trends and promising perspectives. *J. Environ. Manag.* **2021**, *279*, 111547. [CrossRef] [PubMed]
13. Rambabu, K.; Thanigaivelan, A.; Bharath, G.; Sivarajasekar, N.; Banat, F.; Show, P.L. Biosorption potential of Phoenix dactylifera coir wastes for toxic hexavalent chromium sequestration. *Chemosphere* **2021**, *268*, 128809. [CrossRef] [PubMed]

14. Hernández-Cocoletzi, H.; Salinas, R.A.; Águila-Almanza, E.; Rubio-Rosas, E.; Chai, W.S.; Chew, K.W.; Mariscal-Hernández, C.; Show, P.L. Natural hydroxyapatite from fishbone waste for the rapid adsorption of heavy metals of aqueous effluent. *Environ. Technol. Innov.* **2020**, *20*, 101109. [CrossRef]
15. Lee, P.-X.; Liu, B.-L.; Show, P.L.; Ooi, C.W.; Chai, W.S.; Munawaroh, H.S.H.; Chang, Y.-K. Removal of calcium ions from aqueous solution by bovine serum albumin (BSA)-modified nanofiber membrane: Dynamic adsorption performance and breakthrough analysis. *Biochem. Eng. J.* **2021**, *171*, 108016. [CrossRef]
16. Mubashir, M.; Dumée, L.F.; Fong, Y.Y.; Jusoh, N.; Lukose, J.; Chai, W.S.; Show, P.L. Cellulose acetate-based membranes by interfacial engineering and integration of ZIF-62 glass nanoparticles for CO_2 separation. *J. Hazard. Mater.* **2021**, *415*, 125639. [CrossRef]
17. Aigbe, U.O.; Osibote, O.A. A review of hexavalent chromium removal from aqueous solutions by sorption technique using nanomaterials. *J. Environ. Chem. Eng.* **2020**, *8*, 104503. [CrossRef]
18. Varghese, A.G.; Paul, S.A.; Latha, M.S. Remediation of heavy metals and dyes from wastewater using cellulose-based adsorbents. *Environ. Chem. Lett.* **2019**, *17*, 867–877. [CrossRef]
19. Wang, S.-L.; Lee, J.-F. Reaction mechanism of hexavalent chromium with cellulose. *Chem. Eng. J.* **2011**, *174*, 289–295. [CrossRef]
20. Kazachenko, A.; Akman, F.; Medimagh, M.; Issaoui, N.; Vasilieva, N.; Malyar, Y.N.; Sudakova, I.G.; Karacharov, A.; Miroshnikova, A.; Al-Dossary, O.M. Sulfation of Diethylaminoethyl-Cellulose: QTAIM Topological Analysis and Experimental and DFT Studies of the Properties. *ACS Omega* **2021**, *6*, 22603–22615. [CrossRef]
21. Negro, C.; Balea Martin, A.; Sanchez-Salvador, J.; Campano, C.; Fuente, E.; Monte, M.; Blanco, A. Nanocellulose and its potential use for sustainable industrial applications. *Lat. Am. Appl. Res.-Int. J.* **2020**, *50*, 59–64. [CrossRef]
22. Norrrahim, M.N.F.; Kasim, N.A.M.; Knight, V.F.; Misenan, M.S.M.; Janudin, N.; Shah, N.A.A.; Kasim, N.; Yusoff, W.Y.W.; Noor, S.A.M.; Jamal, S.H. Nanocellulose: A bioadsorbent for chemical contaminant remediation. *RSC Adv.* **2021**, *11*, 7347–7368. [CrossRef] [PubMed]
23. Reshmy, R.; Philip, E.; Madhavan, A.; Pugazhendhi, A.; Sindhu, R.; Sirohi, R.; Awasthi, M.K.; Pandey, A.; Binod, P. Nanocellulose as green material for remediation of hazardous heavy metal contaminants. *J. Hazard. Mater.* **2022**, *424*, 127516. [CrossRef] [PubMed]
24. Wang, F.; Hubbe, M. Charge Properties of Fibers in the Paper Mill Environment. 1. Effect of Electrical Conductivity. *J. Pulp Pap. Sci.* **2002**, *28*, 347–353.
25. Khoo, R.Z.; Chow, W.S.; Ismail, H. Sugarcane bagasse fiber and its cellulose nanocrystals for polymer reinforcement and heavy metal adsorbent: A review. *Cellulose* **2018**, *25*, 4303–4330. [CrossRef]
26. Liu, S.; Low, Z.-X.; Xie, Z.; Wang, H. TEMPO-Oxidized Cellulose Nanofibers: A Renewable Nanomaterial for Environmental and Energy Applications. *Adv. Mater. Technol.* **2021**, *6*, 2001180. [CrossRef]
27. Huang, X.Y.; Dognani, G.; Hadi, P.; Yang, M.Y.; Job, A.E.; Hsiao, B.S. Cationic Dialdehyde Nanocellulose from Sugarcane Bagasse for Efficient Chromium(VI) Removal. *ACS Sustain. Chem. Eng.* **2020**, *8*, 4734–4744. [CrossRef]
28. Ossa-Paredes, R.; Bastidas, B.; Carvajal-Barriga, E.J. Remediation of Contaminated Water with Chromium VI by Sorption in Surface-activated-nanocellulose spheroids. *Pollution* **2022**, *8*, 489–500.
29. Ben Hammouda, S.; Chen, Z.; An, C.; Lee, K. Recent advances in developing cellulosic sorbent materials for oil spill cleanup: A state-of-the-art review. *J. Clean. Prod.* **2021**, *311*, 127630. [CrossRef]
30. Zhang, Z.; Sèbe, G.; Rentsch, D.; Zimmermann, T.; Tingaut, P. Ultralightweight and flexible silylated nanocellulose sponges for the selective removal of oil from water. *Chem. Mater.* **2014**, *26*, 2659–2668. [CrossRef]
31. Janik, P.; Zawisza, B.; Talik, E.; Sitko, R. Selective adsorption and determination of hexavalent chromium ions using graphene oxide modified with amino silanes. *Microchim. Acta* **2018**, *185*, 117. [CrossRef] [PubMed]
32. Lee, J.; Park, J.-A.; Kim, H.-G.; Lee, J.-H.; Cho, S.-H.; Choi, K.; Jung, K.-W.; Lee, S.Y.; Choi, J.-W. Most suitable amino silane molecules for surface functionalization of graphene oxide toward hexavalent chromium adsorption. *Chemosphere* **2020**, *251*, 126387. [CrossRef] [PubMed]
33. Jamroz, E.; Kocot, K.; Zawisza, B.; Talik, E.; Gagor, A.; Sitko, R. A green analytical method for ultratrace determination of hexavalent chromium ions based on micro-solid phase extraction using amino-silanized cellulose membranes. *Microchem. J.* **2019**, *149*, 104060. [CrossRef]
34. Saito, T.; Kimura, S.; Nishiyama, Y.; Isogai, A. Cellulose Nanofibers Prepared by TEMPO-Mediated Oxidation of Native Cellulose. *Biomacromolecules* **2007**, *8*, 2485–2491. [CrossRef]
35. Fukuzumi, H.; Saito, T.; Okita, Y.; Isogai, A. Thermal stabilization of TEMPO-oxidized cellulose. *Polym. Degrad. Stab.* **2010**, *95*, 1502–1508. [CrossRef]
36. Balea, A.; Merayo, N.; De La Fuente, E.; Negro, C.; Blanco, Á. Assessing the influence of refining, bleaching and TEMPO-mediated oxidation on the production of more sustainable cellulose nanofibers and their application as paper additives. *Ind. Crops Prod.* **2017**, *97*, 374–387. [CrossRef]
37. Sanchez-Salvador, J.L.; Campano, C.; Negro, C.; Monte, M.C.; Blanco, A. Increasing the Possibilities of TEMPO-Mediated Oxidation in the Production of Cellulose Nanofibers by Reducing the Reaction Time and Reusing the Reaction Medium. *Adv. Sustain. Syst.* **2021**, *5*, 2000277. [CrossRef]
38. Balea, A.; Sanchez-Salvador, J.L.; Monte, M.C.; Merayo, N.; Negro, C.; Blanco, A. In Situ Production and Application of Cellulose Nanofibers to Improve Recycled Paper Production. *Molecules* **2019**, *24*, 1800. [CrossRef]

39. Sanchez-Salvador, J.L.; Balea, A.; Monte, M.C.; Negro, C.; Miller, M.; Olson, J.; Blanco, A. Comparison Of Mechanical And Chemical Nanocellulose As Additives To Reinforce Recycled Cardboard. *Sci. Rep.* **2020**, *10*, 3778. [CrossRef]
40. APHA. Standard Methods: 3500-Cr CHROMIUM. Chromium by colororimetry. In *Standard Methods for the Examination of Water and Wastewater*; APHA: Washington, DC, USA, 2018.
41. Alsaiari, N.S.; Katubi, K.M.; Alzahrani, F.M.; Amari, A.; Osman, H.; Rebah, F.B.; Tahoon, M.A. Synthesis, Characterization and Application of Polypyrrole Functionalized Nanocellulose for the Removal of Cr(VI) from Aqueous Solution. *Polymers* **2021**, *13*, 3691. [CrossRef]
42. Park, S.Y.; Goo, S.; Shin, H.; Kim, J.; Youn, H.J. Structural properties of cellulose nanofibril foam depending on wet foaming conditions in Pickering stabilization. *Cellulose* **2021**, *28*, 10291–10304. [CrossRef]
43. Largitte, L.; Pasquier, R. A review of the kinetics adsorption models and their application to the adsorption of lead by an activated carbon. *Chem. Eng. Res. Des.* **2016**, *109*, 495–504. [CrossRef]
44. Ho, Y.S.; McKay, G. A Comparison of Chemisorption Kinetic Models Applied to Pollutant Removal on Various Sorbents. *Process Saf. Environ. Prot.* **1998**, *76*, 332–340. [CrossRef]
45. Wu, F.-C.; Tseng, R.-L.; Juang, R.-S. Characteristics of Elovich equation used for the analysis of adsorption kinetics in dye-chitosan systems. *Chem. Eng. J.* **2009**, *150*, 366–373. [CrossRef]
46. Andelescu, A.; Nistor, M.A.; Muntean, S.G.; Rădulescu-Grad, M.E. Adsorption studies on copper, cadmium, and zinc ion removal from aqueous solution using magnetite/carbon nanocomposites. *Sep. Sci. Technol.* **2018**, *53*, 2352–2364. [CrossRef]
47. Lim, L.B.L.; Priyantha, N.; Tennakoon, D.T.B.; Chieng, H.I.; Dahri, M.K.; Suklueng, M. Breadnut peel as a highly effective low-cost biosorbent for methylene blue: Equilibrium, thermodynamic and kinetic studies. *Arab. J. Chem.* **2017**, *10*, S3216–S3228. [CrossRef]
48. Patiño-Masó, J.; Serra-Parareda, F.; Tarrés, Q.; Mutjé, P.; Espinach, F.X.; Delgado-Aguilar, M. TEMPO-Oxidized Cellulose Nanofibers: A Potential Bio-Based Superabsorbent for Diaper Production. *Nanomaterials* **2019**, *9*, 1271. [CrossRef]
49. Lu, P.; Liu, R.; Liu, X.; Wu, M. Preparation of Self-supporting Bagasse Cellulose Nanofibrils Hydrogels Induced by Zinc Ions. *Nanomaterials* **2018**, *8*, 800. [CrossRef]
50. Ghiasi, F.; Eskandari, M.H.; Golmakani, M.-T.; Hosseini, S.M.H. Development of highly stable colloidal dispersions of gelled-oil nanoparticles loaded with cuminaldehyde. *J. Colloid. Interf. Sci.* **2019**, *541*, 65–74. [CrossRef]
51. Necula, B.S.; Apachitei, I.; Fratila-Apachitei, L.E.; Teodosiu, C.; Duszczyk, J. Stability of nano-/microsized particles in deionized water and electroless nickel solutions. *J. Colloid. Interfaces.Sci.* **2007**, *314*, 514–522. [CrossRef]
52. Khalil, H.P.S.A.; Davoudpour, Y.; Aprilia, N.A.S.; Mustapha, A.; Hossain, S.; Islam, N.; Dungani, R. Nanocellulose-Based Polymer Nanocomposite: Isolation, Characterization and Applications. In *Nanocellulose Polymer Nanocomposites*; Scrivener Publishing LLC: Beverly, MA, USA, 2014; pp. 273–309.
53. Balea, A.; Fuente, E.; Tarrés, Q.; Pèlach, M.À.; Mutjé, P.; Delgado-Aguilar, M.; Blanco, A.; Negro, C. Influence of pretreatment and mechanical nanofibrillation energy on properties of nanofibers from Aspen cellulose. *Cellulose* **2021**, *28*, 9187–9206. [CrossRef]
54. Kim, Y.; Park, J.; Bang, J.; Kim, J.; Jin, H.-J.; Kwak, H.W. Highly efficient Cr(VI) remediation by cationic functionalized nanocellulose beads. *J. Hazard. Mater.* **2022**, *426*, 128078. [CrossRef] [PubMed]
55. Yuan, H.; Peng, J.; Ren, T.; Luo, Q.; Luo, Y.; Zhang, N.; Huang, Y.; Guo, X.; Wu, Y. Novel fluorescent lignin-based hydrogel with cellulose nanofibers and carbon dots for highly efficient adsorption and detection of Cr(VI). *Sci. Total Environ.* **2021**, *760*, 143395. [CrossRef]
56. Kumar, N.; Kardam, A.; Rajawat, D.S.; Jain, V.K.; Suman. Carboxymethyl nanocellulose stabilized nano zero-valent iron: An effective method for reduction of hexavalent chromium in wastewater. *Mater. Res. Express* **2019**, *6*, 1150f3. [CrossRef]
57. Pourfadakari, S.; Jorfi, S.; Ahmadi, M.; Takdastan, A. Experimental data on adsorption of Cr(VI) from aqueous solution using nanosized cellulose fibers obtained from rice husk. *Data Brief* **2017**, *15*, 887–895. [CrossRef]
58. Huang, H.H.; He, D.L.; Tang, Y.N.; Guo, Y.N.; Li, P.; Qv, W.; Deng, F.; Lu, F.H. Adsorption of Hexavalent Chromium from an Aqueous Phase by Hydroxypropyl Methylcellulose Modified with Diethylenetriamine. *J. Chem. Eng. Data* **2019**, *64*, 98–106. [CrossRef]
59. Wu, F.-C.; Tseng, R.-L.; Juang, R.-S. Initial behavior of intraparticle diffusion model used in the description of adsorption kinetics. *Chem. Eng. J.* **2009**, *153*, 1–8. [CrossRef]
60. Xu, Q.H.; Wang, Y.L.; Jin, L.Q.; Wang, Y.; Qin, M.H. Adsorption of Cu (II), Pb (II) and Cr (VI) from aqueous solutions using black wattle tannin-immobilized nanocellulose. *J. Hazard. Mater.* **2017**, *339*, 91–99. [CrossRef]
61. Saha, B.; Orvig, C. Biosorbents for hexavalent chromium elimination from industrial and municipal effluents. *Coord. Chem. Rev.* **2010**, *254*, 2959–2972. [CrossRef]
62. Qiu, B.; Guo, J.; Zhang, X.; Sun, D.Z.; Gu, H.B.; Wang, Q.; Wang, H.W.; Wang, X.F.; Zhang, X.; Weeks, B.L.; et al. Polyethylenimine Facilitated Ethyl Cellulose for Hexavalent Chromium Removal with a Wide pH Range. *Acs Appl. Mater. Interfaces* **2014**, *6*, 19816–19824. [CrossRef]
63. Singh, K.; Arora, J.K.; Sinha, T.J.M.; Srivastava, S. Functionalization of nanocrystalline cellulose for decontamination of Cr(III) and Cr(VI) from aqueous system: Computational modeling approach. *Clean Technol. Environ. Policy* **2014**, *16*, 1179–1191. [CrossRef]
64. Baral, S.S.; Das, S.N.; Rath, P. Hexavalent chromium removal from aqueous solution by adsorption on treated sawdust. *Biochem. Eng. J.* **2006**, *31*, 216–222. [CrossRef]
65. Nasseh, N.; Taghavi, L.; Barikbin, B.; Harifi-Mood, A.R. The removal of Cr(VI) from aqueous solution by almond green hull waste material: Kinetic and equilibrium studies. *J. Water Reuse Desalination* **2016**, *7*, 449–460. [CrossRef]

66. Omidvar Borna, M.; Pirsaheb, M.; Vosoughi Niri, M.; Khosravi Mashizie, R.; Kakavandi, B.; Zare, M.R.; Asadi, A. Batch and column studies for the adsorption of chromium(VI) on low-cost Hibiscus Cannabinus kenaf, a green adsorbent. *J. Taiwan Inst. Chem. Eng.* **2016**, *68*, 80–89. [CrossRef]
67. Rai, M.K.; Shahi, G.; Meena, V.; Meena, R.; Chakraborty, S.; Singh, R.S.; Rai, B.N. Removal of hexavalent chromium Cr (VI) using activated carbon prepared from mango kernel activated with H_3PO_4. *Resour.-Effic. Technol.* **2016**, *2*, S63–S70. [CrossRef]
68. Mohan, D.; Singh, K.P.; Singh, V.K. Removal of Hexavalent Chromium from Aqueous Solution Using Low-Cost Activated Carbons Derived from Agricultural Waste Materials and Activated Carbon Fabric Cloth. *Ind. Eng. Chem. Res.* **2005**, *44*, 1027–1042. [CrossRef]
69. Dima, J.B.; Sequeiros, C.; Zaritzky, N.E. Hexavalent chromium removal in contaminated water using reticulated chitosan micro/nanoparticles from seafood processing wastes. *Chemosphere* **2015**, *141*, 100–111. [CrossRef]
70. Debnath, A.; Bera, A.; Chattopadhyay, K.K.; Saha, B. Facile additive-free synthesis of hematite nanoparticles for enhanced adsorption of hexavalent chromium from aqueous media: Kinetic, isotherm, and thermodynamic study. *Inorg. Nano-Met. Chem.* **2017**, *47*, 1605–1613. [CrossRef]
71. Yang, R.; Aubrecht, K.B.; Ma, H.; Wang, R.; Grubbs, R.B.; Hsiao, B.S.; Chu, B. Thiol-modified cellulose nanofibrous composite membranes for chromium (VI) and lead (II) adsorption. *Polymer* **2014**, *55*, 1167–1176. [CrossRef]
72. Shao, Y.; Fan, Z.; Zhong, M.; Xu, W.; He, C.; Zhang, Z. Polypyrrole/bacterial cellulose nanofiber composites for hexavalent chromium removal. *Cellulose* **2021**, *28*, 2229–2240. [CrossRef]
73. Jain, P.; Varshney, S.; Srivastava, S. Synthetically modified nano-cellulose for the removal of chromium: A green nanotech perspective. *IET Nanobiotechnol.* **2017**, *11*, 45–51. [CrossRef]
74. Sun, Y.; Yue, Q.; Mao, Y.; Gao, B.; Gao, Y.; Huang, L. Enhanced adsorption of chromium onto activated carbon by microwave-assisted H3PO4 mixed with Fe/Al/Mn activation. *J. Hazard. Mater.* **2014**, *265*, 191–200. [CrossRef] [PubMed]
75. Acharya, J.; Sahu, J.N.; Sahoo, B.K.; Mohanty, C.R.; Meikap, B.C. Removal of chromium(VI) from wastewater by activated carbon developed from Tamarind wood activated with zinc chloride. *Chem. Eng. J.* **2009**, *150*, 25–39. [CrossRef]
76. Liu, S.X.; Chen, X.; Chen, X.Y.; Liu, Z.F.; Wang, H.L. Activated carbon with excellent chromium(VI) adsorption performance prepared by acid–base surface modification. *J. Hazard. Mater.* **2007**, *141*, 315–319. [CrossRef] [PubMed]
77. Zhang, D.; Xu, W.; Cai, J.; Cheng, S.-Y.; Ding, W.-P. Citric acid-incorporated cellulose nanofibrous mats as food materials-based biosorbent for removal of hexavalent chromium from aqueous solutions. *Int. J. Biol. Macromol.* **2020**, *149*, 459–466. [CrossRef]
78. McCabe, W.L.; Smith, J.C.; Harriott, P. *Unit Operations of Chemical Engineering*; McGraw Hill: Boston, MA, USA, 2001.
79. Ayawei, N.; Ebelegi, A.N.; Wankasi, D. Modelling and Interpretation of Adsorption Isotherms. *J. Chem.* **2017**, *2017*, 3039817. [CrossRef]
80. Qiu, B.; Gu, H.; Yan, X.; Guo, J.; Wang, Y.; Sun, D.; Wang, Q.; Khan, M.; Zhang, X.; Weeks, B.L.; et al. Cellulose derived magnetic mesoporous carbon nanocomposites with enhanced hexavalent chromium removal. *J. Mater. Chem. A* **2014**, *2*, 17454–17462. [CrossRef]
81. Rao, D.G.; Senthikumar, R.; Byrne, J.A.; Feroz, S. *Wastewater Treatment: Advanced Processes and Technologies*; CRC Press: Boca Raton, FL, USA, 2012.
82. Dawodu, F.A.; Akpan, B.M.; Akpomie, K.G. Sequestered capture and desorption of hexavalent chromium from solution and textile wastewater onto low cost Heinsia crinita seed coat biomass. *Appl. Water Sci.* **2019**, *10*, 32. [CrossRef]
83. De Andrade Neto, J.C.; Pereira, G.J.; Morandim-Giannetti, A.d.A. Chitosan and corn stover derivative bioadsorbent: Characterization and application in hexavalent chromium adsorption processes. *Cellulose* **2020**, *27*, 6317–6331. [CrossRef]
84. Batool, F.; Akbar, J.; Iqbal, S.; Noreen, S.; Bukhari, S.N.A. Study of Isothermal, Kinetic, and Thermodynamic Parameters for Adsorption of Cadmium: An Overview of Linear and Nonlinear Approach and Error Analysis. *Bioinorg. Chem. Appl.* **2018**, *2018*, 3463724. [CrossRef]
85. Choudhary, B.; Paul, D. Isotherms, kinetics and thermodynamics of hexavalent chromium removal using biochar. *J. Environ. Chem. Eng.* **2018**, *6*, 2335–2343. [CrossRef]

Article

Facile Conversion of Polystyrene Waste into an Efficient Sorbent for Water Purification

Cuizhu Ye [†], Ziyan Pan [†] and Yi Shen *,[†]

School of Food Science and Technology, South China University of Technology, Guangzhou 510640, China
* Correspondence: feyshen@scut.edu.cn
† These authors contributed equally to this work.

Abstract: In this work, we convert a plastic waste, i.e., polystyrene (PS), into a sorbent by a simple sulfonation process. The sulfonation time was optimized and the structures of the resulting sulfonated polystyrene (SPS) was characterized by field emission scanning electron microscopy, energy-dispersive X-ray and contact angle tests. The results showed that the sulfonation time of 7 h can introduce abundant sulfonic groups and preserve the self-standing structure. Additionally, the SPS has a three-dimensional porous structure and hydrophilic surface because of the presence of numerous sulfonic groups, which could serve as effective binding sites for immobilizing varying pollutants. Furthermore, as a proof-of-concept, the adsorption performance of the SPS foams was evaluated using three pollutants, namely Pb^{2+}, lysozyme and methylene blue. The adsorption isotherms were fitted by the Langmuir and Freundlich models, while the kinetics of the adsorption processes were analyzed using the pseudo-first-order, pseudo-second-order and intraparticle diffusion equations. It was found that the adsorption isotherms of Pb^{2+} and lysozyme can be better described by the Langmuir model, leading to maximum equilibrium adsorption uptakes of 10.5 and 15.7 mg g^{-1} for the adsorption of Pb^{2+} and lysozyme, respectively. Importantly, the pollutant-saturated SPS is readily regenerated by acid washing, and the recovered sorbents exhibit outstanding cyclic performance. The abundant availability of feedstock, facile preparation and regeneration processes render the SPS foams a promising sorbent for practical applications.

Keywords: sulfonated polystyrene; water treatment; Pb^{2+}; lysozyme and methylene blue

Citation: Ye, C.; Pan, Z.; Shen, Y. Facile Conversion of Polystyrene Waste into an Efficient Sorbent for Water Purification. *Polymers* **2022**, *14*, 4477. https://doi.org/10.3390/polym14214477

Academic Editors: Marta Otero and Ricardo N. Coimbra

Received: 24 September 2022
Accepted: 20 October 2022
Published: 22 October 2022

Publisher's Note: MDPI stays neutral with regard to jurisdictional claims in published maps and institutional affiliations.

Copyright: © 2022 by the authors. Licensee MDPI, Basel, Switzerland. This article is an open access article distributed under the terms and conditions of the Creative Commons Attribution (CC BY) license (https://creativecommons.org/licenses/by/4.0/).

1. Introduction

With the growth of the global population and the aggravation of environmental pollution, an ever-increasing demand for clean water will be witnessed in the coming decades. Effective removal of pollutants from water is of paramount importance for water security [1–3]. In this context, many water treatment technologies, such as coagulation–flocculation, advanced oxidation, membrane filtration and sorption, have been extensively explored to remove pollutants from water sources [4–7]. Of these techniques, sorption is considered to be one of the most promising approaches because of its simplicity and low cost [8–13]. Since the sorbent plays a key role in the sorption process, developing high-performance sorbents is crucial for the feasibility of applying sorption in water cleanup. Thus far, numerous sorbents, such as carbons [14–18], oxides [19–25], nitrides [26–29] and polymers [30–35], have been reported. For example, Deng et al. prepared thio-functionalized polyacrylonitrile fiber for the selective and enhanced adsorption of mercury and cadmium from water [30]. Moreover, Sapurina et al. expounded the sorbents used for water purification based on conjugated polymers [34]. Unfortunately, most of these reported sorbents suffer a common drawback, i.e., high costs. In particular, the complicated fabrication process of these materials poses a great obstacle for large-scale application. As a result, high-performance and cost-effective sorbents are highly desirable, but unfortunately still lacking.

Our has group devoted many efforts to searching for sorbents for water clean-up [36–39]. We fabricated three-dimensional hierarchical architectures by integrating carbon nanofibers and graphene nanosheets into macroscopic graphite felt supports [36,37]. The resulting composite monoliths showed excellent adsorption performance for the removal of heavy metal ions, dyes and organic solvents from aqueous solutions. In addition, we prepared hierarchical magnetic carbon nanosheet assemblies, which showed remarkable adsorption uptakes of 453 and 724 mg g^{-1} for the adsorption of Pb^{2+} and Congo red, respectively [38].

Recently, we fabricated mono-dispersed sulfonated polystyrene (SPS) nanospheres for water treatment, which could effectively reduce the concentration of heavy metal ions in the solution to ppb levels in several minutes [39]. However, the widespread application of SPS nanospheres is limited because of their cost and the tedious multi-step preparation process. Moreover, SPS nanospheres are always used in powder form and must be collected from aqueous solutions by filtration after the adsorption process, which consumes additional time and manpower, further increasing the overall operating cost of the sorption process.

Polystyrene (PS), a polymer synthesized by the free radical polymerization of a styrene monomer, is extensively employed in plates, utensils (single-use), packing, toys, DVD cases and foam coffee cups [40]. However, the disposal of these products creates environmental pollution because of their nondegradable nature [41]. Additionally, the recycling of PS is currently costly and quite limited, particularly in some developing countries, which leads to serious "white pollution". The growing scientific research elucidates the consequences of PS for wildlife animals and their habitats [42]. Given the concept of changing waste into treasure, it is of great significance to develop valuable products from PS waste [43–45].

Motivated by these critical issues, in this work, we utilized PS waste as a feedstock to fabricate a cost-effective sorbent via a simple sulfonation process. PS foams are widely used in the packaging industry. To this end, herein, we convert spent PS foams into a cost-effective sorbent via a simple sulfonation process. The preparation process is simple and does not involve any sophisticated instruments or complicated fabrication process. The resulting SPS sorbent exhibits a self-standing porous structure, greatly facilitating the separation and regeneration processes, which is highly favorable for practical applications.

2. Materials and Methods

2.1. Materials

Sodium hydroxide (≥96.0%) and anhydrous ethanol (≥99.7%) were purchased from the Nanjing Chemical Reagent Co. Ltd., Nanjing, China. Sulfuric acid (98.0%) and lead nitrate (≥99.0%) were purchased from the Damao Chemical Reagent Co. Ltd., Tianjin, China. Hydrochloric acid (1%) was purchased from the Guangdong Minggu Chemical Technology Co. Ltd., Shanghai, China. Lysozyme and methylene blue (MB) were purchased from the Shanghai Boao Biotechnology Co. Ltd., Shanghai, China. Spent PS foams were collected, rinsed with ethanol and dried at 60 °C in a vacuum overnight.

2.2. Synthesis of SPS Foams

Expanded PS foams were cut into pieces with dimensions of 5 × 5 × 2 cm and immersed in concentrated H_2SO_4. The sulfonation process was conducted at 60 °C with varying reaction times of 3, 4, 5, 6, 7 and 8 h. Subsequently, the resulting SPS foams were taken out from the H_2SO_4 solution, thoroughly washed with deionized water and then dried at 50 °C using a vacuum oven overnight. It is noteworthy that the concentrated H_2SO_4 can be reused in the preparation process, which greatly reduces the consumption of H_2SO_4.

2.3. Determination of Degree of Sulfonation (DoS)

The DoS of the resulting SPS foams was determined by titration. The details of the experimental procedures are presented as follows. A piece of SPS foam was immersed in 80 mL of NaOH solution for 2 days. Afterwards, 20 mL of NaOH solution was withdrawn and titrated using HCl solution. The consumed volume of HCl solution (V_{HCl}) was recorded. The DoS was calculated by following Equation (1):

$$DoS = \frac{80 \times 10^{-3} \times C_{NaOH} - V_{HCl} \times 4 \times 10^{-3} \times C_{HCl}}{M} \quad (1)$$

where C_{NaOH} is the concentration of NaOH and M is the mass of the SPS foam.

2.4. Structural Characterization

The morphology of the PS and SPS foams was observed by field emission scanning electron microscopy (FESEM) (JSM-7600F, JEOL, Tokyo, Japan). The elemental composition of the samples was determined by energy-dispersive X-ray (EDX) analyses. The wetting property of the samples was analyzed by contact angle (CA) tests (JGW-360C, Chenghui Testing Machine Co. Ltd., Chengde, China). Distilled water droplets were used as a probe to study the surface properties of the foams.

2.5. Batch Adsorption Tests

In a typical batch adsorption test, a piece of SPS foam was immersed in a pollutant solution and then agitated at 180 rpm using a mechanical shaker at 25 °C. At given time intervals, a certain volume of aliquot was sampled and filtered through a membrane filter to remove any impurities. The concentration of Pb^{2+} was monitored using an atomic absorption spectrometer (AAS) (Z-2000 Hitachi, Hitachi, Tokyo, Japan), while those of lysozyme and MB were determined by a UV–visible spectrophotometer (North Point Rayleigh UV-1801, Beijing, China). The adsorption uptake of the adsorbent q_t (mg g^{-1}) at time t (min) was calculated by Equation (2):

$$q_t = \frac{(C_0 - C_t) \times V}{W} \quad (2)$$

where C_0 (mg L^{-1}) is the initial pollutant concentration and C_t (mg L^{-1}) is the concentration at time t (min) in the liquid phase, V (L) is the volume of the solution and W (g) is the weight of the sorbent. To determine the equilibrium adsorption capacity q_e (mg g^{-1}), the sorbent was immersed in the pollution solution for at least 12 h to achieve the equilibrium state of adsorption.

2.6. Filtration Adsorption Tests

To study the filtration performance, the SPS foams with a diameter of 1.6 cm and a thickness of 1 cm were fitted into a funnel. MB (initial concentration = 50 mg L^{-1}), lysozyme (initial concentration = 150 mg L^{-1}) and Pb^{2+} (initial concentration = 50 mg L^{-1}) solutions were separately filtered through the foams. The concentrations of the pollutants in the effluents were analyzed.

2.7. Cyclic Adsorption Tests

To evaluate the cyclic adsorption performance, SPS foams were immersed in 50 mg L^{-1} pollutant (MB or Pb^{2+}) solutions. The adsorption process was conducted at 25 °C for at least 12 h. Subsequently, the pollutant-saturated SPS foams were taken out from the solution and rinsed with 100 mL of 1 M HCl solution three times. The adsorption–desorption process was repeated five times. The adsorption uptakes of the SPS foams were recorded as a function of the cycle number.

3. Results and Discussion

3.1. Structural Characterization

Since the number of sulfonic groups in the SPS foams plays a decisive role in the adsorption performance, the sulfonation time was optimized during the preparation process. Figure 1a shows the degree of sulfonation (DoS) of the foams as a function of reaction time.

The DoS values increase with increasing reaction time. Notably, the DoS of the SPS foam reached 0.52 µmol g^{-1} when the sulfonation process was conducted for 7 h. Figure 1b shows the corresponding digital photos of the foams. It reveals that the resulting SPS foams have an intact self-standing structure when the sulfonation time is less than 8 h. When the sulfonation time is over 8 h, the SPS foams start to collapse, leading to many isolated particles with sizes of several millimeters. Although these particles possess large DoS values, they are quite difficult to separate from the solutions, which greatly limits their practical applications. In contrast, the bulky free-standing SPS foams afford a significant advantage during separation. Under this circumstance, the sulfonation time is optimized to be 7 h in this work.

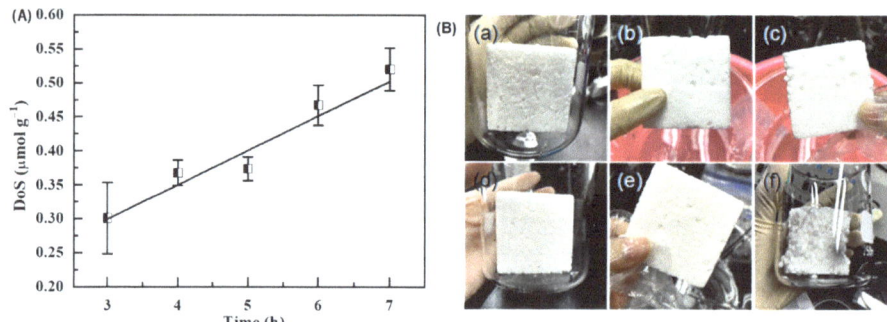

Figure 1. (**A**) DoS as a function of sulfonation time, and (**B**) digital photos of SPS foams with sulfonation times of (**a**) 3, (**b**) 4, (**c**) 5, (**d**) 6, (**e**) 7 and (**f**) 8 h.

The morphology of the foams was observed by FESEM, as shown in Figure 2. The pristine PS foam possesses a macroporous structure, as shown in Figure 2a. The EDX results indicate that the pristine PS foam contains carbon and oxygen elements, as shown in Figure 2b. Figure 2c shows the morphology of the resulting SPS foams. It clearly reveals that the porous structure is well preserved in the SPS foams. Apart from carbon and oxygen, sulfur is also detected by the EDX analyses, as shown in Figure 2d. The insets shown in Figure 2a,c are the digital photos of the PS and SPS foams, respectively. The pristine PS foam has dimensions of ca. $5 \times 5 \times 2$ cm. After the sulfonation and drying processes, the contraction of the foam is noted, resulting in a smaller size of the SPS foam. A close inspection could reveal that the surface of the SPS foams is much rougher than that of the pristine PS foams. This could also be related to the sulfonation process.

The wettability of the foams was investigated by the CA tests, as shown in Figure 3. The CA of the pristine PS foam is determined to be 116° (see Figure 3a), indicating the hydrophobic properties of the PS foam. In contrast, the water droplet readily penetrates the SPS foam (see Figure 3b and the Video S1 in the Supporting Information), reflecting the hydrophilic surface of the SPS foam. The variations in surface properties are attributed to the introduction of hydrophilic sulfonic groups and the increased surface roughness arising from the sulfonation process [46].

3.2. Adsorption Performance

The adsorption performance of the resulting SPS foams was evaluated using three typical substrates, including a heavy metal pollutant (i.e., Pb^{2+}), a protein (i.e., lysozyme) and a dye (i.e., MB). The adsorption characteristics of the SPS foams toward these three substrates were extensively investigated. A piece of SPS foam was immersed in the pollutant solution with different initial concentrations. After reaching adsorption equilibrium, the concentration of Pb^{2+} was analyzed by an inductive plasma emission spectroscope, while those of lysozyme and MB were determined by a UV–visible spectrophotometer. Figure S1 (from the Supplementary Materials) shows the UV spectra of MB. It is noteworthy that several peaks are seen in the UV spectra of MB. The strongest peak, located at ca. 664 nm, was employed to determine the concentration of MB. A broad peak located at ca. 280 nm was observed from the UV spectra of lysozyme, as shown in Figure S2. In the testing concentration range, the peak absorbance is well correlated with the concentration, as shown in Figures S1b and S2b.

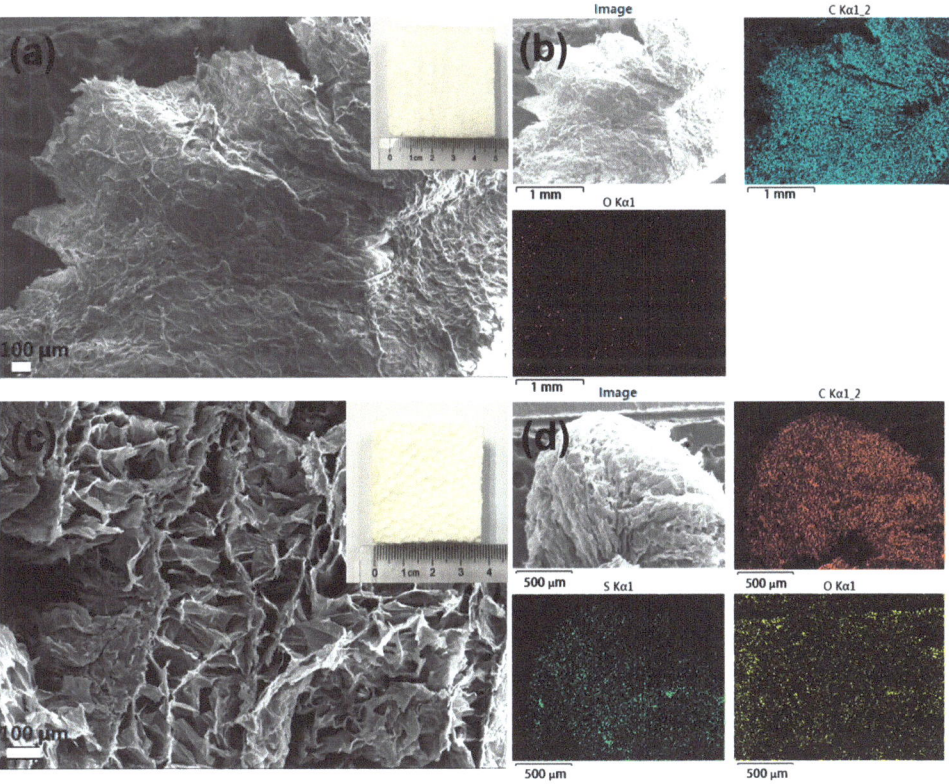

Figure 2. FESEM images of (**a**) pristine EPS and (**c**) SPS foams (the insets in (**a**,**c**) are the corresponding digital photos of EPS and SPS foams); EDX mapping images of (**b**) pristine EPS and (**d**) SPS foams.

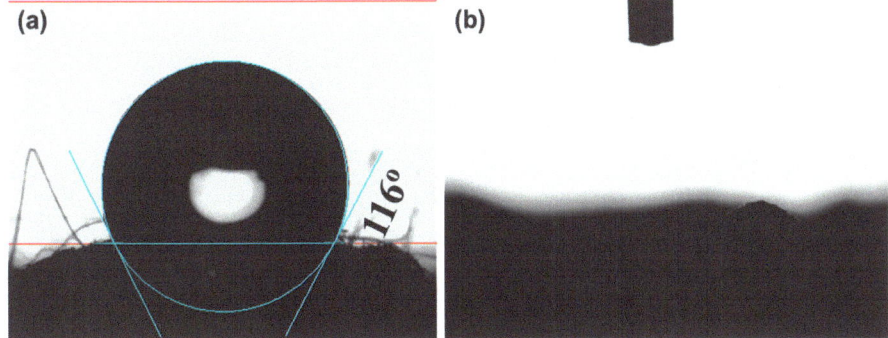

Figure 3. CA tests of (**a**) pristine PS and (**b**) SPS foams.

The adsorption isotherms were first recorded by plotting the equilibrium adsorption uptake (q_e (mg g^{-1})) against the equilibrium substrate concentration (C_e (mg L^{-1})). Figure 4a,d,g display the adsorption isotherms of Pb^{2+}, lysozyme and MB, respectively. To determine the adsorption constant and maximum adsorption capacity, the isotherms were further analyzed by the Langmuir and Freundlich models, as expressed by Equations (3) and (4), respectively.

$$q_e = \frac{q_{max}bC_e}{1+bC_e} \tag{3}$$

$$q_e = kC_e^{1/n} \tag{4}$$

where q_{max} (mg g^{-1}) is the maximum adsorption capacity, b (L mg^{-1}) is the Langmuir adsorption constant, and k (mg$^{1-1/n}$·L$^{1/n}$·g^{-1}) and n are the Freundlich constants associated with adsorption capacity and adsorption intensity, respectively.

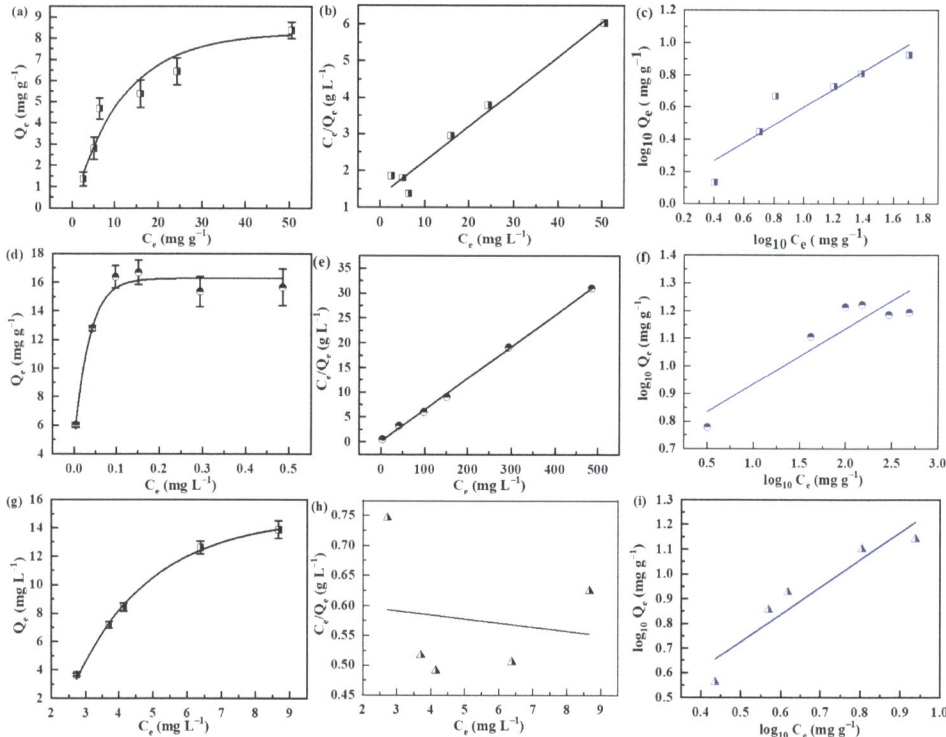

Figure 4. (**a,d,g**) Adsorption isotherms, (**b,e,h**) Langmuir fitting curves and (**c,f,i**) Freundlich fitting curves of the adsorption of pollutants by SPS foams. Square: Pb^{2+}, circle: lysozyme and triangle: MB.

The resulting curves fitted by the Langmuir model are displayed in Figure 4b,e,h, while those obtained by the Freundlich model are displayed in Figure 4c,f,i. The corresponding fitting parameters and coefficients of determination (R^2) are listed in Table 1. Modeling results reveal that the adsorption isotherms of Pb^{2+} and lysozyme can be better described by the Langmuir model, while that of MB is better described by the Freundlich model. The SPS foams possess theoretical maximum adsorption capacities of 10.5 and 15.7 mg g^{-1} for the adsorption of Pb^{2+} and lysozyme, respectively. Since the adsorption of MB cannot be fitted by the Langmuir model, its theoretical maximum adsorption capacity cannot be determined. Notably, the three pollutants used in this work are positively charged at the experimental conditions, while the surface of SPS is negatively charged owing to the sulfonic groups. Under this circumstance, the binding of the pollutants to the SPS foams is mainly attributed to the electrostatic interaction. However, since the DoS value of the SPS foam is low in this study, the π–π stacking interaction could also possibly occur in immobilizing MB to SPS foams. This could be the reason for the different adsorption isotherm of MB as compared with those of Pb^{2+} and lysozyme. It should be pointed out that although the resulting maximum adsorption capacities of the SPS foams are smaller than those of PS-based sorbents reported in the literature [47,48], the prominent advantage of the SPS foams lies in the bulky self-standing morphology, which is highly beneficial for separation but compromises their adsorption capacities.

Table 1. Fitting the equilibrium data using the Langmuir and Freundlich models.

Sorbates	Langmuir Model			Freundlich Model	
	R^2	q_{max}	b	R^2	k
Lysozyme	0.9981	15.7	0.48	0.8079	5.42
MB	−0.3031	−146.8	−0.011	0.8608	1.50
Pb^{2+}	0.9633	10.5	0.071	0.9311	1.11

The adsorption kinetics of the SPS foams were also studied. The instant concentrations and adsorption uptakes of the pollutants were recorded as a function of contact time, as shown in Figure 5. The concentrations of the pollutants (the red lines) rapidly decrease while the adsorption uptakes (the blue lines) increase during the initial period of contact time and then reach an equilibrium state. To study the kinetics of the adsorption of metal ions, three kinetic models, including the pseudo-first-order, pseudo-second-order and intraparticle diffusion models, were applied to fit the experimental data. The pseudo-first-order kinetic model can be expressed by Equation (5).

$$\log(q_e - q_t) = \log q_e - \frac{k_1}{2.303} t \qquad (5)$$

where k_1 (min^{-1}) is the adsorption rate constant of the first-order kinetic model. The pseudo-second-order kinetic model is expressed by Equation (6).

$$\frac{t}{q_t} = \frac{1}{k_2 q_e^2} + \frac{1}{q_e} t \qquad (6)$$

where k_2 (g (mg·min)$^{-1}$) is the adsorption rate constant of the second order kinetic model. The intraparticle diffusion model is expressed by Equation (7).

$$q_t = k_p t^{1/2} + C \qquad (7)$$

where k_p (mg (g h$^{1/2}$)$^{-1}$) is the intraparticle diffusion rate constant and C (mg g^{-1}) is a constant related to the thickness of the boundary layer.

The fitting curves obtained by the pseudo-first-order equation are shown in Figure 5b,d,f, while those by the pseudo-second-order and intraparticle diffusion equations are displayed in Figures S3 and S4, respectively. The resulting fitting parameters are summarized in Table 2. It reveals that among the three equations, the pseudo-second-order best describes the kinetic data. Based on the pseudo-second-order equation, the equilibrium adsorption uptakes are 33, 26.4 and 2.4 mg g^{-1}, while the adsorption constants are 0.021, 0.078 and 0.0043 g (mg·min)$^{-1}$ for the lysozyme, MB and Pb^{2+}, respectively. The equilibrium adsorption uptake could be related to the molar mass and charge of the pollutants. Pb^{2+} ions have a smaller molar mass than those of lysozyme and MB. In addition, a Pb^{2+} ion has two positive charges. In contrast, only one positive charge could be found for the lysozyme and MB molecules. Thus, the binding of Pb^{2+} ions to the SPS foam could involve more SO_3^-

groups as compared with those of lysozyme and MB. These two aspects could result in the lowest equilibrium adsorption uptake of Pb^{2+}.

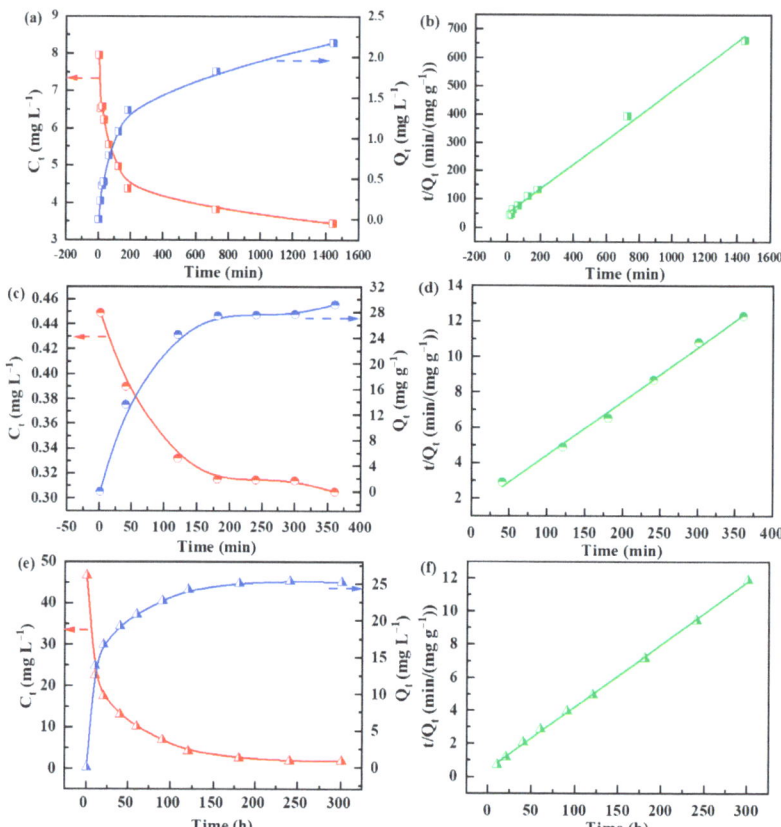

Figure 5. Adsorption kinetics of (**a**) Pb^{2+}, (**c**) lysozyme and (**e**) MB by SPS foams, and (**b**,**d**,**f**) corresponding fitting curves obtained by the pseudo-second-order equation. (The red arrow points to the left axis, the blue arrow points to the right axis.)

Table 2. Kinetic fitting results of the adsorption processes.

Sorbates	Pseudo-First-Order			Pseudo-Second-Order			Intraparticle Diffusion	
	R^2	q_e	k_1	R^2	q_e	k_2	R^2	k
Lysozyme	0.8747	21.29	0.01	0.9937	33.0	0.021	0.9003	1.55
MB	0.9745	18.42	0.025	0.9992	26.4	0.078	0.7243	1.22
Pb^{2+}	0.8967	1.74	0.002	0.9930	2.4	0.0043	0.9126	0.057

One prominent feature of the resulting SPS foam lies in its three-dimensional self-standing structure, which affords a significant advantage for fix-bed filtration adsorption. The filtration adsorption performance of the SPS foams was evaluated. Figure 6a shows the breakthrough curves for the filtration of the pollutants. The concentrations of the pollutants in the effluents present a clear step profile, indicating that the major pollutant molecules were immobilized in the sorbents in the initial filtration stage, but directly penetrated through the foams when they were saturated by the pollutants. At the breakthrough point, 5.2 mL of MB, 3.7 mL of lysozyme and 3.0 mL of Pb^{2+} solution were filtrated through the foams. Figure 6b shows the concentration profiles of MB in the effluents with different initial pollutant concentrations. With an increasing initial concentration, the volume of the effluent decreases at the breakthrough point. For instance, at the breakthrough points, the effluent volumes are determined to be 5.2, 6.3 and 12.5 mL at initial concentrations of 50, 25 and 10 mg L^{-1}, respectively.

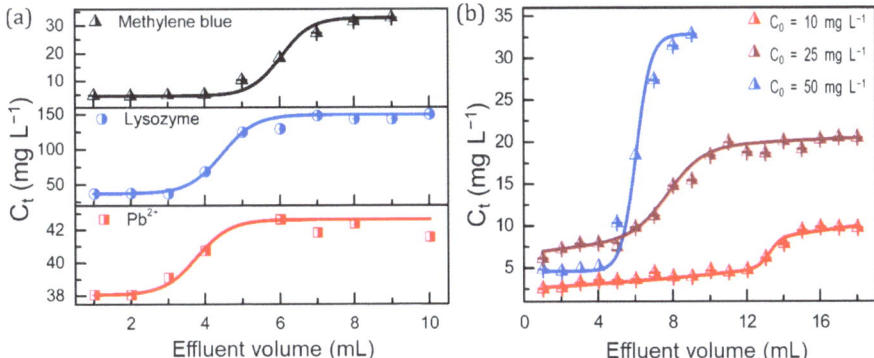

Figure 6. (**a**) Breakthrough curves for the penetration of the pollutants through the SPS foams and (**b**) dependence of breakthrough curves on the initial concentration of MB.

For practical applications, the cyclic performance of the sorbents is also critical. To this end, the desorption of the Pb^{2+}- and MB-saturated SPS foams was conducted by immersing them in 1 M HCl. The regenerated SPS forms were re-utilized to remove the pollutants from aqueous solutions. This adsorption–desorption process was repeated five times and the adsorption uptakes were recorded. Figure 7a,b show the adsorption uptakes of Pb^{2+} and MB, respectively, as a function of cyclic number. It reveals that the adsorption uptakes are quite stable, indicating the outstanding cyclic performance of the SPS foams. Figure 6c–e show the digital photos of SPS before and after the adsorption of MB and after desorption, respectively, which vividly reveal the adsorption and desorption processes. It is noteworthy that the regeneration process is quite easy and does not involve any separation process because of the bulky three-dimensional structure of the SPS foams.

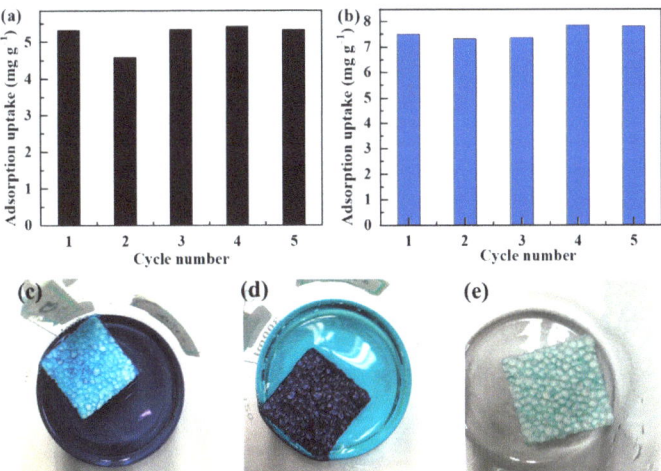

Figure 7. Cyclic adsorption of (**a**) Pb^{2+} and (**b**) MB, and digital photos of SPS (**c**) before and (**d**) after adsorption of MB, and (**e**) after desorption.

4. Conclusions

In summary, we employed PS waste as a feedstock and converted it into bulky SPS sorbents via a facile sulfonation process. Using Pb^{2+}, lysozyme and MB as representative substrates, the characteristics of the adsorption processes were studied. The adsorption isotherms of Pb^{2+} and lysozyme can be better described by the Langmuir model, leading to maximum equilibrium adsorption uptakes of 10.5 and 15.7 mg g^{-1} for the adsorption of Pb^{2+} and lysozyme, respectively. In contrast, the isotherm of MB is better described by the Freundlich model. For all three substrates, the kinetics of the adsorption processes were best described by the pseudo-second-order equation. The electrostatic interaction was identified as the main mechanism of the adsorption process. The Pb^{2+}- and MB-saturated SPS foams can be readily regenerated by immersing them in HCl solution and the recovered sorbents show excellent cyclic adsorption performance. The most prominent features of the SPS sorbent reported in this work lie in the bulky three-dimensional structure, which avoids the tedious separation process. Additionally, the simple preparation process and abundant availability of PS waste as a feedstock indicate SPS foams to be a promising sorbent for practical applications.

Supplementary Materials: The following supporting information can be downloaded at: https://www.mdpi.com/article/10.3390/polym14214477/s1, Video S1 showing the penetration of water droplets into SPS foams. UV spectra of methylene blue and lysozyme and resulting calibrated curves are shown in Figures S1 and S2. Adsorption fitting curves of Pb^{2+}, lysozyme and methylene blue by the pseudo-first-order equation and intraparticle diffusion equation are shown in Figures S3 and S4.

Author Contributions: Conceptualization, Y.S.; data curation, C.Y.; writing, Z.P.; visualization, C.Y. and Z.P.; supervision, Y.S.; funding acquisition, Y.S. All authors have read and agreed to the published version of the manuscript.

Funding: This research received no external funding.

Institutional Review Board Statement: Not applicable.

Informed Consent Statement: Not applicable.

Data Availability Statement: Not applicable.

Conflicts of Interest: The authors declare no conflict of interest.

References

1. Baig, N.; Ihsanullah; Sajid, M.; Saleh, T.A. Graphene-Based Adsorbents for the Removal of Toxic Organic Pollutants: A Review. *J. Environ. Manag.* **2019**, *244*, 370–382. [CrossRef]
2. Shi, C.; Wang, X.; Zhou, S.; Zuo, X.; Wang, C. Mechanism, Application, Influencing Factors and Environmental Benefit Assessment of Steel Slag in Removing Pollutants from Water: A Review. *J. Water Process Eng.* **2022**, *47*, 102666. [CrossRef]
3. Darabdhara, J.; Ahmaruzzaman, M.d. Recent Developments in MOF and MOF Based Composite as Potential Adsorbents for Removal of Aqueous Environmental Contaminants. *Chemosphere* **2022**, *304*, 135261. [CrossRef] [PubMed]
4. Bruno, P.; Campo, R.; Giustra, M.G.; De Marchis, M.; Di Bella, G. Bench Scale Continuous Coagulation-Flocculation of Saline Industrial Wastewater Contaminated by Hydrocarbons. *J. Water Process Eng.* **2020**, *34*, 101156. [CrossRef]
5. Pelaez, M.; Nolan, N.T.; Pillai, S.C.; Seery, M.K.; Falaras, P.; Kontos, A.G.; Dunlop, P.S.M.; Hamilton, J.W.J.; Byrne, J.A.; O'Shea, K.; et al. Review on the Visible Light Active Titanium Dioxide Photocatalysts for Environmental Applications. *Appl. Catal. B* **2012**, *125*, 331–349. [CrossRef]
6. Aldalbahi, A.; El-Naggar, M.; Khattab, T.; Abdelrahman, M.; Rahaman, M.; Alrehaili, A.; El-Newehy, M. Development of Green and Sustainable Cellulose Acetate/Graphene Oxide Nanocomposite Films as Efficient Adsorbents for Wastewater Treatment. *Polymers* **2020**, *12*, 2501. [CrossRef]
7. Cevallos-Mendoza, J.; Amorim, C.G.; Rodríguez-Díaz, J.M.; Montenegro, M.d.C.B.S.M. Removal of Contaminants from Water by Membrane Filtration: A Review. *Membranes* **2022**, *12*, 570. [CrossRef]
8. Dinu, I.A.; Ghimici, L.; Raschip, I.E. Macroporous 3D Chitosan Cryogels for Fastac 10EC Pesticide Adsorption and Antibacterial Applications. *Polymers* **2022**, *14*, 3145. [CrossRef] [PubMed]
9. Fu, L.; Huang, Z.; Zhou, X.; Deng, L.; Liao, M.; Yang, S.; Chen, S.; Wang, H.; Wang, L. Ferrous-Oxalate-Modified Aramid Nanofibers Heterogeneous Fenton Catalyst for Methylene Blue Degradation. *Polymers* **2022**, *14*, 3491. [CrossRef] [PubMed]
10. Liu, Z.; Ling, Q.; Cai, Y.; Xu, L.; Su, J.; Yu, K.; Wu, X.; Xu, J.; Hu, B.; Wang, X. Synthesis of Carbon-Based Nanomaterials and Their Application in Pollution Management. *Nanoscale Adv.* **2022**, *4*, 1246–1262. [CrossRef]

11. Shayegan, Z.; Bahri, M.; Haghighat, F. A Review on an Emerging Solution to Improve Indoor Air Quality: Application of Passive Removal Materials. *Build. Environ.* **2022**, *219*, 109228. [CrossRef]
12. Chen, W.; Zhang, H.; Zhang, M.; Shen, X.; Zhang, X.; Wu, F.; Hu, J.; Wang, B.; Wang, X. Removal of PAHs at High Concentrations in a Soil Washing Solution Containing TX-100 via Simultaneous Sorption and Biodegradation Processes by Immobilized Degrading Bacteria in PVA-SA Hydrogel Beads. *J. Hazard. Mater.* **2021**, *410*, 124533. [CrossRef]
13. Biswal, B.K.; Vijayaraghavan, K.; Tsen-Tieng, D.L.; Balasubramanian, R. Biochar-Based Bioretention Systems for Removal of Chemical and Microbial Pollutants from Stormwater: A Critical Review. *J. Hazard. Mater.* **2022**, *422*, 126886. [CrossRef]
14. Chu, G.; Zhao, J.; Liu, Y.; Lang, D.; Wu, M.; Pan, B.; Steinberg, C.E.W. The Relative Importance of Different Carbon Structures in Biochars to Carbamazepine and Bisphenol A Sorption. *J. Hazard. Mater.* **2019**, *373*, 106–114. [CrossRef]
15. Saleh, T.A.; Gupta, V.K. Processing Methods, Characteristics and Adsorption Behavior of Tire Derived Carbons: A Review. *Adv. Colloid Interface Sci.* **2014**, *211*, 93–101. [CrossRef] [PubMed]
16. Guo, M.; Wang, J.; Wang, C.; Strong, P.J.; Jiang, P.; Ok, Y.S.; Wang, H. Carbon Nanotube-Grafted Chitosan and Its Adsorption Capacity for Phenol in Aqueous Solution. *Sci. Total Environ.* **2019**, *682*, 340–347. [CrossRef] [PubMed]
17. Chen, H.; Wang, X.; Li, J.; Wang, X. Cotton Derived Carbonaceous Aerogels for the Efficient Removal of Organic Pollutants and Heavy Metal Ions. *J. Mater. Chem. A* **2015**, *3*, 6073–6081. [CrossRef]
18. Sharma, V.K.; McDonald, T.J.; Kim, H.; Garg, V.K. Magnetic Graphene–Carbon Nanotube Iron Nanocomposites as Adsorbents and Antibacterial Agents for Water Purification. *Adv. Colloid Interface Sci.* **2015**, *225*, 229–240. [CrossRef]
19. Chen, D.H.; Cao, L.; Hanley, T.L.; Caruso, R.A. Facile Synthesis of Monodisperse Mesoporous Zirconium Titanium Oxide Microspheres with Varying Compositions and High Surface Areas for Heavy Metal Ion Sequestration. *Adv. Funct. Mater.* **2012**, *22*, 1966. [CrossRef]
20. Sendão, R.M.S.; Esteves da Silva, J.C.G.; Pinto da Silva, L. Photocatalytic Removal of Pharmaceutical Water Pollutants by TiO_2—Carbon Dots Nanocomposites: A Review. *Chemosphere* **2022**, *301*, 134731. [CrossRef] [PubMed]
21. Chang, R.; Wu, X.; Cheung, O.; Liu, W. Synthetic Solid Oxide Sorbents for CO_2 Capture: State-of-the Art and Future Perspectives. *J. Mater. Chem. A* **2022**, *10*, 1682–1705. [CrossRef]
22. Wang, L.; Pan, Y.; Li, Y.; Sui, Z.; Li, J.; Xu, X. Destructive Sorption of NF_3 as a Novel Greenhouse Gas over Al_2O_3@Mn_2O_3 Sorbents with High Surface Area. *Process Saf. Environ. Prot.* **2022**, *162*, 1082–1090. [CrossRef]
23. Maiti, D.; Mukhopadhyay, S.; Devi, P.S. Evaluation of Mechanism on Selective, Rapid, and Superior Adsorption of Congo Red by Reusable Mesoporous α-Fe_2O_3 Nanorods. *ACS Sustain. Chem. Eng.* **2017**, *5*, 11255–11267. [CrossRef]
24. Liang, J.; Li, X.; Yu, Z.; Zeng, G.; Luo, Y.; Jiang, L.; Yang, Z.; Qian, Y.; Wu, H. Amorphous MnO_2 Modified Biochar Derived from Aerobically Composted Swine Manure for Adsorption of Pb(II) and Cd(II). *ACS Sustain. Chem. Eng.* **2017**, *5*, 5049–5058. [CrossRef]
25. Wang, J.; Wang, P.Y.; Wang, H.H.; Dong, J.F.; Chen, W.Y.; Wang, X.X.; Wang, S.H.; Hayat, T.; Alsaedi, A.; Wang, X.K. Preparation of Molybdenum Disulfide Coated Mg/Al Layered Double Hydroxide Composites for Efficient Removal of Chromium(VI). *ACS Sustain. Chem. Eng.* **2017**, *5*, 7165–7174. [CrossRef]
26. Lei, W.W.; Portehault, D.; Liu, D.; Qin, S.; Chen, Y. Porous Boron Nitride Nanosheets for Effective Water Cleaning. *Nat. Commun.* **2013**, *4*, 1777. [CrossRef]
27. Shahsavari, R.; Zhao, S. Merger of Energetic Affinity and Optimal Geometry Provides New Class of Boron Nitride Based Sorbents with Unprecedented Hydrogen Storage Capacity. *Small* **2018**, *14*, 1702863. [CrossRef]
28. Sun, Y.; Ha, W.; Chen, J.; Qi, H.; Shi, Y. Advances and Applications of Graphitic Carbon Nitride as Sorbent in Analytical Chemistry for Sample Pretreatment: A Review. *TrAC Trends Anal. Chem.* **2016**, *84*, 12–21. [CrossRef]
29. Zhang, W.T.; Shi, S.; Zhu, W.X.; Huang, L.J.; Yang, C.Y.; Li, S.H.; Liu, X.N.; Wang, R.; Hu, N.; Suo, Y.R.; et al. Agar Aerogel Containing Small-Sized Zeolitic Imidazolate Framework Loaded Carbon Nitride: A Solar-Triggered Regenerable Decontaminant for Convenient and Enhanced Water Purification. *ACS Sustain. Chem. Eng.* **2017**, *5*, 9347–9354. [CrossRef]
30. Sapurina, I.Y.; Shishov, M.A.; Ivanova, V.T. Sorbents for Water Purification Based on Conjugated Polymers. *Russ. Chem. Rev.* **2020**, *89*, 1115–1131.
31. Alnaqbi, M.A.; Al Blooshi, A.G.; Greish, Y.E. Polyethylene and Polyvinyl Chloride-Blended Polystyrene Nanofibrous Sorbents and Their Application in the Removal of Various Oil Spills. *Adv. Polym. Technol.* **2020**, *2020*, 4097520. [CrossRef]
32. Ge, Y.Y.; Li, Z.L. Application of Lignin and Its Derivatives in Adsorption of Heavy Metal Ions in Water: A Review. *ACS Sustain. Chem. Eng.* **2018**, *6*, 7181–7192. [CrossRef]
33. Duan, Y.Q.; Freyburger, A.; Kunz, W.; Zollfrank, C. Lignin/Chitin Films and Their Adsorption Characteristics for Heavy Metal Ions. *ACS Sustain. Chem. Eng.* **2018**, *6*, 6965–6973. [CrossRef]
34. Deng, S.; Zhang, G.S.; Liang, S.; Wang, P. Microwave Assisted Preparation of Thio-Functionalized Polyacrylonitrile Fiber for the Selective and Enhanced Adsorption of Mercury and Cadmium from Water. *ACS Sustain. Chem. Eng.* **2017**, *5*, 6054–6063. [CrossRef]
35. Huang, M.L.; Mishra, S.B.; Liu, S.Q. Waste Glass Fiber Fabric as A Support for Facile Synthesis of Microporous Carbon to Adsorb Cr(VI) from Wastewater. *ACS Sustain. Chem. Eng.* **2017**, *5*, 8127–8136. [CrossRef]
36. Shen, Y.; Li, L.; Xiao, K.J.; Xi, J.Y. Constructing Three-Dimensional Hierarchical Architectures by Integrating Carbon Nanofibers into Graphite Felts for Water Purification. *ACS Sustain. Chem. Eng.* **2016**, *4*, 2351–2358. [CrossRef]

37. Li, L.; Shen, Y.; Wang, Z.M. Synthesis of 3D Iron and Carbon-Based Composite as A Bifunctional Sorbent and Catalyst for Remediation of Organic Pollutants. *Mater. Res. Exp.* **2017**, *4*, 7. [CrossRef]
38. Shen, Y.; Li, L.; Zhang, Z.H. Scalable and Environmentally Friendly Synthesis of Hierarchical Magnetic Carbon Nanosheet Assemblies and Their Application in Water Treatment. *J. Phys. Chem. C* **2016**, *120*, 6659–6668. [CrossRef]
39. Gong, B.; Peng, Y.T.; Pan, Z.Y.; Chen, W.M.; Shen, Y.; Xiao, K.J.; Zhang, L. Gram-Scale Synthesis of Monodisperse Sulfonated Polystyrene Nanospheres for Rapid and Efficient Sequestration of Heavy Metal Ions. *Chem. Commun.* **2017**, *53*, 12766. [CrossRef]
40. Heddagaard, F.E.; Møller, P. Hazard Assessment of Small-Size Plastic Particles: Is the Conceptual Framework of Particle Toxicology Useful? *Food Chem. Toxicol.* **2020**, *136*, 111106. [CrossRef]
41. Maharana, T.; Negi, Y.S.; Mohanty, B. Review Article: Recycling of Polystyrene. *Polym.-Plast. Technol. Eng.* **2007**, *46*, 729–736. [CrossRef]
42. Smith, M.; Love, D.C.; Rochman, C.M.; Neff, R.A. Microplastics in Seafood and the Implications for Human Health. *Curr. Envir. Health Rpt.* **2018**, *5*, 375–386. [CrossRef] [PubMed]
43. Fonseca, W.S.; Meng, X.H.; Deng, D. Trash to Treasure: Transforming Waste Polystyrene Cups into Negative Electrode Materials for Sodium Ion Batteries. *ACS Sustain. Chem. Eng.* **2015**, *3*, 2153–2159. [CrossRef]
44. Mangalara, S.C.H.; Varughese, S. Green Recycling Approach to Obtain Nano- and Microparticles from Expanded Polystyrene Waste. *ACS Sustain. Chem. Eng.* **2016**, *4*, 6095–6100. [CrossRef]
45. Ramanan, V.; Siddaiah, B.; Raji, K.; Ramamurthy, P. Green Synthesis of Multifunctionalized, Nitrogen-Doped, Highly Fluorescent Carbon Dots from Waste Expanded Polystyrene and Its Application in the Fluorimetric Detection of Au^{3+} Ions in Aqueous Media. *ACS Sustain. Chem. Eng.* **2018**, *6*, 1627–1638. [CrossRef]
46. Shen, Y.; Qiu, X.P.; Shen, J.; Xi, J.Y.; Zhu, W.T. PVDF-g-PSSA and Al_2O_3 Composite Proton Exchange Membranes. *J. Power Sources* **2006**, *161*, 54–60. [CrossRef]
47. Sułkowski, W.W.; Nowak, K.; Sułkowska, A.; Mikuła, B.; Wierzba, P. The Conditions of Cationic Exchange with the Use of Recycling Polystyrene Derivative, the Product of Sulfonation by Silica Sulfuric Acid. *J. Appl. Polym. Sci.* **2013**, *128*, 2611–2617. [CrossRef]
48. Mahmoud, M.E.; Abdou, A.E.H.; Ahmed, S.B. Conversion of Waste Styrofoam into Engineered Adsorbents for Efficient Removal of Cadmium, Lead and Mercury from Water. *ACS Sustain. Chem. Eng.* **2016**, *4*, 819–827. [CrossRef]

MDPI AG
Grosspeteranlage 5
4052 Basel
Switzerland
Tel.: +41 61 683 77 34

Polymers Editorial Office
E-mail: polymers@mdpi.com
www.mdpi.com/journal/polymers

Disclaimer/Publisher's Note: The title and front matter of this reprint are at the discretion of the Guest Editors. The publisher is not responsible for their content or any associated concerns. The statements, opinions and data contained in all individual articles are solely those of the individual Editors and contributors and not of MDPI. MDPI disclaims responsibility for any injury to people or property resulting from any ideas, methods, instructions or products referred to in the content.

www.ingramcontent.com/pod-product-compliance
Lightning Source LLC
LaVergne TN
LVHW072333090526
838202LV00019B/2412